Inspections and Reports on Dwellings

Inspecting

Ian A Melville FRICS

Ian A Gordon FRICS

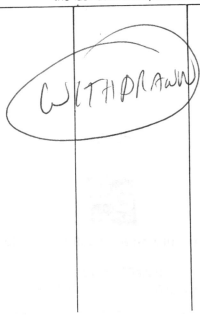

© Ian A Melville and Ian A Gordon 2005
ISBN 0 7282 0448 7
Series: 0 7282 0451 7
Reprinted in 2007

Design and Typesetting by Ted Masters
Printed in Malta by Progress Press Limited

Preface

This series of books, produced at a time of change, cover the work of assessing the age of all types of dwelling, inspecting what is visible of the structure, services and environs and reporting upon the condition to prospective buyers, lenders and to owners about to put a dwelling on the market. They are being published consecutively, the first, *Assessing Age*, having been published in September 2004.

The books follow the investigative practice in England, Wales and Scotland as exemplified in the current editions of the Conditions of Engagement for inspections to produce the Homebuyer Survey and Valuation Report, the Model Conditions of Engagement included in the Specification for the Valuation and Inspection of Residential Properties for Mortgage Purposes and the Guidance Note for Surveyors on Building Surveys of Residential Property, all produced by the Royal Institution of Chartered Surveyors.

The passing of the Housing Act 2004 has also enabled full account to be taken of the Conditions of Engagement and the Requirements for Home Inspectors on the production of Home Condition Reports. These will be required for inclusion in the Home Information Packs to be prepared for sellers before a dwelling is marketed for sale in England and Wales from 1st January 2007. Account has also been taken of the Conditions of Engagement for inspections to produce Single Survey Reports to be made available in advance to those intending to bid for dwellings in Scotland.

The Conditions of Engagement for the five Reports envisaged as being covered by this series of books, require all visible and readily accessible parts of a dwelling to be inspected. This is, in effect, no more than buyers and sellers can see for themselves but the inspector has to bring a great deal more to his task than merely what his eyes can see in order to fulfil his obligations. He is, of course, aided by instruments and equipment such as a damp meter, torch and ladder but otherwise has to rely upon his own physical abilities to obtain access to all the accessible parts of the dwelling, to investigate dark areas and to tap, pry and peer at the numerous components presented to him by buildings through the ages in order to make an accurate diagnosis of their differing characteristics. Most important of all, however, is his ability to diagnose what could be major defects from minimal surface indications since speculation and conjecture are, and must be, necessary elements of an inspection where no opening up forms part of the operation.

Inspecting, as the lynchpin of the series, gives inspectors the detailed information required to provide that added value. How to ascertain and assess construction, materials, exposure, durability, defects and dangers, along with advice on preparing for the task, coping with clients, and, vitally necessary, following the guidance which has been handed down from the Courts over the years on the path to follow when all the information required is not ascertainable. All types and ages of dwelling are covered from the very old to the near new, including those system built in the latter half of the 1900s.

How, after reflective thought, the information gathered at the dwelling will be presented by way of the different forms of report will be the subject of the remaining two books in the series. However, valuations both for open market and insurance reinstatement purposes will be included in both and buyers will, in many cases, wish to have advice on matters outside the provenance of the Home Condition Report, the possibility of alterations for example.

Throughout the books, surveyor and inspector are considered interchangeable and where a pronoun is unavoidable the masculine and feminine equally so.

Ian A Melville
Ian A Gordon

London, July 2005

Contents

A paged list of contents is provided at the beginning of each Section

List of Tables

Acknowledgements

The authors are grateful to the following for permission to reproduce the copyright material set out below:

The Timber Research and Development Association
Figures: 1-11 inclusive and part 13.

The Building Research Establishment
Photographs: 180, 181, 265, 266, 272, 273, 304, 316 and 348.

The Northern Consortium of Housing Authorities
Photographs: 13, 34, 35, 176, 193, 194, 261, 262, 263, 264, 300, 301, 307-311 inclusive, 313, 317, 319, 320, 322-325 inclusive, 327-330 inclusive, 333-340 inclusive, 346, 347, 349-351 inclusive and 386.

The Office of the Deputy Prime Minister
Photographs: 57, 58, 59, 206 and 226.

The National Assembly for Wales
Photographs: 195, 271, 274, 305, 306, 314, 315, 318, 321, 326, 392, 393, 396, 403 and 405.

Domino Books Limited
Figures: 45-58 inclusive and 63.

Section 1 Inspectors

Attributes

The physical attributes of the inspector relate neither to age or sex but to health and the capacity to do the work. Tramping around dwellings, going up and down stairs, crawling around in confined spaces and peering intently can be exhausting. Few are blessed with good health at all times. Working with a sniffle is one thing but aching limbs and a high temperature are liable to cloud the judgement. Surveyors differ in height and girth so that some parts of the inspection may be easy to some and difficult to others. It would seem that earlier generations in this country were smaller than today judging by the restricted size of trapdoors to the lofts of earlier dwellings. Good eyesight is an important attribute since it is the inspector's eyes that take the responsibility for the outcome of the inspection, not those of an assistant whose role should be limited to note taker and bag carrier.

Since it is residential buildings that are going to be examined, a knowledge of their construction must be a basic attribute. This can be acquired through study, attending lectures, examining drawings and specifications, but it is absolutely necessary to look at them, both in their totality and at the details. The study should not only be inclined towards the construction of new properties but should also embrace the study of older dwellings, their forms of construction and the likely defects. An assiduous reading of books, journals, pamphlets and the like relating to buildings and their defects, so as to keep up to date with ideas and research, forms part of this attribute. It is only by maintaining a keen interest in buildings and their defects throughout a career, that experience is gained and consolidated. This experience should be reflected in recommendations from satisfied clients, which is undoubtedly the best way to obtain a continual stream of work.

Knowledge should be extended to a study of the particular characteristics of the residential properties in the area. From what period do they mainly derive and does the topography have any effect on their durability and condition? A keen interest in what is going on in the area at the current time, as well as an interest in its historical development, is a highly desirable attribute. Time should be found to look at the local newspapers as well as the nationals, so that the policies of local as well as central government are kept in mind.

It is easy to think that the cost of building repairs is a matter to be left entirely to contractors, but a necessary attribute is to be able to put an approximate figure to many of the common repair items. It need not be quoted, unless specifically requested, and then only if the full extent of the repair is ascertainable and with appropriate caveats, but it will need to be kept in mind for its possible effect on the valuation figure. Being a member of a firm that manages residential property, or specifies and supervises repair work, is a help and there are publications for this purpose which are invaluable when used with care.

While the finer points of the law can be dealt with by the legal profession, that profession's principals and staff cannot be expected to see the property and the inspector is likely to be the only person to view the property and its site who can appreciate the possibility of legal problems. A knowledge of all the law relating to residential property is therefore an essential attribute, covering environmental, planning, conservation and energy matters as well as easements and rights in relation to party walls and boundaries.

Writing skills come more naturally to some

than others, but the end product of the work under consideration is a report which however it may appear, on paper, screen or disc, nevertheless has to be put together in words. Some are better at stringing words together into sentences and paragraphs to provide clear non-ambiguous advice, but all need to hone their skills in this regard. The attribute of being able to do so is now more necessary than ever, since many reports require information to be presented in defined spaces, if at all possible. The art of being able to turn out a concise meaningful précis from a collection of notes or a number of rambling disjointed sentences is even more essential now than before.

The verbal communication skills used in exercising tact and understanding, as well as the patience to listen to client's problems and concerns, also deserve continual development. If the individual carrying out the work is of mature years and the clients are younger, advice can be given in a 'fatherly' way without being patronising. If both are young, they may be able to share similar concerns. If the roles are reversed, the younger can be an inspiration by demonstrating vim and vigour in dealing with concerns that may seem puzzling. By listening and talking it is sometimes possible, without seeming to be nosey, to find out the client's occupation. It is not good to be caught out in a report labouring a point about a legal matter or the electrical installation when the client is either a lawyer or an electrical engineer.

Tenacity at some cost in time is sometimes an essential attribute during the inspection itself when it is necessary to follow up some minor indication of a defect, 'following the trail' as one Judge put it, so that not only a description of a defect but also, if possible, a diagnosis of its cause can be given together with any categorisation or rating as to 'urgent' or

'significant' for example. Suggesting a remedy, and putting an approximate cost to it is possible with growing experience and can be done on occasions, without suggesting that the client brings in a 'specialist' to do what amounts to the surveyor's job.

Qualifications, Insurance and Client's Expectations

Should the individual putting himself forward to carry out the work have a qualification? In theory, if he possesses the attributes and the necessary abundance of experience, qualifications are irrelevant but, before dismissing them entirely, regard must be had to two important aspects. One governing England and Wales, is that any individual carrying out Inspections for Home Condition Reports must be a member of an approved Certification Scheme. Members have to be qualified by their education, training and experience so that qualification in one or other of the recognised professional institutions would seem to be a desirable pre-requisite for membership of a certification scheme although Assessment Centres are being established to train inspectors up to standard from scratch. Another current requirement for membership of any of the certification schemes approved by the Secretary of State, is that every member has in force professional indemnity insurance. Such insurance was a requirement for members of the professional institutions before any of the certification schemes were set up and its continued requirement for members ensures that those carrying out inspections and preparing reports for residential properties other than for the requirements of the Housing Act 2004 throughout Great Britain are also covered.

Professional Indemnity Insurance is essential when carrying out the work of inspecting and reporting on residential property. It protects both the individual surveyor and the client and the reason why it is compulsory for members of professional organisations may be to protect the profession's standing by way of providing compensation for clients who are let down by inadequate work. Such cover, however, is not available to all and sundry. For example, one collective policy requires that those carrying out inspections and reporting on property be either:

a) a Fellow or Member of the Royal Institution of Chartered Surveyors or a Fellow or Associate of the Faculty of Architects and Surveyors or a Fellow or Member of the Royal Institute of British Architects or a Fellow or Member of the Royal Institute of Architects of Scotland or

b) by anyone who has not less than five years experience of such work or such other person nominated by the Insured to execute such work subject always to supervision of such work by a person qualified in accordance with (a) above.

It is not, therefore, suggested that only members of the RICS are suitable to carry out Home Condition Inspections and Reports. Other organisations, such as the Institution of Civil Engineers and the Institution of Structural Engineers as well as the Chartered Institute of Builders, also require their members to be covered by professional indemnity insurance and, accordingly, it is to be expected that members of these Institutions will also be interested in undertaking the work, providing that they are members of a certification scheme approved by the Secretary of State.

Some individuals strongly object to paying vast premiums for indemnity insurance and, furthermore, being compelled to do so. As there is no law preventing an individual

calling himself a 'surveyor', some practice as such without membership of a professional organisation or indemnity insurance. These usually make sure that they have no assets so that if they fail to spot defects and are sued they are found to be 'men of straw' and the aggrieved is unable to recover any loss. If asked whether they are insured they will, probably, say that they are such good 'surveyors' they do not need it. That may or may not be true, of course, and might also be the attitude of a practitioner wealthy in his own right who while confident of not making mistakes has sufficient resources to pay compensation if necessary.

Unqualified and uninsured 'surveyors' are not allowed, within the law of England and Wales to undertake Home Condition Inspections and Reports for sellers under the Housing Act 2004. However there is nothing to prevent them accepting instructions and carrying out inspections for purchasers anywhere in Great Britain, always provided the report subsequently prepared is not on a copyright form of one of the professional organisations.

The Consumers' Association, in its monthly magazine *Which*, provides occasional articles based on the public's actual experience of dealing with surveyors and on the quality of the reports they produce for those buying dwellings. It has been the only consistent source of information whereby surveyors can gauge the public's reaction to their work and should certainly be read by all practitioners, not only for the comments by the buyers but also for the advice the magazine gives to those contemplating engaging a surveyor, what they should require from him and what they might reasonably expect to get. The existence of unqualified and uninsured 'surveyors' is no doubt one of the reasons why the magazine provides advice to its readers when buying a dwelling is being considered on how to go about choosing a surveyor. First, however, it sensibly indicates that there is no point in even considering the choice of surveyor if from the purchaser's own inspection walls are out of plumb, mortar is soft and weathered, slates or tiles are missing from the roof, there are cracks in the brickwork, damp stains on ceilings of top floor rooms or low down on walls in ground floors or basement and that these would cause problems if the necessary finance was not available to deal with them. If, however, the presence of some, or all, of these items does not put the prospective purchaser off and he needs advice on their cause and what to do about them, then *Which* sets out a number of points which should be considered. It suggests that, if possible, the surveyor should preferably be: recommended, qualified and insured, a member of an Institution which operates a compulsory arbitration scheme and practices in the area in which the property is situated.

Which suggests, next, that the surveyor's Conditions of Engagement be thoroughly examined to see whether they are reasonable. It does not, for example, think very much of disclaimers of responsibility if the purchaser does not have the specialist tests carried out which are recommended in the report. It is not alone in believing that many surveys include such recommendations far too frequently and that, to an extent, surveyors are inclined to duck out of doing the job which they are hired to do. But such a disclaimer is part and parcel of the Conditions of Engagement which clients are required to accept in order to obtain a Homebuyer Survey and Valuation. Such a disclaimer is not considered unreasonable by surveyors on the obvious grounds of cost alone, but surveyors should accept that the recommendation to obtain specialist advice is only used when considered really necessary and not made indiscriminately. Surveyors who spatter their

reports with such recommendations have been said to suffer from 'white van syndrome' as so many firms of specialists run around in white vans.

Which suggests that the prospective purchaser should require the inspector or surveyor, if he makes any recommendations to engage specialists, to say why he does so and what the risks would be if the recommendations were not followed. The magazine found it quite reasonable and indeed thought it necessary, for example, for surveyors to recommend commissioning a specialist report on the electrical installation where the visible wiring looked old, a test on the drains where there were trees nearby and a report on dampness where there were high moisture readings. It is considered that a surveyor should, of course, be able to say a good bit more in these situations than merely to recommend the commissioning of a specialist. Where the visible wiring looks old, most surveyors for example would sensibly recommend a rewire. It is when the wiring looks in order that a specialist's test and report is needed.

While *Which* magazine does not actually suggest a meeting with the surveyor, it is almost implied by other suggestions, such that the prospective purchaser should ensure that the surveyor is prepared to do what is requested by way of commenting on those concerns which viewing the property may have raised, perhaps similar to those already mentioned. He should also say what he will do to comply with whatever terms are set out in written or printed Conditions of Engagement for an agreed fee, which needs to be itemised to cover fee, VAT and, if appropriate, expenses. The Surveyor should also be required to say when he will be able to carry out the inspection, barring undue difficulties and, approximately, how long it will be before the report is submitted. It is further suggested that for the avoidance of doubt, and possible difficulties later on, all the above be set out and agreed by both prospective purchaser and surveyor in writing.

The above advice from the Consumers' Association has been reiterated over the years and derives from member's experience on moving home. From the other side of the fence, so to speak, surveyors possessing the attributes set out in this section will have no difficulty providing the service which Consumers' Association members could well be seeking. If the clients in such instances know what they want in contrast to many prospective purchasers, particularly first time buyers, who do not know, then the surveyor must use that attribute of patience and understanding and talk to them about the alternatives, so as to establish what would suit their particular circumstances.

Of course, advice of this nature is intended for the benefit of prospective purchasers and although derived from experience in the pre-2004 era continues to be valid throughout the United Kingdom until such time as the proposed changes come into effect. In England and Wales prospective buyers will be well advised to view the Home Condition Report in the light of who prepared it. Would it have been prepared in the same way as it would have been for a purchaser? The characteristics of both sellers and purchasers will be considered in the other two books in the series but if a prospective buyer has doubts about the origins of a Home Condition Report supplied by the seller, he would be wise to follow the Consumers' Association advice about choosing a surveyor to provide a report tailored to his specific requirements. While a seller seeking an inspector to prepare a Home Condition Report has the path to follow set out by the Housing Act 2004 and

such protection as provided by the mandatory Certification Schemes approved by the Secretary of State, it would be sensible for him to have regard also to the Consumers' Association advice. If he is what might be thought of by surveyors as an ideal seller, for example sensible, above board, keen to have a quick sale at a reasonable price, a character to be met again later, he will not wish to be disappointed in his choice of inspector.

There is another aspect about surveyors which the Consumers' Association does not mention but might well have done so if known about at the time, since it is designed to provide a significant benefit to clients. As previously mentioned the Consumers' Association advises seeking a surveyor who is qualified but, since 2000, surveyors have been able to add another layer to their level of qualification, in addition to the distinction between Members and Fellows which operates purely as a matter of length of time since initial qualification. The Surveyors and Valuers Accreditation Scheme is voluntary and is for Surveyors working in a variety of specialisms. Benchmarks of competence for carrying out inspections and preparing reports and valuations on residential property have been established to the standard required for the Homebuyer Survey and Valuation Report operated by the RICS. Those members who satisfy the standards set down achieve accredited status. The benefit to clients is that surveyors who are accredited, can offer automatically the advantages of hidden defects insurance in respect of the properties which they inspect. This means that if defects which were not visible at the time of the inspection become apparent later, the cost of the necessary repairs can be recovered through insurance. This can be achieved without having to invoke the surveyor's professional indemnity insurance.

The fact that it was considered necessary to

introduce the Surveyors and Valuers Accreditation Scheme shows that there was a need to monitor the performance of RICS members over the years in a field where their work affected far more members of the general public than any other of their operations. That the necessity was justified was provided by the abundant evidence from the Courts, insurance companies, consumers' organisations, the press and television that the public were being let down by many surveyors. The situation has probably stemmed to a great extent from the time when the Institution decided to change the emphasis of surveyor's education away from the practical to the more academic, abandoning its own examination system in favour of degree qualification through the universities, in an attempt to elevate the status of surveying to a learned profession, held in high regard by the public. Unfortunately, the opposite seems to have happened and, judging from media and television exposure, surveyors are held in less respect now than they ever were. The hurried introduction of Surveyors and Valuers Accreditation was an attempt to claw back some of the lost ground in advance of the Government's introduction of Certification Schemes for Home Inspectors and, no doubt, hopefully to influence their development and operation. In furtherance of this aim SAVA was detached from the Institution to become a wholly independent limited company so that it could take part in the development of standards for the Government's proposed Home Condition Report.

Around the same time as the Surveyors and Valuers Accreditation Scheme was introduced for qualified members, the RICS approved other means of achieving chartered membership as alternatives to either obtaining a degree and completing a two year Assessment of Professional Competence or via Technical membership. These involved the taking of National Vocational Qualifications at different levels coupled with practical experience over periods appropriate to the level of NVQ obtained. Completion of any of these alternatives permitted entry to the final assessment to become a Technical Member and in turn entry on to the pathway to full chartered membership. With an approved Level 4 NVQ, entry is immediately gained to the final assessment and it is at this level of qualification that Home Inspectors are likely to be approved for certification. Even chartered surveyors will have to demonstrate a degree of practical competence with the submission of a series of ten sample reports, at least three in the new Home Condition format, on a varied range of dwellings. It may be that this welcome increased interest in the practicalities of the surveyor's work will have a beneficial effect on the quality of reports in the future.

Of course, if offending surveyors do not change their ways by taking more care over establishing their instructions, more time on site and ensuring that their intentions are expressed clearly and concisely there may not be much difference in the quality of work. If qualifying by degree examination (at one time it was possible to qualify by experience alone) an Assessment of Professional Competence and a Continuing Professional Development programme has not even succeeded in maintaining standards for surveyors, it is hard to see that a further layer of qualification in the form of an Accreditation and Government Certification for Home Inspectors will automatically do so. Insurance, which hardly ever compensates the aggrieved for the misery and upheaval caused by bad reports, is not necessarily the answer since the profit to be made from cramming in as many surveys a day as possible appears, to some organisations, to amply cover the cost of indemnity premiums or even the cost of compensation payments. It does the profession's reputation in the public eye no good, however, to accept the rate of failure that such hurried work and carelessness inevitably produces. However, instilling a sense of pride, refusing to be rushed and insisting on proper payment even if that charge to the client is included in a package set off against receipt of commission later, is not going to be as easy as is generally supposed.

Charges

Most of the attributes believed to be necessary for individuals carrying out inspections and reporting on residential property have been considered but there is one other. It is not too far a stretch of the imagination to say that a further necessary attribute is to know one's worth. In days of severe competition a response that a charge cannot be quoted to an enquiry for this sort of work until the surveyor sees the property is unlikely to produce much work, if any. The enquirer will, most likely, go elsewhere.

Of course, instructions to inspect and carry out a mortgage valuation will invariably be coupled with the amount of fee, based on a percentage of the purchase price, which the lender is prepared to pay to find out the value of the security. Fee quotation will not be involved and, since the instruction may be one of a number received from the same source, the surveyor may well be prepared to accept the rough with the smooth. He will, undoubtedly, have found that such an arrangement bears little relation to the size or condition of the property but will be prepared to accept the situation in return for a steady stream of work.

In other cases, most potential clients will have some information on the property on which they require advice, such as its address,

whether house or flat, its size by way of the number of rooms at least and in the case of a proposed purchase, the asking or agreed price. Having established initially what form of report the client requires, Home Condition, Homebuyer Survey and Valuation or Building Survey, with or without valuation, and the extent of any special instructions, the next step to consider will be the amount to charge for the work. The fee that each surveyor will charge will differ, even for the same work. One surveyor's charges will be found to vary from those of another by the value he places on his time and will have regard to his overheads, all that it costs him to run an office so that he has an organisation behind him when he is away, irrespective of whether he is a sole principal or part of a large office, itself part of a large conglomerate. The Consumers' Association magazine *Which* made the facetious comment that charges for a Building Survey depended more on the surveyor than the property, perhaps not realising how true that was.

Basically there are three methods for assessing charges which can be used according to the circumstances. There are advantages and disadvantages to each.

1) The first method is by rule of thumb of so much per habitable room, such as bedrooms and living rooms. This method would be suitable for instructions to prepare any of the forms of report and is easy to apply. It has the disadvantage, however, in that no two properties with the same number of rooms are exactly the same in character and condition and can take quite different times to inspect. It might be possible to use the method if properties in a particular area are much the same and the client is able to give some idea of the condition. Alternatively, the surveyor could decide that it takes him a minimum time to inspect a room whatever the condition.

Of course it will need to be applied according to a rate of how much per room for either a Home Condition, Homebuyer Survey and Valuation or Building Survey Inspection and Report.

2) The second method is by a percentage basis on the asking price or the disclosed agreed purchase price. A disadvantage here is that, in theory, it is not available for use in connection with the preparation of Home Condition Reports, since no asking or agreed purchase price will have been determined at the time the query is raised. The Home Inspector could, of course, ask the seller what he anticipates receiving for the property and quote a charge based on that figure, which might be acceptable to the seller and be agreed. For a Homebuyer Survey and Valuation Report or a Building Survey Report and Valuation, a disadvantage would be that property in an exclusive area, where demand is strong, would attract a high fee, whereas a property in a poor area would be muchless, yet the two might be quite similar. It could be argued, however, that the higher fee was justified in that if something was missed, the damages awarded would reflect such a difference in property values. As a method it is not thought of too highly and was criticised by the Consumers' Association since, in their view, there was no reason why a higher fee should be charged when the amount of work undertaken could well be much less. Again, if used a series of rates will need to be determined for the different types of report.

3) The third method of a charge of so much per hour spent on the work is acknowledged by most to be the fairest, both to the client and the surveyor. Ideally the surveyor would agree the rate with the client and on completion submit a bill itemising what he has done and the hours spent doing it. Such an arrangement might be acceptable where there is a continuity of instructions and where both are well known to each other and there is an element of trust between them. When surveyor and client are at arm's length, it is unlikely to be acceptable because of being too open ended. Unless some idea was provided in advance of the likely total cost, arrival of the account could well induce, at least, a disinclination to pay. Most clients need to be quoted a fixed fee on the assumption that the information given to the surveyor about the property is correct. For the surveyor not to be able give a quotation would be considered very un-business like.

Every surveyor in practice needs to be able to quote an hourly rate of charge. It can be based on the overall costs of running the organisation, large or small, of which he forms part, including either what he is paid or his expectations as a principal. For a medium sized office, in reasonable premises with appropriate back up staff, it has been suggested that surveyors might be charged at a daily rate 1% of their gross annual salary or expectation. To pluck figures from the air and purely as an example, assuming their productive day comprises six hours, and if salary or expectation is £42,000 pa, the daily rate would be £420 and the hourly charge £70. The percentage figure would probably need to be adjusted, fractionally, for larger or small organisations, but this method has the merit of simplicity when used in conjunction with the surveyor's estimate of how long it would take to complete the work in relation to a Home Condition, Homebuyer Survey and Valuation or Building Survey Report on properties of various periods, sizes and types. Obviously, in addition to estimating the time it will take to carry out the inspection, the surveyor needs to estimate the time needed for travelling, preparation of the report and a reasonable amount of time for discussion with the client, either at the property, in the office or on the telephone.

It is probably best to quote charges exclusive of VAT, as this can be varied by the Chancellor at any time. It is also important that the work should be remunerative. In the case of the Home Condition Reports, it is to be expected, that in the hope of obtaining instructions to sell, the charge for the sellers survey will be driven down. Low fees have been the bane of the surveying profession over many years and has resulted in fees undoubtedly being held down at unrealistic levels for this type of work with the result that clients have never become reconciled to paying for the work at a reasonable rate. The upshot, however, is that whatever the fee, or whoever does the work, the responsibility is the same and the temptation to cut corners in the hope of saving time and being more competitive can rebound to damaging effect.

The surveyor after discussion with a prospective buyer may also receive instructions to carry out tests of drainage, electrical or heating installations or to take up fitted carpets and floorboards for a closer examination of under floor construction, on the assumption that permission will be granted by the seller. It will have to be decided, there and then, how such instructions are to be met. Will the surveyor, provided he has the necessary experience, be responsible for carrying out the tests and reporting on the results or will the work be put out to contractors, in whole or in part, under his supervision or will they be employed by the surveyor on behalf of the client? In the first case, the additional fee for the work can be agreed, exclusive of VAT, and the results incorporated in the report. In the second case, the surveyor may need to ascertain the charges likely to be incurred and obtain the client's agreement to meet them. In the third case, it may again be necessary to ascertain charges on the client's behalf but it would be better, in many cases, for it to be made clear that the contractors are employed by the client, report to him and render their account accordingly, the surveyor merely arranging the timing of the inspections, hopefully, to coincide with his own.

A range of specialist firms has sprung up in recent years to advise buyers on matters extending beyond those of the normal building services consultants and of whose existence the surveyor should be aware. One organisation will do a check on the neighbours to see if they might cause problems. Enquiries and visits to the neighbourhood will be made at hours other than those when the surveyor is likely to be making his inspection. Another organisation will carry out a check on the locality for matters such as airborne pollutants, contamination, subsidence risk, crime rates and the like.

Surveyors and their clients can also take advantage of the services of the woodworm and dry rot eradication and damp prevention firms, many of whom, will carry out a free inspection, prepare a report and provide an estimate for work, if ascertainable, or advice on any further investigations they consider necessary. Obviously, they hope to obtain work as a result but, while their presence provides another useful pair of eyes, the surveyor's responsibility is in no way diminished.

He is the one paid to give the disinterested opinion and advice and, this might well extend to commenting on such firm's reports if asked to do so.

In order to provide its members with some advice on the charges they would be likely to incur on commissioning inspections and reports when considering purchasing a dwelling, the Consumers' Association found in 1999 that for a mortgage valuation on a three bedroom semi-detached house, charges ranged up to £200, depending on price, though some lenders waived the charge if they did not allow the borrower to see the report. On the same sort of house 56 surveyors, thought to be in South Yorkshire although the area was not specifically stated, quoted fees of between £230 and £300 for a Homebuyer Survey and Valuation, but for a Building Survey quoted charges were far more wide ranging from £350 to £600.

In the London area research carried out for the Nationwide Building Society, at about the same time, found on a sample of 10 surveyors carrying out a Homebuyer Survey and Valuation on a two storey mid-terrace Victorian house with a back addition of lower overall height, that fees averaged £380 with a top figure of £520 and the lowest at £270.

Included in a report, prepared by consultants for the Department of Environment Transport and the Regions, now the Office of the Deputy Prime Minister, and published in early 2000, on initiatives to improve the home buying and selling process were fees ranging from £350 for smaller properties to £650 in East Anglia for Homebuyer Survey and Valuation Reports on executive style properties with purchase prices of £100,000 upwards. In South London, where property prices where higher, equivalent charges for a Homebuyer Survey and Valuation were from £375 rising

to £725 for properties over £250,000 in price, but these reports included an energy rating and advice on security. The consultants concluded that the likely cost of a Home Condition Report based on the Homebuyer Survey and Valuation format but excluding the valuation would be to the order of £250 to £350, but clearly this took no real account of area having regard to other findings and even, to some extent, of their own sources.

Equipment

This is not the place to discuss the finer points of the Conditions of Engagement for the five different types of report envisaged as being covered by this series of books. Details of three types of report will be dealt with in *Reporting for Buyers* and the remaining two in *Reporting for Sellers*. However, just as there is a requirement for an assessment of the age of the dwelling being inspected in all reports, so all five require an inspection of all visible parts of the dwelling which are readily accessible. This is expressed by slightly different wording in each set of Conditions but it amounts to the same thing. In other words a non-invasive, non-destructive, visual inspection from the vantage points of ground and floor levels within the site or adjacent communal or public space without the need to move any obstructions, exposing any parts and without carrying out any tests.

The surveyor's eyes can, nevertheless, be supplemented and there is equipment available for this purpose which it is essential for the inspector to have available at the dwelling. As will be seen from the next section 'Inspection', the Courts have, in the past, determined that there is no difference in the scope of the physical inspection where the terms of the surveyor's engagement require him to carry out the type of inspection described in the previous paragraph. Accordingly, there can be no doubt that for whatever report the inspection is being carried out, the surveyor should have with him the same items of equipment.

Clients will expect their surveyor to be appropriately equipped. Some might not give it very much thought but others might expect him to turn up with all manner of sophisticated tools, for example some capable of seeing more than the eye alone can see. They might be somewhat disappointed to find that the only item they will find mentioned in the Conditions of Engagement, presented for signature, with any degree of sophistication is a moisture meter and even that is not mentioned in the Model Conditions for a Building Survey Report. The other items he will see mentioned in some of the other Conditions are nothing more sophisticated than a ladder and a pair of binoculars and in one instance also a torch.

While the 2003 RICS Specification for the Valuation and Inspection of Residential Property for Mortgage Purposes, the 2005 RICS Practice Notes for the Homebuyer Survey and Valuation Report and the 2004 notes accompanying the Single Survey Report for Scotland make no further mention of equipment, advice to surveyors preparing the two types of other report do.

The 2004 RICS Guidance Note for Building Surveys sets out two lists, the first of 21 items of equipment which the surveyor is recommended to have available on the inspection and the second a list of 11 items of optional equipment. While surveyors are not required to follow the recommendations in the Guidance Note, the RICS points out that such recommendations would carry considerable weight if ever the surveyor's

work was called into question. This Guidance Note is available to the public and therefore the list of recommended equipment ought to be considered as essential, indeed mandatory in fact, for the surveyor to have with him on such inspections. Yet it can be criticised, as can the items on the optional list along with those on the two lists in the 2004 Requirements for Home Inspectors which is also in the public domain. The first list for Home Inspectors is of eight mandatory items, with examples of their use and comments, which are considered necessary for the inspection, while the second, also with examples of use and comments, comprises 11 optional items said to widely and commonly used and will 'assist' the inspection. In fact all four lists are a muddle, mixing items of equipment appropriate and useful for completing the task in hand and items for the personal safety of the surveyor.

Accordingly all the items on the lists, together with a couple of others missed by all four, will be considered and contrasted so that a sensible list of essential items to have available when carrying out inspections for any one of the five reports covered by this series of books can be assembled. There should be no list of 'optional' or other useful items. If they are useful they should be on the list of those considered essential. Items for the surveyor's personal safety can be set out quite separately and whether they are taken on inspections can be a matter for the individual surveyor to decide.

In the following paragraphs the letters 'BS' indicates that the item is one of the 21 appearing on the list of 'recommended' equipment contained in the RICS Guidance Note for Building Surveys while 'HI' indicates a comparative item from the mandatory list of tools and equipment from the Guidance for Home Inspectors or its absence.

BS: Equipment for recording findings, e.g. paper, pens or pencils

HI: Writing paper, pen/pencils

There can be no dispute here. The Home Inspector is advised, sensibly, to take enough spares and that pencils are better for damp/moist weather. Neither mention an eraser which can be useful or a clipboard, essential for holding the paper. Of course, many surveyors dictate site notes into a hand held recorder. This is not only quicker than writing but also enables the surveyor to keep an eye on what is being inspected. However, site notes are private to the surveyor and such recorders must not be used in the presence of any others, apart from colleagues, otherwise unpleasant arguments can arise. Any dictated notes must, of course, be transcribed and elaborated if necessary by sketches, plans and photographs, so that a clear legible and unambiguous record of the inspection can be retained without any reliance on memory.

BS: Small tape measure or measuring rod

HI: Measuring tape

A surveyor inspecting on his own needs, primarily, a measuring rod, 2 metre is the usual size, folding handily in to six sections. A long tape measure, also metric, is required for site measurements, when needed, and essentially with a hook on the end. The contemporary use of electronic measuring aids does not obviate the need for the above.

BS: Binoculars or telescope

HI: Binoculars

No difference here and Home Inspectors are told to make use of binoculars regardless of eyesight, for viewing chimneys and roofs and any part of the dwelling from a distance.

BS: Compass

HI: -

A compass for Home Inspectors is only included in the list of commonly used items even though it is essential, not only for orientating the position of the dwelling but also for describing its component parts.

BS: 3m ladder

HI: 4m ladder

The contrast here highlights a discrepancy which ought to have been apparent to the RICS all along. To enable access to a roof void where the hatch is 3m above the adjacent floor level, as required by all reports, a ladder longer than 3m is required, otherwise it would have no support from the side of the opening. As the Guidance for Home Inspectors points out, a 4m ladder can be used to push open the hatch and must be considered an essential item of equipment. It also suggests putting old socks, clean one hopes, on the ends to avoid marking decorations.

BS: Hard hat

HI: -

This is omitted from both the lists in the guidance for Home Inspectors but as it is an item for the personal safety of the surveyor it can be extracted and placed in a separate list to cover Health and Safety aspects. Surveyors inspecting dwellings where building work is in progress, a possibility for the preparation of all types of report, will no doubt be required by the site manager to wear a hard hat to comply with Health and Safety regulations.

BS: Spirit level
HI: Spirit level

Both Guidance Notes and Requirements agree at the inclusion of a spirit level as mandatory. Whereas those for Home Inspectors says its use is only for when following the trail on defects in floors, flat roofs and walls it can and should be used at any time for checking levels and, over short lengths, the verticality of walls. Long and pocket size spirit levels are essential equipment.

BS: Plumb bob
HI: -

Strangely this useful item is only included in the commonly used, not the mandatory, list in the Requirements for Home Inspectors. Even though measurements would not necessarily be taken as they would be on a fuller investigation, hung from an upper floor window a plumb bob can provide a better check on the verticality of a wall than the mere eye. It should be included as an essential item.

BS: Mask, particularly for loft spaces and inspection chambers where there is a greater likelihood of contaminants
HI: -

The RICS Guidance for Building Surveys elaborates on this item to include a description of possible use, as above, but the item only appears in the commonly used list in the Requirements for Home Inspectors. It is not essential for the inspection and as an item for the personal safety of the surveyor can appear in a separate list for that purpose.

BS: Crack gauge or ruler
HI: -

Strictly speaking any item that measures, divided into millimetres, will suffice for measuring crack widths and it could be said that both mandatory lists have already included such an item. Nevertheless a crack gauge is vastly superior for the purpose, providing as it does a vivid illustration of how cracks of various widths appear on plaster or rendered surfaces and even on rougher surfaces such as brick work. It warrants inclusion in the essential list for all inspections and its use avoids sole reliance on subjective descriptions such as 'slight', 'moderate' or 'severe'.

BS: Electronic moisture meter
HI: Electronic moisture meter

Mandatory and essential for all five types of report covered. Home Inspectors are to ensure calibration and testing is done regularly and that they have spare batteries. This also applies to surveyors on building surveys. In a list of essential items the bare title of the instrument can be elaborated with a description of the accessories as set out later.

BS: Torches
HI: Torch

A surveyor without a torch to see into poorly lit areas, is unthinkable and for inspections for both Building Surveys and Home Condition Reports they are mandatory. The carrying of spare batteries and bulbs is mentioned for Home Inspections but not for Building Surveys except as an optional item. Perhaps it was meant to be taken as read but the failure of the torch at a crucial moment in the roof space with out the spares being handy, besides making the surveyor look foolish, can be more than just irritating. A torch that stands up on its own as a lantern is highly desirable as well as the shock proof hand held variety. Better viewing can be obtained in a roof space by means of a protected lamp and an extension lead to a lamp or socket on the landing below. The owner, or occupier should, of course, be asked before use as some can be touchy about electricity consumption.

BS: Camera with flash
HI: -

There is no requirement for photographs to be included in any of the reports considered in these four books. Nevertheless photography is now such an integral part of surveying practice that carrying out an inspection without its use would seem inconceivable. Photographs can be of vast assistance, particularly as an aid to the reflective thought indicated as a requirement before the surveyor compiles both a Building Survey and a Home Condition Report. While it has to be admitted that photographs do not always show a defect as dramatically as the surveyor would wish they can sometimes help to clarify the written description in a Building Survey Report or be added as a supplement to a Report in a prescribed format and, of course, they should always be kept with the site notes as a record of the inspection. Therefore, a camera with flash should be included with those items of an essential nature. The zoom capability on cameras is effective for bringing the likes of chimney stacks into closer view.

BS: Claw hammer and bolster and also, separately, lifting equipment for standard inspection chamber covers
HI: Lifting equipment/crowbar

Something to provide a better element of leverage than the fingers, and to save the nails in the process, is required for loose floorboards and drainage inspection covers, always provided it can be used without causing damage. The Guidance Note for Building Surveys makes a meal of this by including two separate items in its effectively mandatory list without, for the second item, saying what are 'standard', as distinct from 'non-standard' inspection chamber covers. The idea of inspection chamber covers is that the occupier of the dwelling should be able to lift them if required. Domestic covers, even of cast iron, are designed to be lifted by one person, preferably by hooks round the handles but if these are broken, in itself a defect, then with the aid of a bolster, claw hammer or crowbar. The latter provides the best leverage and is also useful for clearing the mud which tends to accumulate around the lifting handles and the edges of covers even when they are in good condition. It is those three tools which provide the essential equipment for gaining access to drainage inspection chambers.

BS: Screwdriver, bradawl or hand held probe
HI: -

No mention of such tools is made on either of the lists in the Requirements for Home Inspectors, no doubt because of the emphasis on the non-destructive nature of the inspection for a Home Condition Report. Phrases in the Requirements such as 'inspector not to cause any permanent damage even if permission given' and 'do not pierce timber frames for rot' seem to ignore the fact that the probes on the moisture meter, which the inspector is required to use, make holes. It is unrealistic, and defeats the objective of the exercise, to deprive the inspector of the judicious use of a finely pointed bradawl, which makes the same size holes as the probes on the moisture meter, which can be useful, not to say essential, for detecting rot in recently painted timber. It is reasonable to expect the surveyor not to do the heavy handed probing shown on 405, which by the looks of it was not necessary anyway, and to show the owner what he wants to do, explain why and ask permission.

If permission is not granted then that should be reported along with the reason behind the surveyor's request. A bradawl probe should therefore be considered an essential tool to be taken on inspections for use as necessary. Also very useful, and accordingly essential, and which can sometimes be more effective as a probe, is a small penknife. This can also double for a little judicious scraping if required but again with the owners prior permission. As for the screwdriver, its use as a probe would be far too drastic and as the surveyor is not to remove screwed down panels because of the risk of damage its only conceivable use would be by the seller if, for good reason, invited to remove a panel and he maintained he hadn't got a screwdriver. The surveyor's could be offered and if the offer was turned down that could be reported along with the reason for the request. As such, worth including as an essential item.

BS: Mirror
HI: -

Only listed among the commonly used items in the Requirements for Home Inspectors despite a comment 'may be coupled with a torch' and its use for 'voids with very restricted access/rear of drainpipes', presumably a misprint for 'downpipes'. It is an essential item for those very reasons and accordingly should have also been included in the mandatory list for Home Inspectors.

BS: Meter cupboard key
HI: -

Surveyors must not go around with their own keys for use in other peoples dwellings. If access is required to any sort of cupboard, or a room for that matter, the owner or occupier must be approached. If a necessary key is not available that should be reported. This is all picked up better in the Requirements for Home Inspectors and no mention of such an item appears in either the mandatory list or the list of commonly used items. Accordingly a meter cupboard key should have no place in the RICS list.

BS: Means of determining inclination of roof pitch
HI: -

An essential mandatory item, of what ever it might consist and surprisingly omitted from either of the lists in the Requirements for Home Inspectors despite being reminded elsewhere that a 'common' defect is 'pitch too shallow for type of cover'. An adjustable set square used with a pocket spirit level is one means of determining inclination, another is with a folding carpenter's rule with protractor like division on the brass hinge or using the same type of rule and a clear celluloid protractor.

BS: First aid kit
HI: -

Included only in the commonly used items list in the Requirements for Home Inspectors and even there commented upon as for the Home Inspector's personal safety, the item has nothing to do with the task in hand and should be extracted from both the RICS and the Requirements for Home Inspector's lists for inclusion elsewhere.

BS: Means of personal identification
HI: -

Here again this item has nothing to do with the task in hand. It is not even mentioned in either of the lists contained in the Requirements for Home Inspectors though, of course, it is stressed elsewhere and is obviously an essential requisite for a surveyor to have with him on every inspection for whatever purpose, not just for these reports. It is not unknown for responsible neighbours to call the police thinking there is a suspicious character mooching around.

Of the 21 items on the list of recommended, in effect mandatory, equipment set out in the RICS Guidance Note for Building Surveys, one is effectively duplicated, four are for the personal safety of the surveyor and one, a key to other people's cupboards, ought not to have been included, leaving 15. From the optional list of 11 items, one is for the protection of the surveyor, five are extensions of items in the recommended/mandatory lists and ought to be included with those, as should be two items not mentioned in that list. These are a marble or golf ball, useful for checking whether floors are level and also the direction of the fall on flat roofs and a

magnifying glass for looking into cracks in plaster, rendering or brickwork, or the exit holes of insects which have been consuming timber. Two further items are also included in the optional list, a hand held metal detector and a boroscope, which could well be needed on further investigations but are in no way essential for a visual inspection as are the additional screwdrivers included in the list. These adjustments raise the number of mandatory items so far to 17.

All eight items on the mandatory list of equipment for Home Inspectors are included in the above 17, but from the list of 11 commonly used items only the marble or golf ball mentioned in the previous paragraph need to be promoted to essential and mandatory, six are for the personal safety of the surveyor and the remaining four are already included among the 17.

To complete a practical and sensible list of mandatory essential items for surveyors to take with them on inspections, two further small items should be added, neither of which are mentioned in any of the four lists already discussed. These both concern airbricks for the essential ventilation of suspended timber ground floors and any other similar concealed areas, for example flat roofs. These are a length of stout wire to check whether air bricks are blocked, perhaps by injected cavity insulation subsequent to construction, and a box of long household matches. The matches are useful for checking whether a current of air is drawn into the air brick from the windward side of the dwelling and whether there is a through current of air when used on the leeward, sheltered, side. These bring the total of mandatory items of equipment to 19 as follows:

1: Equipment for recording findings with spares, eg clipboard, paper, pens or ball points, pencils and erasers

2: Long and short metric tapes, the long with a hook, metric measuring rod

3: Binoculars or small telescope

4: Compass

5: 4 metre sectional lightweight ladder

6: Long and short pocket size levels

7: Plumb bob

8: Crack gauge

9: Properly maintained and calibrated electronic moisture metre and spare battery with accessories and deep insulated probes

10: Lighting equipment with spare batteries and bulbs, eg hand held torches, lantern light and extension lead with protected lamp

11: Camera with flash and zoom lens

12: Bolster, claw hammer and crowbar

13: Bradawl, penknife and screwdriver

14: Mirror

15: Adjustable set square or other item(s) for determining the inclination of roof slopes

16: Marbles or golf ball

17: Magnifying glass, watch repairer's eye glass or magnifying lens with illuminator and spare battery

18: Length of stout wire

19: Box of long household matches.

As well as necessary equipment for the inspection surveyors should pay more attention to items for their own personal health and safety than perhaps they have done in the past. Those items for that purpose mentioned in the RICS Guidance for Building Surveys and in the Requirements for Home Inspectors which are well worthy of consideration are as follows:

1: Means of personal identification

2: Hard hat

3: Protective overalls

4: Latex and heavy duty gloves and hand cleaning wipes

5: Face mask

6: First aid kit

7: Umbrella and rubber boots

8: Personal alarm and mobile phone.

Most of the items have already been discussed where they form part of the four lists but where not previously mentioned their possible use is apparent from the wording. For obvious reasons, the last two items should be kept with the surveyor at all times for use in case of emergency.

Section 2 Inspection

Arranging

Unless instructions are received to inspect an unoccupied dwelling it has to be admitted that the surveyor is not always a very welcome visitor. If acting for a buyer or lender, nearly always the case up to now, his inspection could well bring to light defects not known about previously and accordingly likely to lead to a reduced offer. The welcome might be a little warmer when the surveyor is acting for the seller, since a speedy inspection and a good report will further the objective. Even here though, discovery of defects not known about before will be annoying and either mean a reduction in the amount the seller had been anticipating or the disturbance and cost of the work involved in putting them right.

The surveyor, therefore, has a fairly hard task on his hands to overcome what may range from a slight coldness to almost downright hostility. Seldom will he be welcomed warmly and, such is life, if he is he will immediately be a little suspicious of being wheedled into missing signs of defects. A reasonable formality is probably the best reception, particularly if accompanied by a remark such as 'I am off for the day. Tea and biscuits are in the kitchen and don't forget to lock up and put the spare keys through the letterbox when you go'. Unfortunately this doesn't happen very often.

One of the important aspects in dealing with arrangements for an inspection is forming an estimate of how long it will take. The information given by a prospective buyer not only on what he wants, assuming he knows, but on the dwelling he is thinking of purchasing can be less than generous. It is here that the matter of settling instructions in a precise and structured way comes to the fore and is an essential and vital preliminary to any inspection.

Assuming that the enquiry is not from a client already well known to the surveyor, he or she is likely to say simply 'I am thinking of buying a house, how much would you charge to do a survey'. Bearing in mind the advice that it is considered very un-business like not to be able to respond to such an enquiry, the surveyor must rapidly establish some key facts regarding the dwelling concerned so that he can supply the estimate requested. Vital information at this stage could be limited to location, type of dwelling, detached, semi-detached or terrace house or should the enquiry relate to a flat, on what floor, approximate age, size in the form of the number of living rooms, bedrooms and bathrooms and condition, tenure and if leasehold the unexpired term, occupied or unoccupied plus the asking or accepted offer price. Depending on the replies further questions might be posed, such as, is there a garage, are there outbuildings and if so are they to be included. If the dwelling is more than a hundred years old is it listed or in a conservation area. It is also necessary to ask if the enquirer is aware of the two types of report currently available for buyers. If he is aware, which one does he require? If not, then the surveyor can explain that the inspection is the same for both but that the reports are in a different form, one with a valuation. Of course, the enquirer may have provided details of the type of dwelling on which the surveyor would only entertain carrying out a building survey so that an alternative would not arise. If it transpires that only a building survey will be suitable, nevertheless, does the enquirer require market and fire insurance valuations.

With the above information to hand, and suitably retained as a record of the conversation, it should be possible in most instances for the surveyor to provide an estimate for what the enquirer wants. That, of course, could be the last the surveyor hears from that particular enquirer and if that happens too often, it may be that the surveyor needs to look at the level of his charges.

Should, however, there be an eventual return call with a request to proceed then the opportunity should be taken to have a longer conversation or perhaps arrange a meeting. This is so that the instructions can be fully discussed in order to leave no doubt as to what the surveyor has to do and, hopefully, no doubt in the prospective client's mind as to what he can expect for his money. The enquirer should also be asked to send or bring along any further information he has about the dwelling, agent's particulars for example.

At the meeting, or on the telephone, the opportunity should be taken for the surveyor to put himself into his enquirer's shoes. He should do what he can to take on board some of the apprehensions which assail most people about to embark on probably the most expensive purchase of their lives particularly if they are first time buyers. Trading up to a much more expensive dwelling or downsizing from a family home to a flat, much smaller house or cottage, can produce similar anxieties.

Most prospective buyers by this stage have spent a lot of time looking at dwellings within their price range and have settled on one that more or less meets their requirements and, although an element of compromise usually enters into the decision, it could be at the top end of their financial resources and they are no doubt desperately anxious for a favourable report. However, the surveyor tactfully needs to enquire whether the dwelling has been looked at sufficiently, preferably more than once, for the clients to be satisfied on its visible condition and that they have the resources, with the aid of a loan if necessary, available to

deal with any obvious defects. It is quite surprising how viewers, cheerfully bundled around by owners and their agents, can have a very uncertain idea of what they are proposing to buy and further questions by the surveyor, put with tact, can be vitally necessary for both parties to put matters into perspective. However, given some reassurance that the clients have decided that this is the dwelling for them, the surveyor can then go on to enquire whether there any aspects to which the surveyor should give particular attention and does the client wish to carry out any minor alterations on which the surveyor can comment as to feasibility, for example the removal of a partition without the clients necessarily incurring an increased fee. Furthermore do the clients wish to attend at the end of the inspection to discuss the surveyor's findings or again, is their financial situation so critical that consideration should be given to curtailing the inspection if a major problem requiring urgent work is unearthed, thus saving a proportion of the fee? Does the client wish to be summoned to the dwelling for an explanation should this transpire?

During a discussion of this nature the surveyor will be able to gauge quite a bit about the character of the client. Has he been through all this before and knows precisely what he wants from his surveyor or does the hand need to be held at all times? The surveyor will no doubt react accordingly, but he should seek to avoid an 'I know it all' domineering attitude to the client.

Having established what the client wants, it is important for the surveyor to make quite clear in discussion what he can and cannot do. Some clients have difficulty in recognising that surveyors do not have x-ray eyes and that they are not able on a visual inspection for any of the five types of report envisaged in these four books to pull things apart to see what is behind or take up fitted carpets, laminate flooring, floorboards or decking to see what is underneath. All the surveyor can do if he has grounds for thinking that there is something nasty behind the woodwork is to advise opening up and a fuller investigation. Although all the things a surveyor will do and what he cannot do in someone else's property will be set out in the Conditions of Engagement sent or handed to the client and which he will be expected to sign as having seen and accepted, it is important to go over them verbally, since the significance of the written word is not always fully appreciated and certainly not always read as closely as it might be.

During the discussion, the surveyor will need to find out as much as he can of additional information about access to the dwelling for his inspection. Precise address, the name, address, postal and e-mail, telephone and fax number of the owner, name and the same particulars of any agents involved are required. How is access to be arranged, whether through the owner or the agents? If the dwelling is unoccupied, where can keys be obtained, are all the services connected and turned on, usually necessary if the Conditions of Engagement are to be fully fulfilled. For unoccupied premises, it is important for the surveyor to insist that security systems are deactivated before his arrival. Nothing is worse and more wasteful and time consuming than for such systems to go off with vast noise and some-times the police to arrive. It would be wise for him to stress and confirm in writing that he will not be responsible for reactivating such systems or for draining down services on completion of his inspection of unoccupied premises.

Further discussion will no doubt ensue on a provisional day for the inspection, depending on the state of the surveyor's diary, and the likely day for the delivery of the report with the required number of copies. Towards the end of the discussion details of to whom and to where the report is to be sent and eventually also to where the account should be sent can be taken.

Should the telephone enquiry be along the lines of 'I am thinking of putting my house on the market, how much would it cost to have a report on its condition', much the same procedure would need to be followed by the surveyor irrespective of whether legislation was in force requiring such a report. In the light of current thinking there would be little point in contemplating the inclusion of a 'for your eyes only' clause in a report to an owner thinking of selling. It would have to be assumed that it was the intention to show the report to prospective buyers, and dealings with the enquirer should be conducted on that basis.

Having elicited the same initial information about the dwelling as the surveyor would do in response to an enquiry from a prospective buyer, he would need to consider what form of report would be appropriate. There is no reason whatever why the form of a Homebuyer Survey and Valuation would not be entirely suitable for the purpose. Both the Home Condition Report from 1 January 2007 and the Single Survey Report are intended to be provided under very similar terms and conditions to the Homebuyer Survey and Valuation Report. As to the valuation this could be retained in Scotland as intended for the Single Survey Report, but in England and Wales it could be deleted as for the Home Condition Report but retaining all the other terms and conditions. Of course, as with enquiries from buyers the surveyor may conclude depending on the age, type and construction of the dwelling, that a Building Survey with or without valuation but on the same or similar Model terms and conditions as set out in the Guidance Note would be the only type suitable. Either way the surveyor should be able to provide an estimate.

Should the surveyor's estimate be acceptable a discussion is needed so that the precise terms and conditions can be agreed and accepted in writing and the unique number for the report obtained. Arrangements for access and the inspection should perhaps be more straightforward but, since the object of the exercise is to produce a satisfactory report, it would be even more sensible to have an arrangement to curtail the inspection should a problem be encountered which would necessitate the surveyor advising opening up for a closer investigation. If that in fact revealed there was no problem, the surveyor could then conclude his inspection and submit his report. A problem revealed, however, and the seller would then have the opportunity to decide whether to resolve it to the surveyor's satisfaction or to have estimates obtained and for these to be attached to the report which would contain the surveyors description of the defect as ascertained from the opening up.

Fictional prospective buyers and sellers asking for estimates in these examples are all envisaged as being happy to accept terms and conditions and accordingly appropriate fees for 'proper' inspections. Sometimes, however, a surveyor will be asked, perhaps by a client well known to the surveyor who provides him with an amount of other profitable work or, worse, by a relation or friend, 'Would you mind having a quick look at a house I am thinking of buying and let me know if its OK?' It is highly unwise to take on such a request, even if done for a friend or relation for no fee in the surveyor's spare time and the outcome to be reported either orally or in writing. Negligence cases in tort way back in the 1950s and 1960s, *Sincock* v *Bangs* 1952

and *Sinclair* v *Bowden Son & Partners* 1962, decided that even on a quick brief look inspection the surveyor was expected to indicate the presence of woodworm, dry rot and settlement however slight the manifestations. Even the bait of 'If it seems OK, I will ask you to do a detailed survey' shouldn't be taken. Experience shows that the second instruction never comes and the buyer proceeds headlong to purchase, leaving the surveyor at considerable risk. Far better to point out to friends and relations that their interest would be better served elsewhere by having a 'proper' inspection carried out by a surveyor taking full professional responsibility for his opinions. In the case of valued clients, who really ought to know better, pressing for full instructions to do the work properly is probably better than a flat refusal.

Another request by a potential buyer which needs to be treated with considerable care could be phrased thus, 'The present owner has handed me this report on his house with a view to speeding up its sale. Can you check it over for me?' Checking another's work in these circumstances amounts to transferring the responsibility for the accuracy and the opinions expressed in the report from the person who prepared it to the person checking it and accepting such an instruction should not be countenanced in any circumstances. Since the Contracts (Rights of Third Parties) Act 1999, even if there is no legislation requiring a report to be prepared for a seller, the surveyor preparing such a report is liable in negligence to any buyer acting on the contents and this liability is extended for the benefit of any lender who takes the dwelling as a security for a loan. There is no way of excluding this liability and this is as it should be.

Different, however, would be a request from a prospective buyer to carry out an inspection and report on a dwelling on the basis of the terms and conditions already discussed for this work. Such an instruction could well be forthcoming when there is a Sellers Survey, Home Condition or Single Survey Report available and even though no major alterations were contemplated, although much more likely if they were. In such circumstances, the purchaser's surveyor would probably not wish to have sight of the report prepared for the seller as it might, conceivably, colour his own views. For an additional fee he might be prepared subsequently to make a comparison of the two reports for the prospective buyer although most surveyors would view this as an invidious and unenviable task and prefer to leave it to the prospective buyer to do it himself. If pressed, there can be no real objection to such a comparison despite the possibility of it degenerating into a battle royal with the surveyors on both sides acting as knights in shining armour for their respective clients. Exactly the sort of adversarial confrontation the Government's proposals are intended to curb. The ghost of a third surveyor, acting as referee or arbitrator, can almost be seen looming on the horizon as in the case of party wall disputes.

The discussion concluded and on receipt of the agreed and signed terms and Conditions of Engagement, the surveyor will be in a position to make his arrangements for access to carry out his inspection. He will have all the information he is likely to get in advance on the dwelling and, hopefully, this will be sufficient for him to be able to make a reasonably accurate estimate of the time it is likely to take. He will need to add a margin to cover unavoidable hiccups, undue chit-chat with the vendor and a bit of time for reflective thought on site. Often the vendor will say 'Goodness, I had no idea it would take so long', the first intimation that life might be made difficult. However, the surveyor should stick to his guns and not accept, say, an

appointment for 4pm if he anticipates a three hour inspection even more so if it is winter time. Weekday morning appointments are ideal but, unfortunately, it has to be accepted at times that the odd awkward appointment might have to be entertained, as long as the surveyor's capability to do the work is not impaired in consequence.

It is necessary to stress when the appointment is being made that access to all parts of the dwelling is required and not to assume that this is understood by the occupier. Keys for locked rooms and cupboards need to be produced and if, for example, the basement, top floor or an individual room is sub-let, the occupant must be notified by the owner so that the accommodation can be inspected.

Once at the dwelling the surveyor must be careful not to cause any damage but if he accidentally does so, must tell the owner and be prepared to pay for the reinstatement. The idea of fudging the issue should not be entertained. A particularly relevant Court case, since these inspections require the covers to drainage inspection chambers to be lifted where possible, concerned a surveyor, *Skinner* v *Herbert W Dunphy & Son* 1969, who in doing just that could not replace the cover as he found it. The frame and surround had broken on lifting the cover and he was later sued by the owner's wife who had tripped over it some days later and who it was held had not been warned of the hazard. Damages and costs of £1,500 were awarded for the injuries sustained and, as the case was over 30 years ago, it can well be imagined what the damages would be if the case had arisen in this more litigious age. Even if the damages were paid by the surveyor's professional indemnity insurer it would, nevertheless, amount to a claim which could easily have been avoided with the exercise of more transparency and could well affect the amount of the renewal premium. Even leaving the front door on the latch while trotting in and out can be dangerous. One contractor was held liable for the loss of valuables removed by an unnoticed intruder in just such circumstances and it could easily happen to a surveyor. The National Occupational Standards for Home Inspectors has words to say on the security aspects of homes being inspected.

The surveyor should also bear in mind his own safety. He is only a 'visitor' as far as the Occupiers Liability Act 1957 is concerned. As such the occupier owes no special duty of care to him other than to warn him of any hazards of which he is aware, rotten floor boards, missing sash cords, treacherous trap-doors needing special care come to mind. Sensibly, he should have personal accident insurance cover but, equally sensibly, he should not take unnecessary risks. The temptation on an inspection to stretch just that little bit farther, perhaps while on a ladder, to see something is often very great but should be resisted. Most of the Conditions of Engagement even indicate that flat roofs will not be walked upon by the surveyor. They have been known to collapse.

The RICS publishes a few pages on Surveying Safely which may be downloaded from http//www.rics.org/builtenvironment/ surveying_safely.htm or as quoted in the Requirements for Home Inspectors www.rics.org/resources/surveying_safely/index. html. Although not as comprehensive as might have been expected, it should be read and followed by all surveyors. Among those items listed as of relevance to inspections of dwelling are the need for care when inspecting a dilapidated dwelling where asbestos sheets might have become degraded, releasing fibres, being wary of extending loft ladders which have been known to come down with a rush because of defective catches, and to

expect the unexpected when poking one's head through a loft hatch panel. One surveyor encountered an angry dog, put there to be out of the way when the inspection was in progress, falling off his ladder in the process. Squirrels in the roof space could have the same effect. Dogs in general need to be considered something of a hazard, since they all seem to have a rooted objection to measuring rods. An inspection of a dwelling where work is in progress requires even more care, it should never be carried out without notifying the foreman or charge hand and the site precautions and warning notices, such as the requirement for wearing hard hats, must be strictly followed.

Apart from personal safety aspects to be taken by the surveyor at the site there are the elementary precautions of ensuring that the surveyor's office knows where he is and roughly when expected back. On one inspection the surveyor fell down an inspection pit, was severely injured and left for a whole night unattended, because he wasn't expected back that afternoon. This misfortune highlights the need for the surveyor always to carry a mobile phone in case of emergencies and not just leave it in his briefcase. The ritual adopted by estate agents since the disappearance of one of their number after meeting a stranger at a dwelling has a lot to commend it and similar precautions should be taken where there is considered to be any degree of risk.

Possible compromising situations may necessitate an agreed appointment for inspection being cancelled. An owner leaving a minor to let the surveyor in to look round is a case in point and an owner being too forward should immediately indicate that a situation of possible danger could develop and that the surveyor should find a good reason to delay entering the dwelling or leaving it if already inside.

Scope

A term such as 'inspecting all the visible parts of a dwelling' can conjure up very diverse views of what could be involved in the minds of different people. To some, it could consist of a quick scoot around all the rooms inside and a glance at the outside, taking perhaps no more than a quarter to half an hour in time. To others it could comprise a close methodical examination of every square inch of every conceivable part of the dwelling, taking quite a long time to complete. Both inspections could, strictly speaking, be said to satisfy the requirements of the Conditions of Engagement for all the five types of report envisaged as being covered in this series of books, whether in Model or Draft form or as set out as part of the contract documents for a Homebuyer Survey and Valuation Report.

The finer points of the Conditions of Engagement will be discussed in the next two books, *Reporting for Buyers* and *Reporting for Sellers*, but as far as the inspection is concerned it all amounts to the same thing, even although all use a slightly different form of words. Words and phrases such as, 'visible', 'all', 'visual', 'accessible', 'readily available', 'practicable' and 'surface' are used. All except those for a mortgage valuation mention a ladder but this essential item of equipment needs to accompany the surveyor at all times for every purpose covered by this book since the inspection of all accessible areas is his inescapable duty.

However, it must be pretty obvious that the two forms of inspection mentioned in the first paragraph above could produce widely differing results for the same dwelling, depending on its condition. If in poor condition, important defects could be missed on a quick look round, unless they were blatantly obvious. On an exhaustive inspection, it could be difficult to separate the wheat from the chaff, major from minor defects.

1: There may not be time for a preliminary drive past look at the dwelling before formulating instructions but there are occasions when the surveyor would wish he had taken the opportunity. It also seems unbelievable that a client would not have mentioned the vegetation but this is perfectly possible if he considered it merely as a decorative adjunct. The surveyor is left with no option but to indicate and give warnings about the problems likely to occur and provide a few comments on the little he can see close up of the roof and walls by looking out of the windows.

The reason that there are no requirements and little guidance on how long the inspection of a dwelling should take and how it should be conducted should also be pretty obvious since the possible variations are limitless. It could be said that a small dwelling in good condition might be inspected quickly while a large mansion in poor condition, conversely, much longer but it would be unwise to speculate further than this. The important words in all the conditions add up to 'all visible parts' and provided all such are inspected it is their condition which will form the subject matter of the report. Consequently the length of time an inspection takes needs to be 'appropriate' to the size and condition of the dwelling concerned. Following this principle can cause difficulties if successive appointments are made to carry out inspections on the same day. The advice provided by the RICS Guidance Note 'Building Surveys of Residential Property' that the surveyor is advised 'not to limit the time for inspection but should take the time required for the property in question' is, accordingly, very apt.

However, suspicions that the time required is not always being taken are rather justified by the poor reputation of surveyors in the public eye, findings of numerous Consumers' Association surveys and the high level of professional indemnity insurance premiums which have to be paid to obtain cover. Evidence of inadequate time being taken came from a covert operation conducted on behalf of the Nationwide Building Society in the late 1990s which revealed widely differing levels of inspection adopted by surveyors carrying out Homebuyer Survey and Valuation Reports. A random selection of ten surveyors, all qualified, were anonymously commissioned to prepare such reports on the same mid-terrace, late Victorian two storey house with a back addition, offshot or extension, again of two storeys but of lesser height than the main building, containing all told the equivalent of six rooms and with six visible defects estimated to cost in excess of £10,000 to put right. Of the 10 surveyors commissioned, three spent an average of one hour and seven minutes at the property, producing markedly the worst reports, while the remaining seven spent an average of two hours and twelve minutes. It was considered that only just 20% of the reports were adequate and that 60% at least failed to reach an adequate standard, one surveyor failing to spot any of the defects and five spotting no more than two of the six present at the property. The researcher considered that a minimum of two hours and thirty minutes was necessary to be at the property to identify all the defects and only one surveyor was at the property in excess of this amount of time.

Since the Homebuyer Survey and Valuation Report is said, publicly, by the RICS to be a 'mid-range' service between a mortgage valuation and a Building Survey, it is a bit alarming to speculate how long these same surveyors would have spent at the dwelling if they had only been instructed to carry out a valuation for mortgage purposes based on a 'limited inspection' as said publicly again by the RICS and the report on the inspection had been disclosed and sent to the prospective buyer.

Instead of accepting that the same 'standard', 'scope', 'degree', 'level' or 'intensity', whichever word is chosen to be used in this regard, applies for every inspection which a purchaser either commissions or is shown and for whatever type of report that is required, misconceptions continue to abound among surveyors. These stem from the time between around 1950 and 1980 when the expansion in home ownership really got under way. First local authorities and then Building Societies, who until then had only lent on new dwellings, began to grant mortgages on older dwellings. At that time, there were only two sorts of reports prepared by surveyors on dwellings. For mortgage purposes the Building Society commissioned an inspection and valuation for its eyes only, but collected the modest fee, usually a percentage on the agreed price, from the applicant. If the applicant received an offer of a loan he would, in most cases, proceed with the purchase without having a clue as to what the mortgage report might have said, apart from any sum which might have been retained until certain specified works had been carried out. If later, defects were found and the applicant complained to the Building Society he would be referred to the mortgage deed where he had agreed that he would keep the property in repair, or if pressed, offer to lend more money.

More prudent purchasers would commission a

structural survey without a valuation, but these were fairly few in number, but it was expected that floorboards would be taken up, roofs inspected from long ladders and drainage and electrical systems tested. Most sellers would be understanding and give permission for such an inspection and tests. In consequence because of the detailed work undertaken often taking some hours, there were comparatively few cases of surveyor's negligence in the Courts, apart from those already mentioned in regard to dealing with requests for 'quick look surveys' when it was found that even on a quick look inspection, the Courts expected surveyors to detect defects such as dry rot, dampness, woodworm and settlement.

The situation changed dramatically when, around 1980, one Building Society succumbed to pressure from the press and public but also probably to steal a march on its rivals and recklessly began to disclose mortgage valuation reports to purchasers, without a thought for the consequences to surveyors, and the other Societies rapidly followed suit. Applicants had argued that as they had had to pay for the report they were entitled to see it.

In the days before disclosure, the joke among the general public that the Building Society surveyor dashed through the property in a mere ten minutes or so was not only true in most cases, but was justified, in the residential field, by the fact that that was all that was required to provide a valuation and give a general description of the property, often amounting to little more than ticking boxes in answer to questions. All that mattered was that a property existed, was of conventional construction, not about to collapse and its value. That continues to be the case today where the lender is prepared to give a written undertaking that the report will not be disclosed to the buyer. Valuation being an art rather than a science, the Courts have

generally allowed a 10% margin to reflect possible differences of opinion. Thus, if the surveyor missed the slight signs of a defect, say, for example, one that would cost £9,000 to put right in a property valued at £100,000, the lender would find it well nigh impossible to prove the surveyor negligent, and probably wouldn't even bother to try. Whereas the same defect missed and if the report is disclosed to the borrower there would be a much stronger case against the surveyor. There is no margin for differences of opinion here, the surveyor either sees those slight signs of a defect or he fails to do so and, if that is the case, may have to pay up if found negligent.

Inspections for mortgage valuations have been increasing for the last 50 years in line with expanding home ownership and for over 20 years the results of those inspections have in many cases, been disclosed to borrowers. There is little evidence to show, however, that surveyors have changed their ways over the years to take account of disclosure and the consequences have been a considerable number of actions for negligence against surveyors. It might be thought that the Council of Mortgage Lenders and the RICS between them might have tried to do something more about it than merely making recommendations to prospective purchasers to have a 'proper' survey, which has so clearly been ineffective. The failure of both to do anything really positive has undoubtedly contributed to the general public's poor perception of surveyors and, probably, also to the pressure for the changes about to be introduced by the legislation of 2004.

Those early decisions on 'quick look' surveys should have alerted practitioners to the dangers of considering that on a 'less comprehensive inspection' you could look either less extensively, or in a shorter amount of time, for the signs of major defects than a surveyor would do on a structural, now a building survey. To discover the slight signs of what might be major defects, the level of inspection has to be the same. Limitation may apply to the Report, governing the amount of detail included, but not to the inspection.

To put matters into perspective, it is worthwhile considering the comments made by Judges since the disclosure of reports in cases involving mortgage valuations. In one case, *Smith v Bush* 1987, the Judge made it clear that the surveyor 'owes a duty of reasonable skill and care not to miss obvious defects which are observable by careful visual examination' going on to say that 'the fairly elementary degree of skill and care in observing, following up and reporting on such defects – surely it is work at the lower end of the surveyor's field of professional expertise'. Put this way, unflattering as it may sound, it is not asking too much. If surveyors often fail in this fairly elementary skill, it is hardly surprising the general public has little confidence in their abilities.

In another case, *Roberts v J Hampson & Son* 1988, the Judge said 'if there is a specific ground for suspicion and the trail of suspicion leads behind furniture or under carpets the surveyor must take reasonable steps to follow the trail until he has all the information which it is reasonable to have before making his valuation'. What could be considered reasonable in these circumstances? Irrespective of what it says in the Specification for the Valuation and Inspection of Residential Property for Mortgage Purposes about not moving furniture or lifting carpets, if the answer can be found by moving a small, comparatively lightweight, chest of drawers or lifting a loose carpet, it would be unreasonable, not to say laughable, to maintain that you could not find the answer because the Specification stated you did not move furniture or lift floor coverings. An obstructive

heavily loaded wardrobe, nailed down fitted carpets or laminate flooring are a different matter and could mean that the trail could not be followed and, therefore, further investigations would need to be recommended, as they would have to be in a Homebuyer Survey and Valuation or Building Survey Report. The same Judge also commented on surveyors having to take the rough with the smooth as to fees 'if you have to take two hours to do a mortgage valuation of a £19,000 house, so be it'.

Two other mortgage valuation cases, which went to the Court of Appeal, did not involve matters of principle but are included here to illustrate the degree of care required for this work, clearly not provided on these occasions. The first, *Sneesby v Goldings* 1994, involved the surveyor failing to spot the lack of support to a chimney breast which he could have seen if he had opened a cupboard door. The surveyor argued that a mortgage valuation involved only a limited inspection. This argument was not accepted by the Judge and the surveyor was found negligent. In the other case, *Ezekiel v McDade and others* 1994, the surveyor, viewing the roof space from the hatch, said he could not see a 40 mm gap at the end of a purlin. On this occasion the trial Judge went to see for himself and found that he could see the gap from the hatch opening. Finding the surveyor negligent, he awarded damages of £36,000 and for the distress caused £6,000. The appeal Court upheld the judgement but reduced the damages to £27,000 and for the distress, to £4,000.

Confusion was compounded by the introduction of the Homebuyer Survey and Valuation Report by the RICS as a copyright Standard Service in 1981 first as the House Buyers Report and Valuation, followed shortly after by the Flat Buyers Survey and Valuation. These early versions were brought out in something of a rush to meet a believed demand for some-thing between a Mortgage Valuation, the contents of which were by this time being disclosed, in many instances, to borrowers, and a Structural Survey (as Building Surveys were then called) which would be acceptable to both Lenders and prospective purchasers. It was trumpeted at the time as being attractive when 'time is short and economy important, involving a less comprehensive inspection and a concise report' coupled with a valuation, which it was pointed out did not automatically come with a structural survey.

Prior to the introduction of the RICS Service the Building Societies themselves had seen what they thought to be the advantage of a mid-range product having been heavily criticised earlier for not disclosing mortgage valuation reports to borrowers, who had been forced to pay for them, even though they might, or might not, have been having a structural survey carried out at about the same time. Some had begun to offer more detailed reports with a valuation as an alternative to a plain mortgage valuation on an abbreviated form. Following the introduction of the RICS Service more Building Societies began to offer prospective borrowers the three alternatives of Mortgage Valuation, Homebuyer Survey and Valuation or Structural Survey, although to satisfy Society requirements this latter would have had to be provided with a valuation as well, signed by an appropriately qualified individual to satisfy the requirements of the Building Societies Act 1986.

In most cases, these alternatives were offered with no or little explanation to a public that often failed to appreciate the differences. Continuing to prevail was the view that if the Lender came up with the offer of a loan, there could not be much wrong with the property. Most prospective purchasers, about 80%, therefore continued to plump for the cheapest, the mortgage valuation, while of those who listened to the advice to have something better, about 18% chose to have either the House or Flat Buyer Survey and Valuation while only 2% or so opted for a Structural Survey. In the case of the middle range option, it is stated that the use of the RICS Homebuyer Survey and Valuation Service has declined somewhat in recent years in the face of competition from similar services offered by groups of surveying practices producing their own versions with different features and on different terms. Structural Surveys in Scotland were said to be fairly rare because of the speed at which purchases of residential property normally take place. Either way, the limitations and conditions which originally applied to the RICS Homebuyer Survey and Valuation generally became overlooked in the rush to carry out the work in boom times, and because instructions often came second hand from lenders, not direct from borrowers, so that there was no opportunity for discussion or explanation. The fixed scale charges which originally applied also went under pressure from the Monopolies Commission, although charges quoted were and are still often influenced by purchase price.

The public were not the only ones confused by the differences in what was being offered. For the new mid-range product, many practitioners expressed doubts about the extent of the duty and what the reaction of the Courts would be when something went wrong. The early decisions mentioned at p34 on 'quick look' or 'general opinion' inspections should have yet again altered practitioners to the dangers of considering that on a 'less comprehensive inspection' you could look either less extensively or in a shorter space of time for the signs of major defects than one would do on a Structural Survey. Instead, it ought to have been considered that the words could only mean that you need not report in detail or comprehensively on comparatively minor matters observed on the inspection, such as split door panels, broken locks, corner cracking to glass, small holes in the plaster and the like. By adopting that practice the surveyor could save both note taking time on the site and space in the report without compromising the intention and still fulfil his duty. There was, however, no clear advice along these lines given to RICS members at the time of introduction although the Joint General Practice and Building Surveyors Division Working Party on Structural Survey Advice to the Profession on Residential Property said as much in July1984.

However, it was the Courts who again brought some semblance of reason out of the confusion which persisted through the 1980s. The Judge, in what is considered now the leading case on the subject, *Cross* v *David Martin & Mortimer* 1989, thought the conclusions of the Working party well founded *viz* 'We are convinced that the same level of expertise is required from the surveyor in carrying out a House Buyer Survey and Valuation as that for a Structural Survey'. Applying the conclusions to the facts of the case before him, he found the surveyor negligent in failing to relate the visual evidence of sloping floors and doors out of alignment, to the serious degree of settlement in a house on a sloping site and the undermining of stability in the roof by the alteration of roof trusses to enable the loft to be converted into a room.

In another case, *Howard* v *Horne & Sons* 1990, at about the same time, the Judge said the surveyor should have been able to overcome the difficulties of bad light and dust to see that on a board, where several electrical junction boxes were mounted, the wiring was a mixture of old and new and instead of saying 'electrical wiring is in PVC cable', he should

have said that it was mixed and to have recommended a test.

Quoting the Court's views on what the surveyor should do to fulfil the Conditions of his Engagement for both a mortgage valuation and a Homebuyer Survey and Valuation should dispel any doubts or confusion in respect of what is required of him as far as the inspection is concerned. It is to be expected that surveyors will accept different fees according to the type of report required, particularly if it is to include a valuation. That difference, however, should be reflected by the extent of the amount of work involved in the preparation of the report and the degree of skill employed in the analysis of the information gathered, whether by expressing an opinion on an observed defect or assessing a value from comparables.

Given the recklessness of lenders in disclosing reports when they were never intended to be seen by borrowers, it would have been better for the RICS to have taken a more robust line at the time or at least in the wake of the first negligence cases which subsequently came about after disclosure. It certainly would have been better, and would have been possible in 1990, to say that as reports were being shown to borrowers then fees needed to be adjusted upwards to allow for the preparation of a more detailed report, at least to the standard of the Homebuyer Survey and Valuation Service available at the time. If that had been done then the situation might have been recovered.

Should lenders have said, in response to the above suggestion, that borrowers would not bear increased fees, the answer to that, surely, is that lenders make their profits out of mortgages and therefore if they want the business it is not unreasonable for them to pay, if not all, then a proportion of the valuation fee. After all, most buyers do not need a separate valuation as, by the time they have got to the stage of applying for a loan, they are happy to pay either the asking price or, at least, somewhere near it. If a borrower decided he did not wish to have anything disclosed, then Lenders could content themselves with charging applicants an administration fee, as some do, to cover some part of their costs in obtaining a valuation in whatever way they considered necessary and as a means of avoiding time wasting applications. If the applicant opted for disclosure they could still have charged their borrower the administration fee towards the cost of the valuation element in the report, but also charge the borrower for the more detailed aspects in the report on the property itself. The difficulties for surveyors over inspections and reports for mortgage valuations will, of course, be resolved by the compulsory requirement for sellers from January 2007, to provide a Home Information Pack with its Home Condition Report which can be relied upon by lenders as well as buyers. The lender will be provided with a copy and it will be up to the lender how the valuation is obtained. If an 'inspection' is required, 'driving past' may be sufficient. Alternatively, accessing available information on local comparables, the 'automated valuation model' (AVM), coupled with an examination of the contents of the Home Information Pack may be employed.

The surveyor will, of course, prepare his valuation after he has carried out his inspection and the aspects to be considered for valuation purposes will be discussed in the next book in the series *Reporting for Buyers*. Nevertheless, unless it is made apparent by an express statement in the Report that he has not done so, the surveyor is required by the Specification for the Valuation and Inspection of Residential Property for Mortgage Purposes to make a considerable number of assumptions, an

incredible 31 are itemised about the security being offered for mortgage purposes, though he is under no duty to verify them. There is little point, however, in returning to the office to prepare both valuation and report without having given thought to those assumptions during the inspection itself. The surveyor must consider from the evidence on site whether it is safe to make those assumptions.

Unless any indications are found to the contrary, the necessary and basic assumptions to be made by the surveyor can be summarised as follows: that vacant possession is provided, all planning and statutory approvals have been obtained, no deleterious or hazardous materials or techniques have been used and there is no contamination, there has been no landfill on the site, there are no onerous restrictions, encumbrances or outgoings and good title can be shown, the value would not be affected by any matter found by local searches, its use is not unlawful, nothing found by further survey or in hidden parts would affect value, the services are connected and normal terms apply to their use, sewers, services and roads giving access are adopted and any lease provides rights of access and egress over estate roads, paths, corridors, stairways and to use communal grounds and parking areas, the construction of new property will be satisfactorily completed by a builder registered with the National House Building Council or Zurich Municipal Mutual, or an equivalent organisation, so as to obtain its indemnity cover.

For flats and maisonettes, it is to be assumed, unless there are contrary indications, that repair costs to buildings and grounds are shared equitably between all, suitable covenants are enforceable against other lease-holders, or through the landlord, and upon free or feu holder, no onerous liabilities are outstanding, over the next five years no substantial defects will require expenditure over the current amount of annual service charge equivalent to 10% or more of the market value.

For other leasehold property, it is to be assumed that there is an unexpired term of 70 years and there are no moves afoot to acquire the freehold or extend the lease, there are no onerous covenants, the lease cannot be determined other than for serious breach, terms and conditions are in the same form for all separate freeholders, head or sub-head leaseholders and mutually enforceable, there are no breaches or disputes, all other leases are materially the same, the ground rent is fixed and payable throughout the term, where there are more than six dwellings, the freeholder manages directly or through a professional, properly bonded agent, where intermixed with commercial property, there will be no significant change of use, where there are a number of blocks terms only apply to the subject block, other than for common roads, paths, communal grounds and services, where ownership is split rights and reservations are reserved, no unusual restrictions on assignment or subletting for residential purposes, there are no outstanding claims or litigation on the lease, there is adequate provision for continued use and maintenance of facilities without exceptional restriction or extra charge.

As to insurance on leasehold residential property, it is to be assumed that the property will be insured under all risks cover, to include subsidence, landslip and heave for current reinstatement cost on normal terms, there are no outstanding claims or disputes, leases make provision for mutual enforceability of insurance and repairing obligations, where separate insurances are allowed, there is an obligation for the landlord to rebuild as may be altered to comply with current regulations.

A reduced version of these assumptions which the surveyor has to take into account on a valuation appears in the published description of the Home Buyer Service and some of those referring to leasehold properties in the appendix to the Report. These are accompanied with the advice that details of the full range are available from the surveyor. Why the reduced versions are included is puzzling when all surveyors are required to comply with the RICS Appraisal and Valuation Manual as a matter of course and where they appear in full. If it was thought necessary, and there is considerable doubt as to whether it is, to include in a report an explanation of what the surveyor does, even if only in part, to arrive at his opinion on value then copies could be taken from the Manual for enclosure with any report which included a valuation.

The compilers of the Requirements for Home Inspectors give the impression of being well aware of the consequences of rushed work as evidenced by the remarks made in the Court cases discussed in these paragraphs. They indicate that while 'there are no mandatory or set timescales governing the production of Home Condition Reports' decisions must be based on 'sufficient information'. 'Home Inspectors must set aside sufficient time for the inspection' and the compilers provide 'a reasonable guideline for a typical house (eg semi-detached three bedrooms from 1960s)' at 'around 1.5 to 2 hours'. 'Following the trail' is stressed at various points throughout. Furthermore there is a 'risk of damage to the reputation of Home Inspectors which may be caused by incomplete or hastily prepared Home Condition Reports produced under competitive timescales that do not allow for proper professional reflection'. What issues should be considered during that period of reflective thought are given a fair amount of attention.

Of course, whether there is an improvement

in the standard of work produced by the surveyors providing Home Condition and Single Survey Reports for sellers over the standards of the past when providing reports for lenders and buyers will depend very much on how much notice they take of the Code of Practice, the National Occupational Standards and the Requirements, the three documents produced for Home Inspectors governing the work of producing Home Condition Reports in England and Wales. It may be that an improvement in the standards of the past might have been achieved if the RICS had produced something similar to the comparatively clear and straightforwardly expressed Requirements for Home Inspectors a long time ago, instead of the highly confusing Practice Notes for the production of Homebuyer Survey and Valuation Reports and which were extant right up to the second edition produced at the beginning of 2005.

Liability

When an individual puts himself forward and agrees to carry out an inspection of a residential property and to provide a report on its condition, he takes on a responsibility and incurs liabilities in two ways. First, to the person with whom the contract to do the work is made, as expressed in its terms, however they may be formulated, either in writing as Conditions of Engagement, or by word of mouth. Second, and quite irrespective of the contract, to other persons who may have been expected to rely upon the work should it have been reasonable to anticipate that they would do so.

If by reason of a failure to exercise reasonable skill and care, damage is suffered, either physical or financial, a claim may arise by way of breach of contract in the law of Contract

of for breach of duty in the law of Tort in England and Wales or the law of Delict in Scotland, negligence in other words, against the individual carrying out the inspection.

The law covering negligence applies to all individuals who exercise their skills and do work or give advice and it is irrelevant whether they are qualified or not or whether they are experienced in the particular field. To take an extreme example, if an individual, totally unqualified and without any experience of inspecting residential property, undertakes the work he takes on the same responsibility as others, who have both qualifications and experience. Going further, if he takes on the work for a low charge to undercut others, or even for no charge at all, perhaps for a friend, his responsibility again remains the same.

For a claim to succeed, however, in contract, tort or delict, the recipient of the report has to have acted on its contents and suffered loss. If the individual carrying out the inspection fails to put into words what he intended to say, that is no defence. The reasonable skill and care requirement extends to how the results of the inspection are expressed in words of advice to the client. It is this aspect that often causes as much difficulty to individuals carrying out inspections as the inspection itself, if not more so. Should a subsequent meeting be necessary to clear up any misconception or misunderstanding in the client's mind, it is most important that the outcome of the meeting is confirmed by letter, with an acknowledgment by the client that he has received it.

It will be noted that reasonable skill and care has to be exercised, not that which an expert might employ in similar circumstances. Indeed plaintiffs who succeeded in a lower court in obtaining a judgement for damages on the evidence of an expert, not of a surveyor but that of a structural engineer, found the decision overturned in the Court of Appeal, *Sansom v Metcalfe Hambleton & Co* 1997. The individual carrying out the work has to exercise the reasonable skill and care of others carrying out the same sort of work at the same time, not some later time. In other words, hindsight does not come into it. This aspect was argued by Lord Justice Brooke in a dissenting judgement in the Court of Appeal case of *Izzard and Izzard v Field Palmer* in 1999. Although he agreed with his colleagues in finding a surveyor carrying out an inspection for mortgage purposes negligent in failing to provide warnings about future liability for charges relating to a flat in a non-traditional block, he disagreed with them on the amount of damages awarded. Other valuers at about the same time had apparently taken a not too dissimilar view and he considered comparison with those was the proper measure rather than with a figure produced with hindsight.

The responsibilities undertaken do not subsist for ever. If actions are to be brought arising out of reports, then sellers or purchasers in England and Wales have to do so in contract within six years of the date of the report to comply with s5 of the Limitation Act 1980, three years if a claim involves personal injury or death or within 12 years if the contract is under seal, very rare in this type of work. In tort, the 1980 Limitation Act in s2 prescribes a different time scale, in that the six or three year periods run from the date when the claimant acted in reliance on the report by entering into a contract to purchase or sell. This action is considered to be when the damage is suffered, not as might be thought when defects are found in the property. To overcome this anomaly for purchasers who did not know they might have a cause of action until defects were found, the Latent Damage Act was passed in 1986. Under this Act actions, which can only be brought in tort

not contract, have to be brought within a long-stop period of 15 years from the date of the report, but within that period, six years from the discovery of defects. There is provision for a further three years to be added to the six from a 'starting date', which is defined as the earliest date on which the claimant had the knowledge required to bring the action and the right to do so.

There was no need for an Act to cover latent damage in Scotland because the anomaly had already been dealt with by way of the two periods for bringing actions laid down in the Prescription and Limitation of Actions (Scotland) Act of 1973. This provides for a five year limitation period, three years if involving personal injury or death, from the time when the claimant could with reasonable diligence have become aware of the damage and a long stop period of 20 years within which all claims must be brought.

As mentioned previously, in brief, when defects costing considerable sums of money to put right are not detected and a claim against the surveyor is successful the award of damages by the Court is often a source of annoyance to the claimant, to put it mildly. Often it is at a sum far below the cost of the repairs, because the law only provides damages to cover the diminution in value of the property. That figure is the difference in value between the property as reported on and its value as it should have been reported if the defects had been known about at the time of the inspection. In the case which established this principle, *Philips v Ward* 1956, the claimant paid £25,000 for a house in 1952 having taken advice. If the advice had been correct he would have paid £21,000. In consequence he was awarded £4,000 although the repairs would have cost £7,000, as estimated at the time of purchase. The view of the Courts is that the claimant should not be put in a better position than he would have been if there had been no breach of contract or negligence ie compensation for the assessed loss, not punishment of the offending individual, which would have been the case if the cost of repairs had been awarded. If that higher sum had been awarded the claimant would have been in possession of a house worth more than the £25,000 he paid for it.

There is a certain element of logic in the Court's way of awarding damages to successful claimants. On the other hand, it is small consolation to hard-pressed claimants without the necessary financial resources to cope. The response to criticism that owners can just sell on to someone else is somewhat facile, since the problem giving rise to the sale will be known about and will have a detrimental effect on the price. Although the Courts have tended to lean at times towards figures more nearly matching the claimants actual costs, there has been much adverse comment. The Consumers' Association over many years has commented unfavourably on the situation and cited one exceptional case arising out of a negligent Building Survey where £5,000 damages were awarded against a £12,000 bill for repairs. The problem was compounded when the surveyor became bankrupt and didn't even pay the damages.

Because of the unsatisfactory outcome of some claims in the Courts and the seeming unlikelihood of any change being made to the principle, since they apply to all aspects and categories of negligence claims, the Consumer's Association investigated various aspects of warranty insurance schemes which it was believed could be tied into seller's surveys and were said to be available in other countries. This idea has been followed up and RICS members who also qualify for the Surveyors and Valuers Accreditation Scheme, now offer hidden defects insurance to back up their reports. This operates to provide the purchaser who after purchase, having relied upon the report, finds defects which were hidden, with the means to claim on insurance for the cost of repairs. It is said that the establishment of defects will be sufficient to invoke the insurance without reference to the contents of the report. This does, however, seem to overlook the fact that the words used are 'hidden defects' and if it turns out that what is being paid out is not for 'hidden defects' but defects which were ascertainable by a more careful visual inspection then either the scheme will be in need of rethinking or, alternatively, premiums will rocket.

Section 3 Roofs

Pitched Roof Structures

The Inspection

As discussed previously the surveyor, on arriving at the property, should extend his initial reconnaissance of the dwelling's surrounding area and its front elevation to a quick but very vital examination of the whole of both inside and outside. From this he can obtain a general feel as to layout, construction and, perhaps, see some of the more obvious flaws, in particular any in connection with the roof.

On his reconnaissance of the inside the surveyor will see whether there are any damp stains to top floor ceilings and whether there is a trapdoor to space below any pitched roof coverings. It is a requirement of the Conditions of Engagement for all types of residential survey, with one exception, that where there is access to a roof space inside the dwelling, and even outside from the common parts in the case of a flat in Scotland, within 3m of the floor below, it should be entered and the interior inspected. The exception is that on an inspection for a mortgage valuation the Model Conditions state that the inspection need only be by 'head and shoulders' from the trapdoor without entering. Whether the inspection of the roof space should be limited in this way is a matter of judgment for each individual surveyor, irrespective of any conflicting advice or instructions, since he is personally responsible and signs the report under the terms of the Building Societies Act 1986 where applicable. Even with the most powerful of torches it is often difficult to see the character of construction and detail of defects from the trapdoor alone without entering and inspecting. How, for example, would the surveyor know whether the dwelling was not one of the 150,000 or so steel framed houses many of which, from the outside and the rooms inside,

look exactly like a conventional house. These crop up all over the country and were built in both the private and public sectors from the 1920s onwards. That they are steel framed can only be ascertained from a close inspection in the roof space. With terrace or semi-detached dwellings presenting the appearance of brick, often the surveyor's first indication of modern timber framed construction will be the non-traditional appearance of the party, or separating, walls seen by entering the roof space Furthermore a situation where a Judge, hearing a claim in connection with a mortgage valuation said, following his own inspection, that he could see a defect but the surveyor maintained he could not is to be avoided, see *Ezekiel* v *McDade* 1994 on p40.

If there is access within range, the sensible advice must be to inspect by entering in all cases. If there is no access or any trapdoor provided is unopenable, then that amounts to a serious limitation on the inspection to which attention needs to be drawn and, if necessary, elaborated on as to the possible consequences of basic structural problems let alone disrepair leading to leaks and other defects such as aged electrical wiring and defective plumbing.

In this and subsequent sections dealing with the detailed inspection, the features of typical properties of different eras will be discussed with a view to describing defects which tend to be characteristic of the age of construction. The points to look for and note will be brought to the fore so that a defect can be described, its cause perhaps given if visible, or surmised if not, and what subsequent action should be taken. Where possible a defect will be specifically rated where this is a requirement of a particular form of report.

Access to the roof space is vital to the surveyor

as it provides him with the opportunity to determine how the roof covering is supported. Firm support is necessary otherwise the covering can be dislodged from its pre-determined position and this could interfere with its proper performance. While most of the coverings for pitched roofs will tolerate and accommodate a small amount of movement before leaking, undue sagging in the support or sideways displacement of timbers causing unevenness of the covering in either direction can lead to intractable problems of water penetration, particularly when rainfall is accompanied by high winds. In most properties built before the 1960s the framing of the roof structure would in turn often require support not only from the external walls but from internal load bearing partitions as well. With access for inspection all these aspects can be checked if there are any doubts about the effectiveness of both covering and structure.

Pitched Roof Structures to Post 1960 Properties

Bolted and Connector Roof Trusses

As over a third of all residential accommodation has been built since the early 1960s, the revolution in roof design and construction which took place around that time will, in the vast majority of cases, be reflected in what is found on site when properties built since then are inspected. Until that time most domestic roof structures were formed on site by carpenters cutting and fitting the timbers together to designs prepared on the basis of the required span, what structural support was available from below and the desired roof covering but, primarily, on the general use of economically sized timbers, almost invariably 100mm by 50mm.

A partial departure from those earlier procedures had come about in the ten years or so before the 1960s with the development of trusses based on bolted joints using double sided toothed plate connectors. Prefabricated, hoisted and spaced at 1.8m centres, site work would be reduced to fixing purlins and ridge plate and cutting and fitting intermediate 100mm x 50mm common rafters and ceiling joists at 450mm centres, as shown on Figure 1.

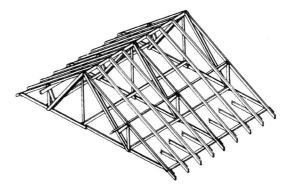

Figure 1: Typical bolted and connector roof truss with purlins. Developed in the 1950s and spaced at 1.8m centres, they first freed designers of roofs of the need to rely on intermediate support from structural partitions in small dwellings.

The use of bolted and connector roof trusses still provided a reasonable amount of storage space between the trusses, a saving in time formerly spent on site, a saving in timber used but primarily the freedom to plan space in the building below unrestricted by the need to include any structural partitions. Depending on the design, sometimes component members would merely consist of two common rafters bolted together and, for some, plywood gussets would be used instead of toothed plate connectors. Their use continues and clear spans of 12m are possible with pitches of between 15 to 45 degrees. Standard details for forming hips are available for some designs.

Trussed Rafter Roofs

In the early 1960s, with the aim of saving even more timber and yet more time on site by further prefabrication, the idea behind the truss with bolted connections was developed so that in effect each 'couple' of rafters could become in themselves a separate truss, a 'trussed rafter'. These were installed generally at 400mm centres or at the very maximum of 600mm centres. Rafters, ceiling joists and the struts together with any other web members, became generally thinner at 35mm of stress graded timber connected together in a factory with thin punched toothed metal plate fasteners of galvanised, or more rarely, stainless steel. Again the alternative of nailed or glued plywood gussets could be used.

Various shapes were developed by manufacturers and British Standard BS5268 Part 3 Code of Practice for Trussed Rafter Roofs, first issued in 1985 and amended in 1988, covers their design, manufacture and installation. It also includes span tables for the 'fink' and monopitch trusses shown in Figure 2, enabling these to be used under specific conditions without further testing although it would still be necessary for the supplier to certify fabrication in accordance with the Standard. The maximum span tabulated for the fink and monopitch trusses shown is 11m when 35mm thick timbers are used, all based on a standard loading, although a maximum

span of 12m is possible when 47mm thick sections are employed.

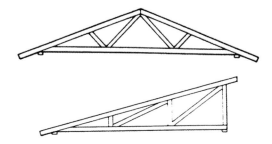

Figure 2: Common types of trussed rafter; fink and monopitch. Generally spaced at 400mm centres, trussed rafters came complete with single section ceiling joists and web members, joined where they meet by steel toothed plate connectors, avoiding the need for much further site work once hoisted into position

Some other available shapes are shown at Figure 3, including an asymmetrical design, but a manufacturer would have to supply a certificate for these to show that their design had satisfied the Code's requirements for structural adequacy, either by engineering calculation or load testing as laid down in the British Standard.

All trusses that satisfy the Standard's requirements should be marked clearly to show the name of the manufacturer, the species and strength class of the timber used and confirmation that design and fabrication

Figure 3: Other trussed rafter designs.

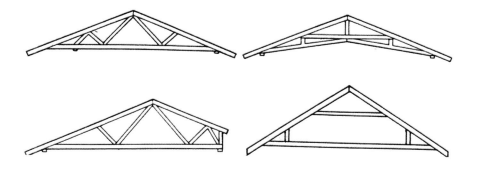

has been carried out in accordance with Part 3 of the Standard. Additionally there may be a TRADA (Timber Research and Development Association) marking if the manufacturer is a member of that Association's quality assurance scheme.

Part 3 of the Code with its span tables provides a useful measure against which existing trussed rafter installations can be checked, particularly those unmarked and the many installed before the Code was published. This checking can be extended to include how the trusses have been arranged and adapted to form openings for access hatches and chimneys and for the support to any cold water storage cistern. Recommended methods for all these adaptations are included in the Code.

Considered individually, trussed rafters are very flimsy and should receive considerable care in storage, handling and loading at the factory and unloading and storage at the site.

Fixing of trusses is normally to wall plates, but initially these need to be securely held down to the inner face of the external walls to prevent subsequent uplift by strong winds. Hoisting and fixing into position on to the wall plates, by either proprietary clips or skew nailing needs further care to avoid damage to the truss joints by distortion. However, when set up and provided with the essential bracing, both diagonally and longitudinally, as shown in Figure 4, in accordance with the current British Standard they provide a firm base for the roof covering.

Additional chevron bracing is required for duopitch spans over 8m and monopitch spans over 5m unless a sarking underlay of boarding, plywood or chipboard is used when it can be omitted. In particular trussed rafters suited the

Figure 4: Standard bracing for duopitch trussed rafters (Chevron bracing, required on spans over 8m, omitted)

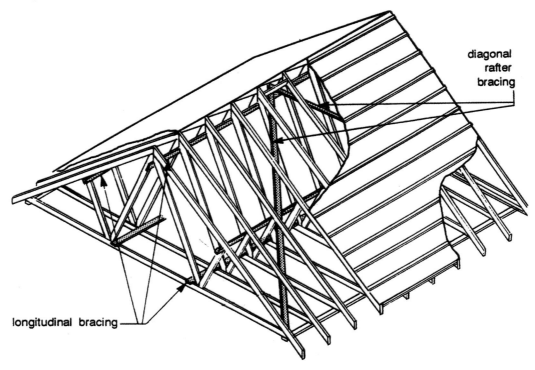

diagonal rafter bracing

longitudinal bracing

2: Typical shallow pitch roof of the 1960s and 1970s

3 and 4: The availability of different shapes of trussed rafters encouraged the use of monopitch and even butterfly roofs in the 1960s and the 1970s.

5: Some idea of the difficulty of inspecting trussed rafter roofs from the 1960s onwards with their pressed steel toothed connectors can be gained from this example. This one has a comparatively steep pitch but many, particularly in the early days, were provided with pitches as low as 20° to 30° and some even lower.

6: Instructions to carry out surveys on newly con-structed dwellings are rare, in view of most purchaser's reliance, sometimes unwisely, on NHBC or other guarantees. If forthcoming on a dwelling forming part of an estate, the possibility of inspecting other incomplete structures, as here, may present the opportunity to assess the quality of details which may, subsequently, be hidden from view, such as fixings of and to wall plates and the presence and fixing of bracing straps and connectors.

then current fashion for clean lined, minimum interest roof shapes of low pitch to utilise the availability of new designs of interlocking tiles made of concrete. The almost universal inclusion of central heating in new dwellings often meant that whole roofs could be covered without cutting a single tile but utilisation of space within the roof became virtually impossible because of the close spacing of timbers and the shallow pitch favoured at the time.

If on occasions, fairly rare in the 1960s and 1970s, a hipped end or pediment with valley gutters was incorporated, these tended to be formed either with truncated trusses or with a girder truss and cut members or, for a pediment, by diminishing trusses fixed on top of the main run as shown on Figures 5 and 6. In either case their use tended to be solely for cosmetic purposes in the prevailing shallow pitch roofs of the day, perhaps when some individuality was required in the private sector.

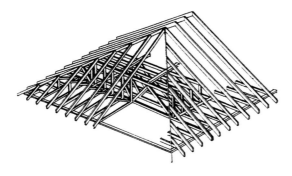

Figure 5: Typical hip construction in trussed rafter roofs using a girder truss.

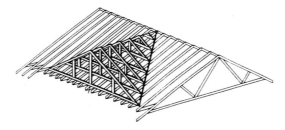

Figure 6: Valley set of diminishing trussed rafters to form gable. These would be used to purely improve appearance by introducing variety of shape to roofs of the 1960s and 1970s.

Later, in the 1980s, when the provision of housing in the public sector was changed from local authorities to housing associations and the styles and fashions of the 1960s and 1970s became anathema in both public and private sectors, hips and featured gables took on much greater importance. Roof pitches increased, often to accommodate rooms in the roof with either roof lights or dormer windows and in an endeavour to provide more interesting and varied configurations as well as increased utilisation of available space.

Attic Trussed Rafter Roofs

A type of trussed rafter design found suitable, when further developed, for the purpose of accommodating rooms in the roof space is shown at Figure 7 and came into use at this time .It is generally known as an 'attic' trussed rafter. Because of the increased height required, transportation becomes more difficult and the type of single unit shown is normally limited to a height of 4.5m. Intermediate support for the floor tie is normally required when spans exceed 9m and the use of single unit attic trussed rafters becomes uneconomic when spans exceed 10m. However, it is possible to design and manufacture larger trusses in sections and for them to be assembled on site, in strict accordance with the designer's instructions, although, of course, this does involve much more site work.

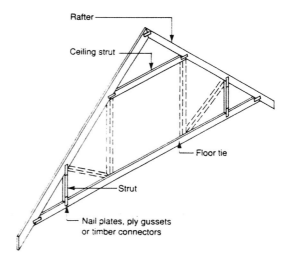

Figure 7: Attic trussed rafter. Introduced in the 1980s to provide rooms in the roof space.

The maximum spacing for attic trussed rafters is 600mm but it is possible to form structural openings for dormer windows or roof lights up to three times the spacing adopted. It is, however necessary for staircases to be

7: *In the 1980s roof pitches became steeper to provide accommodation in the roof space either with dormer windows or with gables or roof lights as here on this block of terrace houses.*

arranged to run parallel to the attic trussed rafters to minimise the number of rafters with cut floor ties. Once in place bracing is

Figure 8: (below)Typical hipped roof construction with lightly constructed attic trussed rafters at a maximum spacing of 600mm.

required in accordance with the British Standard as for other trussed rafters. One way to form a hipped slope with this type of roof is to provide a girder truss as shown on Figure 8 but other ways are possible depending on span and pitch.

Attic Trussed Roofs

More freedom in the location and size of dormer windows and staircases can be obtained by the use of more strongly constructed attic trusses. Whereas the lighter attic trussed rafters are spaced at 600mm centres, attic trusses can be spaced at 3.6m centres with only the common rafters needing to be cut and fitted to purlins at the narrower spacing of 600mm. A typical layout where the trusses are spaced at 2.4m is shown at Figure 9. Account needs to be taken of the heavier

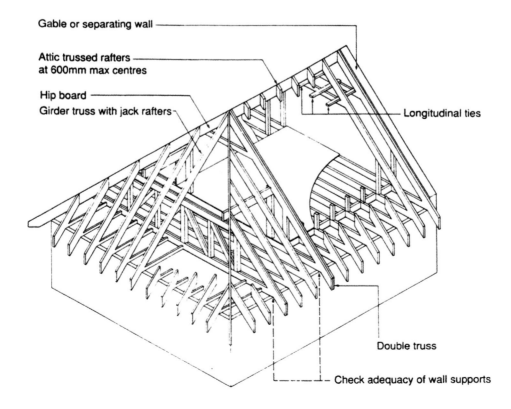

Gable or separating wall

Attic trussed rafters at 600mm max centres

Hip board

Girder truss with jack rafters

Longitudinal ties

Double truss

Check adequacy of wall supports

point loads involved with the use of attic trusses in the design of the supporting walls.

Sheathed Rafter Panel Roofs

Where domestic buildings are still preferred to have simple continuous ridges and separating walls or gable ends, a different form of construction can be used which eliminates all struts, ties or bracing and provides a completely clear triangular roof space. This comprises a structure of sheathed rafter panels as shown on Figure 10. The sheathing can be of any number of different materials such as plywood, chipboard, tempered hardboard or bitumen impregnated softboard, among others, and can be placed on top of

the rafters as in the traditional manner of boarded sarking in Scotland, or below.

The panels can be factory made as in Figure 11 to include all the necessary noggings, insulation panels and vapour barrier and be either hoisted or craned into position. Spans up to 10.5m at roof pitches of 35° to 45° are possible with this method of construction but it is vital for the floor decking to be taken to connect with a base plate at the outer edge and to be securely nailed so that it acts to restrain roof spread.

Figure 9: Typical roof construction with attic trusses at 2.4m centres. Being stronger these provide greater freedom for the arrangement of dormer windows and the staircase, but do require more site work to provide the cut and fitted common rafters at 600m centres.

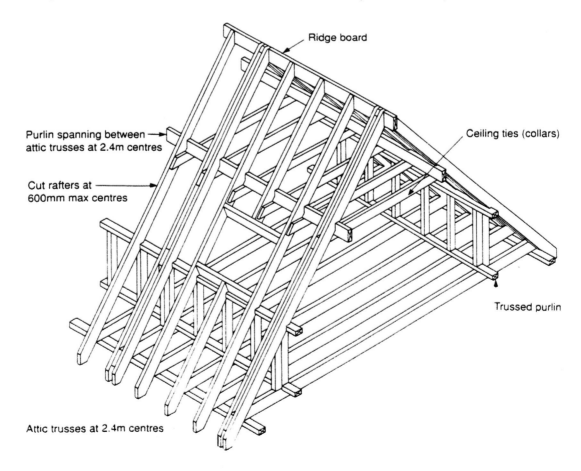

Ridge board

Purlin spanning between attic trusses at 2.4m centres

Cut rafters at 600mm max centres

Ceiling ties (collars)

Trussed purlin

Attic trusses at 2.4m centres

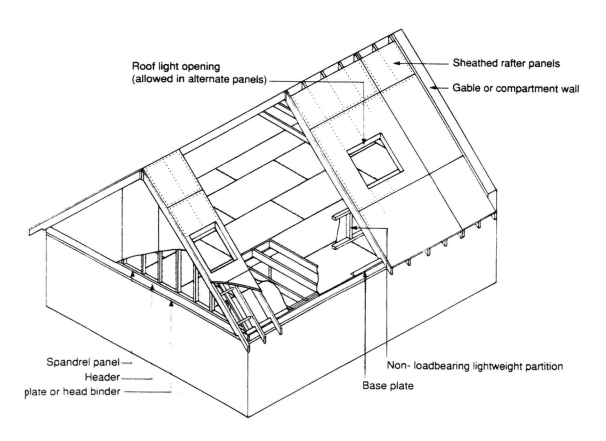

Roof light opening
(allowed in alternate panels)

Sheathed rafter panels

Gable or compartment wall

Spandrel panel
Header
plate or head binder

Non- loadbearing lightweight partition

Base plate

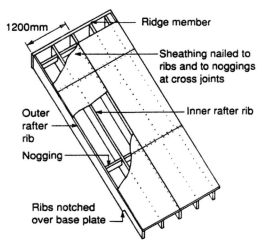

1200mm

Ridge member

Sheathing nailed to
ribs and to noggings
at cross joints

Outer
rafter
rib

Inner rafter rib

Nogging

Ribs notched
over base plate

Figure 10: (above) Typical sheathed roof panel construction. Much strength is gained when rafters are joined together in a panel with a sheathing of plywood or the like. Provided care is taken with jointing to a similarly constructed floor and to walls a sturdy roof can be produced.

Figure 11: (left) Typical sheathed roof rafter panel, factory produced and installed on site.

Roof lights in the larger panels can be formed with this type of roof construction but an imperforate panel needs to be located on either side. Panels with dormer windows are also possible, framed up and using short rafter panels for the roofs of the dormers.

Faults in Post 1960 Pitched Roof Structures

The requirements for design, fabrication, marking and certification of BS5268 mean that trussed rafters, provided they have been correctly selected according to span, pitch and the loading required and have been carefully transported by the fabricator, arrive at the site suitable for their purpose. It is their subsequent treatment at the site which invariably causes any problems that arise. Storage on site, hoisting into position, installation, bracing and packing require care and are just those operations which are easy to bodge and skimp, compounded, perhaps, by less than sound adaptations for service runs, hatchways and possibly the installation of a cold water storage cistern. When the roof structure is insulated, enclosed with sarking and the roof covering, the plethora of light timbers and the often shallow pitch make inspection of the roof space in post 1960s dwellings more than usually difficult. More difficult than in the case of pre-1960 dwellings with their cut and fitted roofs, and the incidence of reported defects is far greater making inspection all the more important.

Disruption to the lie of the roof covering and perhaps damage to top floor ceilings observed on the reconnaissance may well be indications of defects in the structure of the roof and it may be possible on inspecting the roof space to see the cause. This will involve a check on the visible forms of construction This check is an essential requirement for any type of residential inspection as the surveyor needs to be on the look out not only for actual defects but also for faults likely to cause defects within a reasonably short time span, say about ten years.

Installation defects in the roof structure which could cause displacement of the roof covering and which can usually be seen from within

the roof space include a total lack of, or incomplete or inadequately fixed, diagonal and longitudinal bracing. Bracing of this nature prevents racking or lateral movement of trusses which is unlikely to be prevented if reliance is placed on the battens alone which was often the case with the installation of the earliest of trusses.

The bracing as illustrated on Figure 4 is necessary in all cases except where a form of boarded sarking has been used and this is likely to be found only in Scotland where it is traditional, rather than in the South.

Also possibly omitted, either totally or in part or inadequately fixed, even if provided, could be the straps, packing and noggings tying the trussed rafters to separating walls between terraced dwellings or to gable walls. In the case of tying to a gable wall, besides the element of restraint provided to the roof structure as there is in tying to a separating wall, support against wind pressure is provided to the gable itself by transferring that pressure to the front and rear walls. Furthermore instances of gable walls being sucked out by high winds are by no means unknown when such straps, packings and noggings have been omitted. To be effective, straps, packing and noggings need to connect three rafters to the wall in question with adequate screw or nail fixing, not less than three at each point of connection, as against the one or two often found.

Without both bracing and the straps mentioned above, cases have been found where all the trussed rafters were leaning over to such an extent that the last one was resting against either a separating or gable wall. Omissions of both, or either bracing or straps, requires urgent attention in view of the danger of collapse at any time, not just when there are exceptionally high winds blowing. To obtain a design for a repair that will be both effective and practical to install, reference to an engineer is essential.

8: BRE has reported problems with monopitch roof trusses exerting pressure at the top of the upright section where there is differential movement between this and the adjacent wall. There are signs of movement with this truss which needs to be investigated.

9: The increased requirements for insulation coupled with the provision of roofing felt to the underside of the covering have rendered the timbers of roof structures far more liable to attacks by wet rot and sometimes even dry rot should ventilation levels be too low. In the upgrading of this roof substantial provision for ventilation has been made in the soffit of the generously projecting eaves.

Perfection is, of course, too much to expect from installation in all cases and a certain amount of lean is considered permissible. The permitted amount is 10mm for every 1m in height from ceiling tie to apex. Trussed rafters are also permitted to oversail the wall plate to a maximum of 50mm although 50% of the ceiling tie to rafter connecting plate should overlap the wall plate. Any bow on the rafter itself should not exceed 10mm at its maximum.

A combination of internal and external inspection should establish whether any deviations are within these permitted tolerances. Out of plumb in excess of the permitted amount will also require reference to a structural engineer although if limited to under twice the permitted amount is probably correctable on site. Above this amount, rebuilding is more likely to be required.

If storage facilities on site are inadequate and/or installation is not carefully carried out, the connections of trussed rafters can become loosened and trusses distorted rendering them weaker and less able to carry the applied load. If trusses are cut and altered without taking proper advice, the same effect can be produced. It is not unknown for the pressed toothed fasteners to be prized out, timbers cut and altered and the fasteners nailed back on. Any signs of such nailing needs to be viewed with suspicion and a careful check made for any consequential distortion.

All pitched roofs constructed in the last 30 years or so have been required to be insulated to minimise heat loss from the habitable rooms to conserve energy. This adds to the difficulties of inspection. If there are no habitable rooms in the roof space, the insulation is usually placed at top floor ceiling level This leaves a 'cold' roof space which needs ventilation to prevent condensation.

Both insulation and ventilation levels are prescribed in the current Building Regulations. These provide a standard against which the levels of both can be gauged in existing dwellings. Over the years the requirements have been raised and what is found now may be considered woefully inadequate. In particular, there was a time when a concentration on insulation tended to leave roof spaces inadequately ventilated with the resultant high risk of condensation and possible corrosion damage to punched toothed plate metal fasteners. This could be a developing fault and therefore something which must be closely checked in every case along with the possibility of rot in the timbers from this cause.

It is also particularly important to check metal fasteners for signs of corrosion in areas where the Building Regulations require the treatment of all roof timbers against the ravages of the House Longhorn beetle. Use of the wrong type of preservative can cause corrosion in itself. Using the wrong method even with the correct type, if it involves re-wetting the timber and such application is not carefully controlled, can cause distortion of the timbers and this may result in disruption of the toothed metal plate fastener joints.

Trussed rafters are normally fixed on installation to previously secured wall plates with clips or by skew nailing from both sides. By the time the insulation is in place the connections are virtually impossible to see and consequently an opportunity is provided for skimping at this stage of construction. It is probably, however, only when there is no other reason ascertainable for a visible defect that a recommendation for exposure of this detail might, of necessity, be needed.

Being an engineered product, cutting any component member of a trussed rafter, other than strictly in accordance with the manufacturer's specification and drawings, is hazardous. Yet cutting and adapting to form hatchways, or for chimney stacks or the installation of services, is one of the most frequent causes of problems in near new construction. If the dwelling is without habitable rooms in the roof space, it is probably possible to diagnose the cause of any such problems. However, if rooms are present and defects are visible from the exterior, it is unlikely to be possible and the report may

need to advise opening up in the area of the defect to ascertain what is wrong.

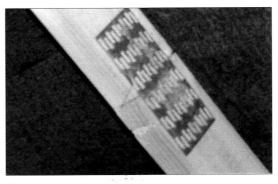

10: It is not unknown for the pressed metal toothed fasteners for trussed rafters incorporated by the fabricator, as shown here, to be prized off, timbers cut and adapted on site to suit changed circumstances and for the fasteners to be nailed back on. A careful check for distortion needs to be made if this is seen to have been carried out.

If there is a cold water storage cistern in a trussed rafter roof access is vitally necessary for any adjustments or replacements necessary to the float or ball valve. If there are signs of distortion in truss members it would be essential to check the position of the bearers carrying the cistern platform. To be effectively supported by the roof structure these need to be as near as possible to the node points (i.e. where the sloping or vertical truss members connect to the ceiling ties) and taken over a sufficient number of trusses to avoid distortion.

Pitched Roof Structures to Pre 1960 Dwellings

Dwellings built before the 1960s, the remaining two thirds of the total stock, will have some form of traditional roof structure, as will those built later not provided with trussed rafter roofs. Apart from the comparatively few with flat roofs, these will be pitched roofs based, in the main, on 100mm x 50mm common rafters, considered over many many years to be the most economic size of softwood timber for the purpose. These would be cut and fitted on site.

Couple Roofs

A 100mm x 50mm common rafter can support, on the slope, a load of typical roof covering comprising slates or tiles, sarking and battens over a length of 2.4m. Joined together at the top they could span between supporting walls 3.5m to 4.5m apart, depending on the pitch adopted. Such a design, a 'couple roof', would, however, be severely flawed as such since the rafters would exert a substantial thrust at the top of the walls. Such thrust could only be resisted if the walls were of massive proportions, such as the stone walls of some medieval dwellings. Consequently couple roofs are generally found only on comparatively ancient dwellings and their use for many years in more recent times has been limited to covering very small spans, such as entrance porches.

Close Couple Roofs

The problem of thrust at the top of the supporting walls is solved by the addition of a tie, usually a ceiling joist, to connect together the feet of the rafters to form a 'close couple roof', illustrated on Figure 12. As indicated above, the span between supporting walls with this type of roof can vary between 3.5m and 4.5m depending on the pitch. Over this distance there is a danger of the ceiling joists sagging if subjected to any load or being walked upon so that they are normally provided with a binder hung from the ridge plate. Properly nailed at each of the connecting points, as circled on Figure 12, the load of the roof structure and its covering will be transferred vertically downwards through the

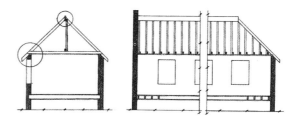

Figure 12: Cross and longitudinal sections of a close couple roof with hangers and binder to support the ceiling joists. Scale 1:200. Capable of spanning roughly one room and accordingly used either gable end facing street or in parallel alongside another, but then needing a structural partition in support.

supporting walls.

Simple close couple roofs of this nature are occasionally found as the main roof over the smallest of dwellings, only one room in depth when the ridge is parallel to the frontage. Of course, such dwellings if turned around so that the gable is facing the street can be much more extensive, 11 is one such of a type popular in the 1950 and 1960s when the availability of trussed rafters enabled a respectable frontage to be achieved.

11: Presenting the gable end of the roof to the street was never very popular because, until the advent of trussed rafters in the 1960s and 1970s, the appearance was mean. Here using the shallow pitched trusses and shape fashionable at the time a reasonable frontage could be provided for this 1960s detached house.

Sometimes, however, two will be found facing the street constructed in parallel, a central structural partition between front and rear rooms being taken to top floor ceiling level to provide support for the lower end of the central rafters. There needs then to be a valley gutter between the two roofs discharging generally to a flank wall in the case of detached or semi-detached dwellings, a frequent source of trouble if the gutter becomes blocked and overflows. Even more troublesome are those cases where such parallel roofs are provided to terrace property involving either an internal rainwater pipe through the whole height of the building or a secret gutter to take the rainwater, either through the roof space or through the floor.

13: The half hipped parallel close couple roofs with rooms in the roof space of this terrace of 1700s houses drain to a central valley gutter with outlets through the flank walls at each end. This is also the method of disposal for the parapet gutter at the front and is a much better arrangement then the secret gutters mentioned at 12 if the terrace is short.

12: This end of terrace house with parallel small mansard roofs clearly shows the outlet from the valley gutter on the flank wall. Elsewhere in the terrace rainwater from the valley gutter is taken to the rear via a secret gutter in the top floor.

14: Secret gutters whether within a floor or within a roof space as here are a frequent source of problems both from leakage and consequential rot.

As previously stated, the principle behind the dual pitch close couple roof is the creation of a rigid triangle which transfers the weight of the roof covering and its own weight vertically downwards through the supporting walls. It relies on the provision of adequately sized timbers which will resist deflection under load, accurate cutting and fitting of the component parts and secure fixing at the joints. A failure in one or more of those requirements can produce unevenness in the roof covering to the extent that it may not be able to fulfil its function of keeping water at bay. Further damage may also be caused by thrust due to deflection or weak jointing. It may be that when such defects are visible, a cause is readily ascertainable and can be included in the report with appropriate recommendations, but it may be that further investigation will need to be advised.

Close Couple Purlin Roofs

The three up two down semi-detached dwelling is the most frequently encountered of the period 1920-1940, each half of the pair being relatively square in plan. Something like 40% of all dwellings were built during this period. Usually two rooms from front to back, the roof would consist of a dual pitch close couple hipped design as on Figure 13, comprising common rafters, ridge plate and purlins, shown circled on the diagram. The purlins divide the length of unsupported rafter into approximately 2.4m lengths. To support the purlin, struts are required and sometimes a collar is provided. For a span of about 8–10m, with a structural partition separating the front and rear rooms taken up through

Figure 13: Cross and longitudinal sections (Scale 1:200) of close couple roof with purlins (circled) and struts where structural partition with plate on top (circled) is parallel to the frontage. The typical square shaped semi-detached house of the 1930s has a roof of this type. The diagram also shows a similar roof over an L-shaped dwelling but with collar omitted for clarity.

the first storey and supporting the struts from a plate at the top, as also circled on the diagram, it could take the form shown. The binder and hangers shown on the cross section would be needed to prevent the ceiling joists from sagging if one room was considerably larger than the other.

Because of the long lengths involved, the 100 x 50mm ceiling joists were often in two lengths intended to be lapped and nailed together on top of the load bearing partition to maintain the tie between the feet of the rafters. It is not unknown for the significance of this to be lost and the nailing to be inadequate. Sometimes it is omitted entirely and the two lengths not even lapped with reliance placed on skew nailing the ends to the plate on top of the partition. Skew nailing in this position is easily pulled out as the roof starts to spread.

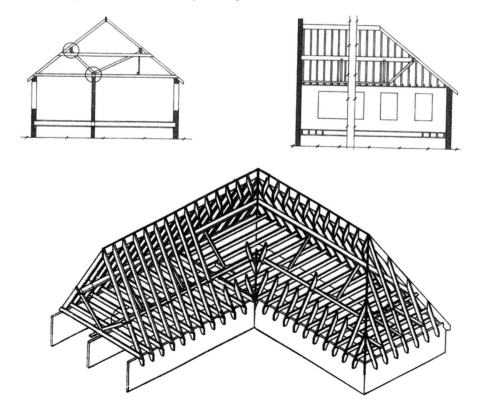

If the dwelling had been designed with a large ground floor through room the two storey structural partition would probably be at right angles to the frontage and the roof arrangement could be as shown on Figure 14 with binders, as circled, spanning from the partition to the separating wall supporting both first floor joists and roof struts.

Smaller dwellings, particularly those built in the early 1920s in a cottage style together with many others going back to the late 1800s were built in the form of long or short terraces. The dual pitch close couple roof with purlins is also the common form of roof for such dwellings. The purlins are usually strutted from a central structural partition parallel to the frontage as shown on the half cross section at Figure 15.

15 (top) and 16 (bottom): Semi-detached houses with typical close couple purlin hipped roofs from the periods 1940-1960 and 1920-1940 (bottom). The earlier pair have a covering of the clay glazed and coloured pantiles often featured in conjunction with rendered and white painted walls, curved corners and windows, the estate builders interpretation of 'modern'. Both glaze and colouring is wearing off the pantiles.

Figure 15 (above): Half cross section of dual pitch close couple roof with purlins and struts supported by a structural partition parallel to the frontage. Scale 1:200. A type of roof much used for late 1800 and early 1900s two storey terrace houses along by-law streets

Figure 14 (below): Cross and longitudinal sections of close couple hipped roof with purlins and struts where structural partition at right angles to frontage. Scale 1:200. This form of close couple purlin roof would be required where a large through room was provided.

If the frontage of the dwelling was not too wide, the purlins could be arranged to span between the separating walls in which case there would be no requirement for supporting

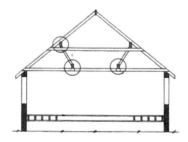

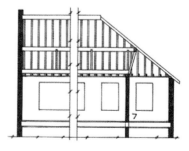

struts or for the partition to be of a structural nature above first floor level. Obviously the purlins would need to be of a substantial nature to carry the load without deflection but such timber would have been available at the time of construction. Later in the 1950s and 1960s, timbers of the required size would be hard to come by for narrow fronted terrace dwellings and lighter timbers were used made up in the form trussed purlins, similar to those shown on Figure 9. Existing purlins can sometimes be turned into trussed purlins and this is an option which is often adopted when the struts are removed in older dwellings so as to provide accommodation in the roof space by way of a loft conversion. Failure to provide a satisfactory alternative to the struts can lead to severe consequences for both roof above and ceiling below.

Collar Roofs

A more problematic form of roof can be produced when the tie, instead of connecting the feet of the rafters, is positioned higher as a collar. The higher it is placed the less effective it will be at resisting thrust at the top of the supporting walls. Nevertheless, a collar roof, as this type of roof is called, will often be found where it is desired to provide rooms half in and half out of the roof space. The arrangement is attractive to designers since the same accommodation can be provided as in a full storey but with a saving in the amount of walling required, as shown on Figure 16. Early local authority dwellings favoured such an arrangement following the examples set earlier by the Garden City and Arts and Crafts movements.

To compensate for the less effective nature of the collar, as against a tie at the feet of the rafters, best practice indicates that the size of the rafters and the collars forming the ceiling

should be increased to 150mm by 50mm and the joint between the two, circled on Figure 16, be made with a dovetailed halved joint, instead of reliance being placed solely on nailing. Best practice is not always followed, however, and if the joint is merely nailed and these pull out, thrust may push out the top of the walls, overturning the wall plate in the process. If standard size rafters only are used these may deflect being unrestrained at their feet in consequence, possibly disrupting the lie of the roof covering.

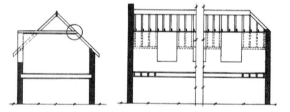

Figure 16: Cross and longitudinal sections of collar roof providing rooms half in and half out of the roof space. Scale 1:200. There is a saving in the wall material when this type of construction is adopted and accordingly it is often encountered on cheaper housing.

17: Rooms half in and half out of the roof space of this pair of early 1900s semi-detached houses necessitating a check on the form of roof construction and whether any roof spread has occurred. None is apparent on the photograph and the distinct hogging of the tiling is due to the separating wall retaining its position while all else has settled slightly.

18: The influence of the Arts and Crafts movement pervaded the dwellings of the garden cities, as here, and the earliest local authority housing at the beginning of the 1900s resulting in many with rooms half in and half out of the roof space. Coupled with gables to the dormer windows they have a cottage like look but nevertheless the construction needs checking.

19: The shape and construction of a mansard type of roof, very common on the terraces of the 1700 and 1800s since it provided the attic rooms for the servants, can be discerned from this example in process of repair. Stout beams spanning from party wall to party wall at the top of the steeply sloping sections front and rear and at the centre carry the load of either a flat topped roof, covered probably in lead, or a shallow pitched dual sloped roof with possibly large slates or pantiles. The top floor acts as a tie. The front elevation has a cornice and parapet, to maintain the classical façade, and a parapet gutter while the rear slope discharges into either an eaves gutter or another parapet gutter.

Mansard and Duopitch Butterfly Roofs

All the roofs described so far, even those of very shallow pitch favoured both in the periods 1960-1980 and 1810-1840, are at least visible. However, over quite extensive periods, particularly in the 1700s, fashion dictated that the roof should be hidden as much as possible from the front of the property. This could be achieved by hiding most of the bulk of a close couple or mansard form of roof and dormer windows behind a parapet, 12 and 13, or sometimes a balustrade, see 21. Later for smaller houses, when there wasn't quite the same need for attic rooms, the height of the parapet could be reduced by the use of a butterfly roof, as illustrated on Figure 17. The

height could be reduced even further if hipped slopes were incorporated at the front, as on 22.

The top end of the rafters of late 1800s butterfly roofs take their bearing from a plate or corbels or may be built into the flank or separating wall, marked C on Figure 17, but the lower end would rely on support from the long beam spanning from front to rear, marked A. Invariably this beam was intended to be provided with intermediate support from

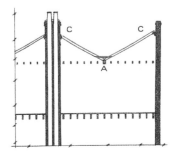

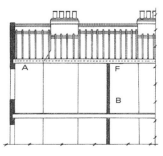

Figure 17: Butterfly roofs were not only a creation of the period 1960-1980 when monopitch trussed rafters became available but were often employed as the roof for tall narrow fronted dwellings towards the end of the 1800s. Cross and longitudinal sections. Scale 1:200.

20: The unified facade was not to be disturbed by too much sight of the roof in the 1700s, yet the servants had to be accommodated somewhere. The parapet capped many a facade, as on this brick built house, along with what could be, at times, troublesome parapet gutters.

22: A butterfly roof, seen from above, re-covered with concrete tiles. The need for a higher parapet has been reduced by adopting hipped slopes at the front.

21: Sometimes an ornamental balustrade would be used to cap the front elevation, as on this stone built terrace.

23: The rear of a terrace of late 1800s dwellings with butterfly roofs. It is not unusual for the defects in one house of a terrace to be repeated elsewhere. A surveyor inspecting adjoining dwellings here would need therefore to consider this possibility.

a structural partition dividing front and rear rooms. A frequent cause of problems often lies in the construction of this partition, allowing sagging in the beam and consequential overflow from the valley gutter. The partition is often of comparatively flimsy construction on either an inadequate, or practically non-existent, foundation. Timber framed lath and plaster was frequently used above ground floor level and subsequent cutting of principal members to open up through rooms can have serious consequences. In particular, the settlement of the front to rear gutter can cause disruption to the bearing of the top of the rafters, break supporting corbels, disturb flashings

and allow water penetration. While this may not reach ceiling level it can cause rot in the timbers, see 24.

Ends of rafters, when supported at the top of a flank wall, can push the wall outwards, since there is no balancing thrust from rafters opposite, as there is on a separating wall. As the flank wall is often inadequately restrained lower down, particularly if the staircase of the dwelling is positioned against it, the movement can be quite dramatic. The result can often be seen in gaps at the edge of floors and ceilings and cracks at the junction of flank with front and rear walls.

24: The provision of roofing felt to the slating of this butterfly roof of an 1800s terrace house reduces the ventilation very substantially and increases the risk of rot in the rafter ends and in the bearing. Flashings have clearly been defective and the inspector needs to check for rot and on the bearing.

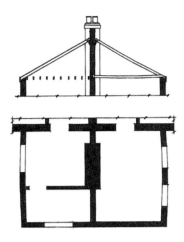

Figure 18: Cross section of typical monopitch back addition offshot roof with plan of possible accommodation below. Scale 1:200. These structures are very commonly found attached to small dwellings of the late 1800s and early 1900s.

Commonly encountered to the rear of many dwellings from all periods, except a fairly high proportion of those built since the 1920s, is the back addition or offshot. This very often has a monopitch lean-to pitched roof. Arranged, as on the right of Figure 18, with proper joints, adequately nailed and with suitably sized timbers, all should be well even though lacking entire compliance with current Regulations. However, the span on slope of the 100mm x 50mm rafters is well over the recognised safe span of 2.4m and accordingly it is not surprising that the plan shows a bow on the flank wall, particularly as this is broken up by two windows. It could be that deflection in the rafters under the load of roofing could be combined with inadequate nailing giving rise to roof thrust.

In cheaper construction, headroom in the back addition was often kept to a minimum and the ceiling joists to provide a tie omitted. If ceiling joists are provided but there are two rooms in the back addition with a separating partition the ceiling joists might be arranged as on the left of Figure 18. The bow visible in this flank wall could again be caused by roof thrust and be manifested to its maximum extent at the centre away from the restraint

provided by the bonding to the rear wall of the main structure and the rear wall of the back addition itself.

Trussed Roofs

Comparatively rarely encountered, but types of roof of which the surveyor should be aware, can be found on larger dwellings, or where larger spans were required up to the earlier part of the 1900s and shown on Figure 19. They do not appear on the cheaper of dwellings and were therefore usually soundly designed and constructed of well-seasoned timber of ample proportions, put together with joints of long standing design.

Nevertheless, the bolts, nuts, plates and screws which are used to strap the joints together can rust when exposed to condensation over long periods. Joints can be disrupted if the support to a truss is disturbed placing the joints under increased strain. This can happen when a pier along a length of wall subsides because the point loading has been

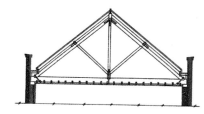

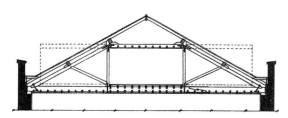

penetration. The confined airless spaces below such gutters are ideal situations for outbreaks of dry and wet rot and are seldom examined when the gutter is repaired and the decorations are made good. Again truss joints can be disrupted and it is often a repeat of severe flooding that will draw attention to the collapse of gutter boards due to the spread of rot and the consequential need for major repairs after opening up. Traversing the gutters, if the opportunity avails itself, can disclose indications of such problems and the need for 'urgent' warnings to be included in the report.

25: Because of the gap between the underside of a parapet gutter and the ceiling below slow leaks can often affect the woodwork setting up attacks of dry and wet rot before becoming apparent on the ceiling. Nevertheless the effects are visible by inspection and it ill behoves a surveyor to miss them. Some replacements have been carried out here but can a clean bill of health be given?

Figure 19: Types of trussed roof. Top, King Post, centre a variant of the King Post where the central timber post is replaced by a metal rod, more efficient in tension than the timber, and the purlins support boarding at a shallow pitch and large slates or alternatively lead, copper or zinc. Bottom, Queen Post, where it was desired to have rooms in the roof space. Scale 1:200. These will occasionally be found on the more expensive dwellings of the 1700, 1800 or early 1900s where there was often a requirement for roofs of a considerable span to cover large rooms. They are made up of substantial sections of soft wood with iron straps to bind the joints together.

miscalculated or the ground has settled. It is usually failures in support of this nature which cause structural problems, not the original design or construction of the truss itself.

Many trussed roofs are provided for dwellings which incorporate parapets. Frequent overflows from the gutter behind the parapet in heavy storms or at times of snow cause damp

Faults in Pre 1960 Pitched Roofs

Among the most frequent causes of defects in pre-1960 cut and fitted timber roofs are:
(a) Inadequately sized timbers.
If suspected, inadequacy can be checked by

measurement of size, unsupported length and spacing, coupled with an assessment of the dead load and a comparison made with the appropriate table for the roof pitch appearing in Approved Document A of the Building Regulations 1991 in the case of England and Wales. In the case of Scotland it is necessary to use British Standard BS8103: Part 3: 1996 Code of Practice for Timber Floors and Roofs for Housing. This latter document covers a wider range of circumstances than Approved Document A and therefore may also be of use in England and Wales when the Approved Document is found to be inadequate. In addition, the Standard contains tables for trimmers and beams, it being pointed out that joist span tables in many cases were erroneously being used for trimmers and beams supporting joists when they should not have been.

26: A purlin unduly deflected. Is this due to inadequate sizing or has the partition supporting the strut undergone settlement? If allowed to continue the roof covering could be disrupted.

In this connection it is useful to be familiar with the self weights which comprise the dead loading figure used in the tables but which neither the Approved Document nor the British Standard provide. Including battens and roofing felt, only thin Welsh slates, shingles, asbestos cement slates and thatch are below 0.5 kN/m^2 dead loading. Thick Welsh slates, manufactured concrete slates, thin Westmorland slates, machine made plain clay and concrete tiles, single lap interlocking tiles

and pantiles of both clay and concrete are between 0.5 and 0.75 kN/m^2 dead loading. Thick Westmorland slates and plain concrete tiles of stone aggregate laid to a 75mm gauge involve a dead loading of between 0.75 and 1.00 kN/m^2. The dead loading from stone slabs would, of course, have to be calculated depending on type of stone, thickness and detail of laying, though typically the weight exceeds 100 kg per m^2 which converts (factor 0.009807) into a dead loading of 0.98 kN/m^2. It could be a fine point therefore whether the Approved Document or the British Standard could be used. It would depend on the circumstances at each location. In every case, however, as fairly old buildings are involved, it is best to assume that the lower strength of timber covered by the tables has been used.

27: One way of providing support to a sagging ceiling. A substantial new beam has been inserted and metal hangers provided to support each ceiling joist. It is to be hoped that the inspector heeds what is in effect a warning to tread warily. He will need, however, to see how the beam is carried on the brickwork of the chimney flues.

(b) Bad cutting and jointing of components. The former will often make the latter difficult to complete satisfactorily since it may reduce the area of overlap and therefore the area where nailing can take place. Consequently the number of nails used could be inadequate. For example, 5 should be used at the joint between ceiling joist and rafter, but more often it is 3, or even 2, that are used with the danger that the joint is weak and could be pulled apart. Cutting square so that the whole surface of one component is in contact with one to be supported is necessary otherwise only a proportional part will be effective. Sometimes overcutting to form joints can be damaging by removing too much material thereby reducing strength.

(c) Damage caused by fungal decay or timber boring insects.

Previous damp penetration can set up attacks of dry rot, serpula lacrymans, or one of the

28: Could this be an entirely DIY effort to provide additional support to ceiling joists by screwing them to steel angles and to the original bearer which was clearly not fulfilling its function. The lack of insulation at top floor ceiling level at least aids the inspection but needs commenting upon in the report.

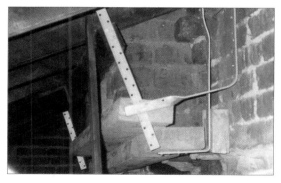

29: It seems an unduly complicated way to replace the missing or defective support for the trimmed rafters around the brick chimney flues in this roof space. An external inspection, usually possible through a trap door in this type of roof, would verify whether it has been successful.

varieties of wet rot, particularly if the timbers are poorly ventilated. Condensation could be another cause of dampness in the timber. As to insect attack, the common furniture beetle, *anobium punctatum*, is unlikely to cause undue structural damage. The house long-horn beetle, *hylotrupes bajulus*, on the other hand can cause severe damage to softwood timber, particularly in those areas where its presence has to be reported to the local authority in accordance with the Building Regulations. The ravaging which the death watch beetle, *xestobium rufovillosum*, can cause to structural oak timbers in old church roofs is well known and it is most essential to indicate the presence of this destructive insect when encountered in the oak framing of both walls and roof of old wholly framed dwellings, (see the 1994 case of *Oswald* v *Countrywide Surveyors Ltd*, reported on p151, where the surveyor in his report failed to distinguish between death watch beetle and common furniture beetle with expensive consequences).

30: Defective flashings are a frequent source of damp penetration affecting here the end of a purlin and the rafter nearest the brick wall with wet rot. The absence of roofing felt enables an underside view of the natural slates to be obtained. This shows that they are near the end of their life by the signs of discolouration and decomposition near the edges while there is also wet rot in the battens. Major replacements and updating are clearly required.

Roofs of Early Timber Framed Dwellings

The roof structure of modern timber framed dwellings built since the 1960s will almost invariably be of the trussed rafter type while the comparatively few built before then will have cut and fitted roofs. Both types will have been made up of softwood sections and perhaps exhibit some of the defects already described. In the case, however, of what are commonly identified as timber framed dwellings from earlier times, built perhaps between 1300 and 1600, the framing extends to both walls and roof and is in hardwood. Oak was the favoured material and is usually found but when that came to be in short supply, black poplar or elm was often used.

Because the roof framing is an integral part of the whole structure, and not just placed on top of a masonry wall, it is important that both wall and roof framing are considered together, since the former is a continuation of the latter from one side of the dwelling to the other. Cutting or removal of one principal member can often have considerable effect at another point in the structure.

Three typical early timber framed roofs are shown at Figure 20 and it will immediately be seen that the top two very much resemble in shape the King and Queen Post roof trusses constructed in softwood in later times, while the one at the bottom is not very different in outline from a close couple roof with collar. There are structural differences however. For example, the centre post is a tension member holding up the tie beam in a King Post roof truss whereas the central vertical member in the Crown Post roof is in compression taking the load from a medial or longitudinal purlin and transferring it down through the tie beam to the side wall framing. The simple collar roof at the bottom is not as simple as it looks and is really a truss. Besides the purlins housed

into the principal rafters, which support the common rafters shown dotted, there will usually be extensive wind bracing between the principal rafters. This type of roof is sometimes varied in that the purlins can be set into the top of

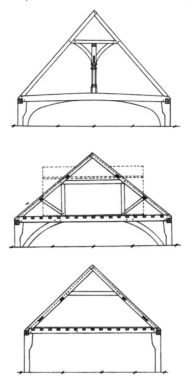

Figure 20: Some roofs of early timber framed dwellings. Top, Crown Post roof with medial purlin and braces, the most frequently encountered type, although the details may vary. The carved Crown Post indicates that it was originally meant to be seen. Very often, however, a ceiling will have been formed subsequently at tie beam level leaving it smoke blackened, unseen and marooned in a newly formed roof space. Centre, Queen Post roof truss, enabling rooms to be formed in the roof space. Bottom, Simple Collar Roof, a truss in all but name, and where rooms can also be formed in the roof space. Scale 1:200.

the principal rafters and the collar provided with bracing from the top of the wall plates producing an attractive arched ceiling from below. Many variations in the details of all these types of roof will be seen and, of course, they were all developed over the years from cruck framed roofs of even earlier times, the shape of one of which can be discerned from the flank wall of the dwelling shown at 33.

31: Early timber Crown Post roof truss with medial purlin and braces.

32: The deformation in early timber framed dwellings often leaves the roof in a very uneven condition for the roof covering. That on the right would appear to have been packed out successfully. That on the left has had some attention but needs more.

It is fairly unusual to find early timber framed buildings in other than a deformed state although this is, of course, part of their attraction to many a purchaser, even if they do not quite realise what they are letting themselves in for. Such dwellings were normally constructed from newly cut down green oak which subsequently shrank. Although iron is occasionally found most joints were oak pegged and the majority consisted of splayed housings which took up the strain when the wood shrank. If the allowance for shrinkage

was insufficient, the timber might split or severely warp. Subsequent alterations to install fireplaces and chimney stacks, and perhaps the rearrangement of rooms, were often ill considered and resulted in the cutting of members with little regard to their importance to the frames overall stability.

All this means that the surveyor needs to look very closely indeed at all visible timber and joints and even if there are no immediate problems of rot, beetle infestation, damp penetration or apparent current movement, make a careful assessment of the existing structural situation.

33: The flank wall of a single storey, with attic, timber framed and brick infilled cottage of about 1500 showing the end pair of the crucks on which both the roof and the wall framing are based.

Steel Framed Roofs

Whereas the prefabricated reinforced concrete framing of the walls of some of the system built dwellings of the 1940s and 1950s merely provided a base for a conventional cut and fitted roof typical of the period, the framing of some of the steel houses also built around this time and earlier was extended to form the basis of the roof structure. While experience indicates that where the framing comprises steel sections, visible corrosion is likely to be superficial, particular attention needs to be paid to dwellings where thin steel sheet has been folded to form the load bearing members. Very often the protection to the steel has gone and corrosion has left little remaining. This necessitates particular attention being paid to dwellings where the metal trussing system is found to be based on fabricated sections, Trusteel and Hills are examples. Generally these types are well passed their original design life but if inspection reveals little wrong may still last for many more years. Obviously comment on the need for eventual replacement must be included in a report of whatever type with a recommendation for close inspection at regular intervals.

34: Trusteel steel framed semi-detached houses of the 1950s. These need particular attention because of their extensive use of thin steel fabricated sections which can suffer greatly from corrosion. BRE lists approaching 300 locations throughout England and Wales where they were constructed in both the public and private sectors.

35: The steel framed houses produced by Dorman Long & Co. Ltd known as 'Dorlonco', in the 1920s and 1930s have steel roof trusses and are characterised by being double fronted. Some are brick clad and some are rendered on the outer leaf.

Alterations to Roof Structures

As stated previously, the main structural functions of a pitched roof structure are to transfer the weight of the roof covering, together with the weight of the structure itself, vertically downwards on to the solid external walls, or frame, and to provide a level, firm base for the components comprising the roof covering.

Less than satisfactory performance in either of the above functions can lead to the components of the roof covering being unable to fulfil their purpose of keeping out damp, not necessarily due to any fault in either the material of the covering or the workmanship. If the surveyor encounters evidence of movement in the structure and suspects that this is a continuing problem, either causing dampness currently or likely to do so later by way of disruption to the covering, then he needs to include a description of what he sees from both the exterior and the interior of the roof space. He also needs to provide advice on what needs to be done next. If a remedy is fairly obvious then that can be included, otherwise it may be necessary to advise further detailed investigation which might include monitoring the situation over time.

Apart from cases of current movement, many roof structures, particularly those of early timber framed dwellings, will have perhaps moved in the past but settled into a new position of repose, leaving the base for the roof covering by no means level. Much of this can be corrected by packing out and firring up timbers to leave the roof structure if not exactly flat at least with undulations which, with a bit of skill, the roofing contractor can provide a satisfactory covering. In this respect, of course, small roofing components will fulfil this function more satisfactorily than, say, large profiled components intended to interlock. Sometimes for example an over hasty decision to change from one type of component to another, without proper regard to pitch and structure, can cause problems and two examples can be cited.

The first concerns roofs of old timber framed dwellings. Many of these, particularly on the eastern side of England and Scotland and in some other places, Bridgewater for example, were originally roofed in clay pantiles at a shallow pitch of around 30 degrees. Undue movement in the frame would render these unable to fulfil their function and the temptation to recover in plain tiles, which fit together much better on undulating slopes would be hard to resist. Unfortunately plain tiles at such a low pitch do not perform satisfactorily and much resort over the years to torching is often evidence of continual efforts to keep out damp and wind blown snow and rain. When either enters, it tends to be soaked up by the torching which is fine for preventing staining on the ceiling below but not so good for its effect on the oak pegs used to fix the tiles, which will eventually disintegrate due to wet rot. A check on the angle of pitch could be useful corroboration if a change from the original is suspected and there are signs of damp and slipped tiles.

The second example relates to the replacement of slates. Towards the end of the 1800s, substantial numbers of the two storey terrace houses lining the streets laid out under the bye-laws of the time, had roofs which were covered with the widely available natural Welsh slates, laid again at a pitch of around 30 degrees. The Welsh slates, in the main, lasted well, around 100 years in most cases. However, by the time their life had become exhausted replacement slates were either very expensive or often said to be unobtainable. Nevertheless, profiled interlocking concrete tiles were available and said by their makers

to give a satisfactory performance at this pitch, unlike plain tiles. Roofing contractors were happy to use them but failed to appreciate, along with the difficulty of cutting and fitting them properly to the often occurring awkward shapes, that their weight was at least twice that of the slates they were replacing.
In consequence, many roofs settled badly and there were reports of some even collapsing. It became a common type of failure in the 1970s and 1980s and necessitated the inclusion of requirements in Approved Document A of the Building Regulations 1991. Calculations now have to be made whenever a roofing material is to be replaced by one which is either heavier or lighter. On inspection, a surveyor should be able to form an opinion of the material likely to have been used originally according to the dwelling's age and quality. Evidence suggesting a change of material, much more likely to be found on cheaper dwellings, will necessitate a check on whether there have been changes to the roof structure appropriate to the circumstances. If not, a thorough examination for signs of movement needs to be made along with, if necessary, a check on timber sizes by reference to the Building Regulations or BS8103: Part 3: 1996.

36: Defects in roof structures or a replacement of lighter weight slates with heavier components, such as concrete tiles, often require additional strengthening as here. Additional struts off a load bearing partition and collars with bolted connections have been provided and the ends of purlins supported by corbels. As a structural alteration the work should have been subject to Building Regulation approval.

A loft conversion to increase the living accommodation of a dwelling has been a popular improvement for the last 40 years or so. Not all roof structures are either suitable or capable of conversion for this purpose. For example, the trussed rafter roofs almost universally used since the 1960s, all comprising a multitude of light timbers at close centres and usually with shallow pitches of 30° or less, do not lend themselves for this purpose. Cut and fitted roofs, to be found on the vast majority of dwellings, on the other hand, can usually be adapted, although it is simpler to do so if the dwelling has gable ends or separating party walls on either side, rather than hipped slopes. It is even more practical if the dwelling has a reasonably steep pitched roof and there are internal structural partitions. What it is always necessary to provide is a load bearing floor for the new living accommodation and, of course, Building Regulation approval for the necessary structural work.

It is unlikely that there would be any difficulty in identifying a property where a loft conversion has been carried out. A comparison with similar neighbouring properties can provide a clue, the presence of roof lights of modern design or unsympathetic dormer windows another, or what seems to be a curiously arranged internal staircase. Once the presence of a loft conversion and its date of construction has been established, depending on the circumstances, a request to the owner for a sight of the Building Regulations approval and the submitted plans and details on which it was based would be a sensible first step. If produced and there are no signs of defects, checking the details may have to be taken on trust as much may be hidden from view. If no Building Regulations approval was obtained and such instances are not unknown, particularly if no alterations were made to the front of the dwelling, then an owner would have to be advised to regulatise the work otherwise

he would find it virtually impossible to sell because a prospective purchaser would be warned not to proceed unless retrospective approval was obtained. In the latter circumstances, it may be possible to see, of course, defects or sub-standard work of a DIY nature, struts removed or an inadequately supported floor. Although not the principal cause of complaint in the case of *Cross and Another* v *David Martin & Mortimer* (1989), alterations to roof trusses to form a room in the loft featured prominently.

Pitched Roof Coverings

The Inspection

Whereas roof structures, given reasonable construction, freedom from damp, the ravages of insects and untoward disturbance from settlement will last for hundreds of years, this is not the case with the coverings. Depending on the type of covering, and there are many, the quality, the fixings and the degree of pollution in the atmosphere, renewal may be required in anything from 30 to 100 years. Even if the covering material itself will last for 100 years or longer, the fixings will probably corrode in perhaps half that time requiring the roof to be stripped, the components sorted and then relaid, possibly with a modification of details and improvements to the underlay.

The prospect of having to incur major expenditure on the roof often provides a reason for moving. Accordingly, it is essential for the surveyor to form an assessment of the current condition and to consider, if necessary, whether entire renewal or recovering using existing material is essential in either the short or long term. This is an important aspect for whatever form of report the inspection is being carried out. If either of these major repairs is assessed as being required within say five years then this will have to be rated appropriately according to the form of report required, or suitable paragraphs included if there is no requirement to provide a rating.

Since no roof covering is likely to have a life of more than 100 years, the surveyor's knowledge of the age of the dwelling could be crucial if the covering is contemporary with the structure, in determining it's life expectancy. For example, it may be thought unlikely that he would find a great deal wrong with the actual material covering the roof of a dwelling built since 1980, although having regard to the pitch, the covering may be used or fixed incorrectly and the detailing may be faulty. One built between 1960 and 1980 could, however, be a different matter. Defects are less likely to be found if the dwelling was built in the private sector which was more inclined, even in those days, to favour more traditional materials and methods. If built in the public sector, however, it was a time of much inno-vation and, although it now seems incredible, much housing was roofed in what was thought of even at that time as temporary material, three ply bituminous felt, for example, on pitched roofs. Most of these dwellings will, of course, have been re-roofed by now but it is still necessary to consider very carefully how this was done. Was it a repeat operation or was a longer lasting material employed and if so was it appropriate to the pitch and was the roof strengthened?

Dwellings built between 1940 and 1960 will probably have traditional roofs in shape, with a steeper pitch, and in consequence, more sharply angled hips but it was a time of shortages and substitutes, in particular concrete products, were much used, some more successful than others. Those covered with tiles, however, may now, along with those built between 1920 and 1940, be showing signs of

the need for recovering depending on the quality of the original clay or concrete tiles used. Some will already have been recovered, presenting the surveyor with a variety of tile shapes representing the full range of what has been available in comparatively recent times, some guaranteed by their makers for periods of up to 60 years.

Dwellings built between 1900 and 1920 if not re-roofed already will definitely need to be re-roofed very soon, while there will not be many built before 1900 that have not already been dealt with.

Such simplification is not, of course, valid for dwellings built, say, before 1870. It is to be expected that such dwellings will have had their roofs radically attended to at least once and possibly twice since the original date of construction.

Ideally, to obtain a complete appraisal of an existing roof covering a close examination of both upper and lower surfaces of the covering should be made but this may not always be possible. As to the upper surfaces the limitations of a 4m ladder indicate that it might only be possible to look closely at the lower courses of the covering material on bungalows or single storey extensions to taller dwellings, unless there is a balcony or dormer window which would enable a closer inspection to be made. If single storey extensions are covered with the same material as the main roof slopes at higher level all to the good, but often they are not. 'Cat slide' roofs provide another opportunity for close inspection as do cases where, with taller older dwellings, there are internal facing roof slopes and safe access to a flat roof from the interior. If none of these facilities are available then the surveyor must have resort to his binoculars, but this is not as satisfactory.

As to the underside of the roof covering

material and the fixing components, close inspection was usually possible until comparatively recently, within the last 20 to 30 years or so, except in the relatively few cases where roofs were boarded in England and Wales, although the use of boarding as an underlay has always been common practice in Scotland. Since then, however, an underlay of roofing or sarking felt, to keep out wind driven rain or snow, has been provided on practically all new or renewal work so that the opportunity to see the underside is now fairly rare. Advantage should be taken, however, of any situation where there is a tear or gap in the felt to take a peek at the underside. This can sometimes be possible where a ventilating pipe is taken through the roof slope for example.

On the assumption that neither surface of the pitched roof covering is available for close inspection, what can the surveyor tell from a sight of the external surface of the roof covering through his binoculars? It is expected that he will be able to go some way at least towards identifying the material and type of covering used. To take, at one extreme, some of the rarer examples first: he is not likely to fail to identify sheet materials where these have been used to roof a dwelling, copper by its green, or brown, patina, lead by its dark grey colour, zinc by its lighter grey shade, each with its characteristic cappings or rolls down the slope. Dark grey material with flat-welted seams ought to suggest to him roofing felt and he should have no problem in identifying where, even more rarely, corrugated asbestos or corrugated iron or thatch have been used. What can be said about condition requires further consideration of each material, but these materials account for but a small proportion, probably only about 5%, of roofs covering dwellings.

It is when it comes to identifying what the smaller components, tiles and slates, are made

37: The characteristic green patina of sheet copper is unlikely to be confused with any other material when used on the slope as here.

38: Thatch is an unmistakeable covering for a pitched roof although to believe that it is only found on old dwellings in country areas is disproved by this suburban example built in the 1920s.

of and which cover the vast majority of dwellings that doubts can begin to form. Tiles, which cover 75% of the total housing stock, can easily be distinguished from slates covering about 20% by their shape, thickness, texture and colour and also one from another by whether they are plain or profiled. Plain tiles, however, can be of hand or machine made clay or machine made if of concrete. From a distance they can appear indistinguishable. In rare cases, tiles might be mistaken for shingles, a long traditional version made of wood. Profiled tiles, again of long tradition, are also hand or machine made of clay and machine made of concrete. Accordingly from a distance, while distinguishable from slates, it is not possible to be certain from which material tiles are made. Nor may it be possible in all cases to give a name to the shape, unless the surveyor is familiar with tile makers' catalogues. These can sometimes reveal different names for similar shaped tiles and the surveyor should also be aware that it is possible to obtain pressed steel tilesheets which when fixed look remarkably like profiled tiles from a distance.

39: At a quick glance these roofing tiles could be mistaken for profiled concrete when in fact they are in the form of coloured pressed steel sheet.

40: The set back two storey house on the right has what appears at first glance a roof covering of clay pantiles, a traditional feature of the area, but which is in fact one of terracotta coloured steel tilesheets with a much shorter life. A potential hazard for an inspector in a hurry.

Similar doubts can be generated when thin flat slates are seen as the covering. Natural slates, split from metamorphic rock, can be copied by moulding resins with powdered

slate waste so that from a distance they are virtually indistinguishable, presenting the same riven texture, irregular edges and colour. Such good copies, from the appearance point of view, were not the first artificial slate to appear on the market. Earlier versions, some made of Portland cement and asbestos fibres, although flat and of a dark grey colour, could usually be distinguished from natural slates by their mechanical, uniform shape, straight sharp thin edges and unnatural very smooth and somewhat shiny appearance. Even earlier, before attempts to actually copy natural slates for replacement work were made, asbestos cement slates were available in colours such as russet brown, blue and the natural grey shade of asbestos. Since the raising of health concerns in the 1970s they have not been used but they often provided the roof covering over the chalet type bungalows fashionable in the 1930s because of their very light weight. When seen they are highly unlikely to be mistaken for slates in any other material and will be showing signs of their age, indicating a need for replacement. Since most artificial slates can now be supplied with matching accessories for ridge, hips, valleys and roof ventilators, the inclusion of these may be the only features distinguishing artificial from natural slates.

Once again the surveyor should be aware that the concrete tile manufacturers produce a flat, lightweight, slate coloured, interlocking tile. When used, these too from a distance and at a quick glance give the impression of a slate roof. A second look, however, will show that they are thicker and too uniform in shape, texture and colour to be mistaken for either a natural or artificial slate covered roof.

So-called slates were, of course, the traditional roof covering in areas where the local non-metamorphic stone formed the main building material. They continued to be so until the

41: *The contrast between natural on the right and artificial slates in the centre after 10-15 years in use is not always as clear as when they are newish and often still fairly shiny. An equally popular replacement for the natural slates originally provided on this terrace of late 1800s houses is profiled concrete tiles suitable for the original pitch and this alternative to artificial slates has been chosen for the house on the left but they are a great deal heavier.*

42: *Asbestos cement slates, available before the health concerns about asbestos, are not likely to be mistaken for slates of any other material and by now are bound to be due for replacement, yet many will still be found.*

product of the massive North Wales mines and quarries with the help of the canals and railways tended to oust them on grounds of cost in the 1800s. Although usually called 'stone slates' they are more like stone slabs and are sometimes called 'tile stones'. Whereas natural and artificial slates normally range from 5 to 8mm in thickness, stone slates at their least are twice that in thickness and can range up to over 30mm. Through binoculars that difference coupled with the

contrast in texture and colour is sufficient to set them apart. However, where there has been sufficient demand, the local stone slate has been copied using Portland cement and crushed stone moulded to give a texture which at a distance makes it virtually indistinguishable from the real thing.

43: *Concrete slates replace the original coloured asbestos cement slates on this 1930s bungalow. They are much heavier and the surveyor will need to check what has been done to strengthen the roof structure.*

44: *Whether stone slates are the real thing or manufactured from cement and stone dust is not possible to tell through binoculars. Newness or the age of the dwelling might be a guide.*

While, therefore, there will be times when a surveyor may be confident in identifying the material used to cover the roof of a dwelling, there will be other times in the absence of facilities for both close external and internal inspection when it would be unwise to be dogmatic. Phrases such as "appears to be ... when viewed through binoculars from ground level" would be quite legitimate for inclusion in a report, if that is the only vantage point available, irrespective of whether the roof looks sound or in need of repair. If the latter happens to be the case, the next person inspecting could well be another surveyor or roofing contractor with a long ladder, able to be certain and possibly embarrassing, at the very least, to the initial reporting surveyor should he have been too dogmatic in his report.

With a lack of any facilities for close inspection, but with reasonable confidence in identifying the material used, the sight of an apparently sound roof with no sign of damp stains internally on surfaces not decorated for some time, the surveyor could report favourably without further ado. Where surfaces immediately below the roof have been redecorated recently, he needs to be much more circumspect since temporary, but adequate in the short term, roof repairs and redecoration could have been carried out specifically for the purposes of sale. He needs then to check whether the pitch and other features are appropriate for the covering and are satisfactory, looking very carefully and using his damp meter on timbers in the roof space at places where there could be a possibility of damp penetration.

In other cases where the roof of a dwelling is covered in either tiles or slates, the sight of many missing, slipped or broken items or in the case of slates, many refixed with "tingles", (the narrow strips of lead used to re-fix slipped slates), is indicative, depending on the number, of the need for either minor or major repairs. A few items on a roof, not particularly old looking, might be due to isolated incidents of wind gusting or the clumsy feet of aerial installers and be straightforward to rectify. On an older roof it may be the beginning of a more general deterioration, likely to accelerate as time goes on and worthy of further

45: A recently recovered roof in plain machine made tiles where, in the absence of any new damp stains below, the surveyor can be reasonably confident in his report. Right to buy consequences are in evidence in the mismatch of roof coverings.

investigation. Circumstances will dictate how the surveyor phrases his report in such a situation, but it would be wise to recommend a closer examination.

On the other hand, the sight of many defective components and, because of the lack of facilities for close inspection, little possibility of ascertaining the likely cause of failure,

46: In the absence of facilities for close inspection, the surveyor cannot do otherwise than suggest a need for close investigation to check that the remaining old tiles on the roof of the 1930s semi-detached dwelling on the left are not going to need replacement shortly. They have all been renewed on the other house of the pair as well as elsewhere on the estate suggesting that that might have been the better course to follow rather than the patching carried out here.

requires the surveyor to indicate, again quite legitimately, that major expenditure will be required. This would follow further investigation to obtain recommendations of what needs to be done and the obtaining of estimates. It will probably not be possible for the surveyor to see, for example, with binoculars from a distance whether tiles or slates are delaminating and that therefore it would be a matter of renewing the covering or whether, when the underside is obscured by felt, it is the fixings which have corroded and the repair will involve relaying rather than renewing. Either way it will be a major expense, but not quite so much in the case of the former.

Given access for close inspection to both external and internal surfaces, but primarily the latter, it ought to be possible for the surveyor to provide more information about the covering whether sound or unsound and the reasons, if applicable, for disrepair.

If the surveyor has full access for inspection and finds evidence of deterioration he may be offered a document of guarantee or possibly an estimate for repairs. Both should be regarded as irrelevant until the surveyor has decided from his own experience what advice he is going to provide. All the parties envisaged as likely to commission an inspection and report would wish to have an opinion expressed on how long the roof covering will last. Obviously if the roof covering seen is at the end of its life, the surveyor should say so. A situation, however, where only a few repairs are needed or, in the surveyor's view, there is nothing seriously wrong, the question could be posed. The surveyor needs, therefore, to keep in the back of his mind received information and that derived from his own experience about longevity and if asked the question apply it with some circumspection because his opinion has to be based on what he sees.

Having concluded to the best of his ability the qualities of the material which has been used, he needs to look closely for any emergent signs of degradation however slight both in the actual covering used and the fixings and battens. Even if he has been told when the roof covering was provided and with what material, either as new on a comparatively recently built dwelling or as a renewal, it is always possible, more likely perhaps on renewal work, that an inferior product and/or inferior fixings have been used, less than ideal practices followed or corners cut. If the surveyor has full access, he ought to be able to identify shortcomings of this nature and, if necessary, include appropriate advice in his report.

Tiled Roof Coverings

By far the most common form of roof covering which the surveyor is likely to encounter is tiling since it is estimated that throughout the UK something like three quarters of all dwellings have tiled roofs. Two thirds of these tiles are of concrete, this being the cheapest form of satisfactory semi-permanent form of roof covering for the last 50-60 years, the remaining third being of clay.

Clay, moulded and fired, has been used to make tiles in the UK for many hundreds of years. Clay tiles lost out in favour of slates for around 140 years from about 1780 to 1920, resurfaced in popularity for a further period until overtaken in the 1950s by the first really successful concrete tiles. Both clay and concrete tiles have always been, and continue to be, made in two forms, plain and profiled.

Unlike the makers of concrete tiles, some of whom currently guarantee their products for 50-60 years, the makers of clay tiles are more reticent, perhaps because the processes for producing both hand made and machine made tiles from natural materials are less capable of fine control. While the makers of clay tiles naturally extol their virtues, pointing out that many seen can be well over 100 years old, guarantees in excess of 30 years have not been encountered. Whereas British Standards for the manufacture of concrete tiles have existed since the 1950s there is no British Standard for the manufacture of clay tiles, only for the size and performance by testing of plain, not profiled, tiles. Some makers are, however, members of a Ceramic Industry Certification Scheme which purports to provide an assurance of quality and there is the possibility of a European Standard for all clay tiles being introduced which will, of course, be applicable here.

47: A roof of machine made plain tiles with bonnet hip tiles in need of a fair amount of attention but given that, should well be sound for some years.
The accumulation of moss due to the proximity of trees is not damaging in itself but interferes with the free flow, encouraging water retention and capillary attraction between the tiles. A growth inhibiting wash should complete the repairs.

Plain Tiled

Plain tiles are rectangular in shape, 265mm x 165mm or thereabouts, with a slight camber lengthwise and sometimes also a cross camber, holed to be secured with wooden pegs or nails at one end and in most cases with nibs for hanging on battens. Being comparatively small, plain tiles are adaptable, unlike the profiled variety, to roofs of relatively small and complicated shapes and can also be hung vertically on walls when tiles shaped at the lower end and bearing such self evident names as arrowhead, spade, club or fishtail are often used. They are laid with a double lap so that there are always three thicknesses of tile over the point of support. Every nibless tile needs twice nailing and those with nibs at least every fifth course below 60° and every course at a steeper pitch along with every tile at the crucial points of verge, eaves, ridge, hips and valleys.

In order to prevent the entry of wind blown rain and snow, a more likely occurrence when, because of their better less mechanical

appearance, tiles with a cross camber are used, the surveyor should look for the underlay of boarding or roofing felt which should be employed nailed to the rafters below the battens on all tiled roofs. Ideally, counter battens should also be provided when boarding is used.

Although it may be limited, the facility for a closer look at the tiles should enable the surveyor to decide by scraping the surface whether they are of concrete or clay. Early concrete tiles had an application of coloured granules to the surface which can usually be scratched off, even if they have not already been partly removed by weathering to give a patchy appearance to the roof. Although unsightly, this weathering does not usually affect the roof's performance. Later concrete tiles are usually through coloured and either smooth or granular faced but can still be distinguished close up from clay.

Hand made clay tiles are usually distinguishable from their machine made cousins by irregularities of shape, both as to thickness and camber as well as to size and the rougher

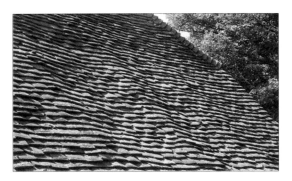

48: Too much disturbance to the lie of these hand made plain tiles, which have a pronounced cross camber, can introduce gaps and, if allowed to become worse, compromise the effectiveness of the covering. Strengthening the structure may be necessary, along with packing out the rafters before the covering can be relaid.

50: Shaped tiles are not only used for hanging on walls, whether timber framed, brick or block. They make an attractive roof as here. When used in this way the run off to the gutter is facilitated by a few courses of plain tiles at the eaves.

49: Patched plain tiling in serious need of general attention particularly adjacent to a dormer window which is equally in need of repair. It is to be hoped that soakers are present below the lead flashing and, with care, the inspector may be able to check by gently lifting the flashing.

51: The regularity of machine made tiles shows disturbance here where they are beginning to delaminate.

texture produced by sand moulding. Though machine made tiles can be provided with a sand faced finish, this too can usually be scraped off, unlike the texture produced by sand moulding. Much more typical of machine moulded tiles is the regular appearance both as to texture and colour and often just the one way camber. Usually they lack that variety of texture, size and colour which make the hand made product considered so much more suitable for use on older dwellings.

52: The underside of a hand made plain tile roof covering secured with pegs. Signs of efflorescence on the tiles can be seen and elsewhere some of the pegs have worked loose, leaving some hanging on one peg only.

Bearing in mind that the pitch of the tile when laid is always less than the pitch of the pitch of the latter for plain tiles whether of clay or concrete should never be less than 40° and preferably should be around 48 to 50° or even more, as on many early tiled roofs. When the basic roof pitch is shallow, there is always a danger that when, for example, sprockets are provided, rainwater may not clear sufficiently rapidly to prevent moss forming or, in winter, to prevent it soaking into the more porous tiles and accelerating frost damage at the eaves, a frequent cause of tiles delaminating.

Efflorescence of salts from within the tile itself has been known to cause the underside of the tile and the nibs, if any, of both clay and concrete tiles to erode and crumble. Unduly porous tiles lacking a cross camber are more prone to this defect as the salts carried in solution to the bottom of the tile migrate by capillary action towards the head of the tile where the solution evaporates, leaving little more than a white powder. It was the failure to spot this white powder which had fallen on to the top of the ceiling joists in an unfelted roof space, which led to a surveyor being found negligent in one of the earliest reported cases of surveyor's negligence, *Last* v *Post* (1952). If the dwelling is one of an estate, a glance at other properties might indicate whether a total renewal of tiling has been found to be generally necessary from whatever of these possible causes of delamination.

Close examination from a balcony or dormer window may confirm that the many slipped tiles are delaminating from one or other of these causes and that those still present are crumbling away and will meet the same fate fairly shortly, thus indicating that total renewal is required.

On the other hand if the tiles, particularly those that have slipped down, show no signs of delamination then it is likely that the fixings have corroded or the battens have rotted. Although it may not be possible, in the absence of full exposure, to be sure which of these possible defects is the cause, the sight, from within the roof space, of roofing felt stretched tightly across the top of the rafters preventing drainage of wind blown rain from the underside of the battens would suggest rotted battens as the possible cause.

Rot could also be the cause of failure in the fixing of old tiles. Very often early hand made tiles were nibless and relied on wooden pegs for their security. An old custom to improve weathertightness, particularly if the lap was less than ideal, was to torch with lime plaster and hair the underside of the tiles, covering the pegs in the process. While torching helped to hold the peg and tile in place, it frequently became wet and led to rot in the peg.

In either of the above cases of fixings corroding or battens rotting, it will be a case of stripping the roof, sorting sound tiles and providing and fixing new felt and battens and relaying the tiles.

Profiled Tiled

Roofs covered in profiled tiles have just as long a history of use in the UK as those covered in plain tiles. S-shaped clay pantiles probably introduced and imported from the continent, were developed as a one piece tile from the flat or curved undertiles and curved overtiles which made up Roman and Spanish tiling of long tradition. Being roughly twice as wide and one and a half times as long as a plain tile and laid at a lower pitch with only a single lap, they have always been much more economical in use. Against this advantage, however, is that they do not lend themselves to roofs of complicated shapes because they

53: *Traditional clay pantiles finished with half round hip tiles. This shows that whenever other than the simplest shapes are adopted there is always awkward cutting and filling in with mortar which has a tendency to shrink and fall out, in this case to the ridge tiles.*

56: *Different types of profiled tiles on a pair of semi-detached local authority houses of the 1920s, replacing plain tiles originally provided.*

54 and 55: *(above) Replacement traditional styled pantiles on a restored early timber framed dwelling and on a 1920s pair of semi-detached houses.*

57, 58 and 59: *(opposite) Shallow pitch roofs of the early 1960s with a variety of the patterns available for profiled tiles at the time.*

60: *Contrasting roofs of adjoining semi-detached houses. On the right, the original green glazed pantiles from the 1930s now losing their colour and glaze and beginning to show signs of delaminating. On the right a roof fairly recently recovered matching the original. Owners should not be fobbed off with a tale of 'you can't get them now', as enquiries can often prove fruitful. It may, of course, be a requirement to match should the dwelling be in a Conservation area.*

61: *Glazed clay Roman tiles coloured both green and blue from the 1930s used as copings to the parapets of these flat roofs to art-deco style semi-detached houses. Unfortunately the curved glass steel framed casement windows, the principal characteristic feature, along with the white walls and flat roofs, of this style have been removed from the bay window of the house on the right and the curved sections of walling crudely filled in with brickwork, totally altering the appearance.*

are more difficult to cut, providing problems in forming satisfactory joints at abutments to walls and at projections through roofs and they are less tolerant of unevenness in the structure below.

As well as the traditional pantile shape, clay tiles also came to be made as single units with either one bold roll or a pair of narrower rolls so that when laid, an impression of Roman tiling was produced. Manufacture in concrete as well as clay began in the early 1900s leading to the development of the wide range of interlocking tiles available today.

62: *To even out the flow and lessen the risk of overshooting the gutter in heavy rainstorms pantiles are sometimes provided, as here, with a few courses of plain tiles at the eaves.*

Finishes can be smooth or sand faced in various shades and also as on clay tiles, when glazed, highly coloured. In shades of green, purple and blue their use was popular on some of the rendered and white painted dwellings of the 1930s. Such tiles are still available from the continent with an extended colour range, but it may be difficult to match both shape and shade if replacements for tiles 60 to 70

years old are required. A few of the shapes of profiled tiles available are illustrated along with some of the colours still to be seen on dwellings from the 1930s.

Profiled tiles because they are made from the same materials, clay and concrete, as plain tiles suffer from the same flaws in manufacture

63: It is perhaps surprising that with so much cutting necessary to the pantiles on a roof of this complicated shape that there are not more problems of leakage but the size and weight of each tile helps to keep them in position.

64: Profiled concrete double Roman tiles. As can be seen each tile has two flat sections separated by a complete roll with the interlocking joint two thirds of the way round the next roll. These are heavily marked by lichen and algae and there is extensive moss growth spoiling the appearance which although harmless in itself requires removal and a surface treatment to inhibit growth because of the encouragement of moisture retention. Attention would also seem to be required to the flashings which look less than satisfactory.

valley gutters and at the junction with hips, the necessary cutting to make them fit is often crudely done and can remove the inter-locking anti-capillary grooves, carefully provided by the makers, which normally help to keep them in place and keep water penetration at bay. Sometimes even the nail hole is removed leaving a section of tile to rely on its weight to hold it in position and vulnerable to high winds or physical disturbance when gutters are cleared.

65: Old, broken and patch repaired profiled clay double Roman tiles which point to a roof covering of limited life.

66: A very old clay pantile roof extensively patched with cement in a very rough and ready manner and now in need of complete renewal.

and the same effects from weathering, particularly frost damage. Even so, the main problems encountered with profiled tiles are their use in circumstances not appropriate to their shape. When they are used over splayed bay windows, over small entrance porches, at the sides of

Furthermore, the very shape of profiled tiles means that when they abut a hip tile or ridge piece, much mortar has to be used to fill in the spaces between the two. If the mortar is too strong it usually shrinks excessively leaving gaps. If galleted, when slips or pieces of tile

67: A wide range of shapes is available these days for profiled concrete tiles, providing choice to designers. Where based on traditional designs such as the double Roman, as here provided on a recent development, they can be described as such by the inspector. The flashings provided against the pre-cast concrete artificial stone block wall are of a curious design and not adequately tucked into the joints and risking dislodgement, while the lead dressed over one roll of the tiles is no doubt effective but inelegant and a neater and better arrangement for both could have been provided. One tile has already slipped.

69: Profiled concrete tiles in non traditional shapes, of which there are many now available, are given proprietary names by their manufacturers which, unless the inspector knows the make, leaves him with little alternative but to confine his description to just those words. The shape of these tiles, or something very similar, on a recent development are given the names of 'Ludlow' by one maker and 'Vanguard' by another. Unless spares have been left on site by the tiler so that a positive identification can be made, it is best that the inspector avoids such descriptions.

68: Profiled concrete tiles to the simply shaped roof of a house on a recent development in the form of economical double pantiles. The verge to the gable has no overhang leaving the detail with little protection and a rather mean look but a least the tiles at the edge are provided with clips to help resist wind uplift.

70: More often than not cement fillets are provided to profiled tiles at abutments but here lead has been used. However, the cutting of a groove in brickwork for the top edge in lieu of stepping into raked out joints is not satisfactory being much more liable to work loose, and neither is the dressing down on to the profiled tiles adequate. The tiles here are of a type much favoured for use as a replacement for slates.

are inserted, they, in consequence, will become loose and slip down. If too weak, the mortar might not hold the galleting and will let the water through. In recent years to help overcome these problems and also to provide ventilation to the roof structure manufacturers have produced a range of fittings, some in plastic, for use as ridge pieces. The plastic versions come in a limited range of colours and while, initially, they achieve their objective, the matching to the tile shade needs care and their longevity is far less predictable.

Slated Roof Coverings

Natural

Even though a roof may be identified as one of the 20% or so of all dwellings covered with natural slate, as distinct from the man made artificial, stone or asbestos cement varieties, the surveyor should be aware that there are good and bad slates. The good can last for hundreds of years but the bad for very few.

For new or replacement work, a British Standard has been available since the 1930s and provides for slates to satisfy tests for water absorption, to cover frost resistance, wetting and drying for differential movement and acid immersion for coping with atmospheric pollution. Presented with a recently provided natural slate roof, the surveyor needs therefore to seek evidence that the slates themselves comply with the Standard. If they do, then a new owner can be reasonably assured of the material's sound performance for many years to come. However, as natural slate roofing has been the most expensive form of pitched roof covering for some years now, encountering such a situation will be comparatively rare. This is so even though imported slates may cost only half as much as their UK counterparts. Failing evidence being produced of compliance with the Standard, the surveyor needs to be cautious as BRE has found that only about a third of imported slates meet the requirements and some have shown signs of failure after only a few years of exposure to the elements.

There is also the factor of second hand slates to be considered. For some years roofing contractors have been replacing the natural slate roof coverings of 1800s and early 1900s dwellings with other materials. While the broken slates end up on the skip as rubbish, those that are not are carefully removed and sorted to appear again on the roof of a dwelling whose owner would rather see his roof recovered with the same material as before, instead of a substitute. He may know nothing of British Standards but be assured by the roofing contractor that what he is getting are specially selected fine quality slates. Yet these slates may well be approaching the end of their useful lives.

Depending on the thickness of the slate, its size in relation to the pitch and careful selection, it may well be that many years of service will be obtained from non-British Standard slates but after very close inspection and, in the absence of any signs of premature failure, the risk that this might not be so is one that a new owner should be expected to take rather than the surveyor. The fact that there is no guarantee needs to be explained with great care in the report to avoid any ambiguity.

More often, however, the surveyor will be presented with a natural slated roof which is neither particularly new or old looking and on which no information is forthcoming from the current owner of the dwelling. If there is no access for close inspection to either upper or lower surfaces the scope of advice which the surveyor can provide in various circumstances has already been covered under the heading of the Inspection. With the opportunity for close inspection of either surface, the surveyor will be able to examine and check a great deal more based on his knowledge of the material and its characteristics when used as a roofing material in the conditions of exposure and pollution relevant to the area in which the dwelling is situated.

Natural slate, locally mined or quarried and capable of being riven ie split as distinct from sawn, has long been a traditional roof covering in many of the areas where metamorphosed rock occurs. Its use expanded mightily in the late 1700s, 1800s and early 1900s when slates from the huge quarries in North Wales could be transported relatively cheaply, first by sea and canal and then by rail, to most parts of the country. Most of the dwellings remaining, built during the Industrial Revolution and in the late 1800s and early 1900s, were originally covered in thin Welsh slates but, because of its high cost, when it is found necessary to renew the covering an alternative is usually selected. There are still many dwellings remaining, however, with their original slate roofs, provided and installed according to the custom of the time, still unfelted and accordingly well ventilated but with the only sop to present ideas of comfort being a layer of insulation laid between the ceiling joists.

The custom of the time looms large when it comes to roofing in natural slate, since there is really no scope for improvement in the basic component. Slates have always been sold in many different sizes, up to 30 or more, and usually in three thicknesses between 4 and 8mm but, apart from being provided with either one or two nail holes at the head or centre, they remain thin and flat with none of the cambering, anti-capillary grooves or inter-locking features of manufactured tiles. As a material slate either complies with the British Standard or it does not.

All slates eventually, the non-complying sooner, those complying with the Standard later, succumb to disintegration even assuming they survive splitting due to strong wind uplift, particularly at verges, hence the preference for slate and a half rather than half slates at this point, and the depredations of those clambering over roofs to install TV aerials and the like. Being a natural material, sometimes the cause of disintegration can be inherent due to the presence of impurities. Iron pyrites for example, present in the slate when mined, oxidise in the atmosphere into sulphur dioxide

which, besides causing a brown stain, when combined with rainwater turns into sulphuric acid, attacking the calcium carbonate component of slate and turning it into mushy calcium sulphate.

Even if iron pyrites are not present in the slate, pollution in the atmosphere, which is increasing all the time from the products of combustion, combines with rainwater to descend as acid rain. Because slates lie flat against each other, the acid is drawn along the underside of the slate by capillary action to continue its attack, particularly in the area of the nail holes, long after it has stopped raining.

The signs of such degradation are a lightening of colour on the edges of the slates and powdering, flaking or blistering of the surface, initially much more pronounced on the under-side. It is essential therefore for the surveyor to keep a sharp eye open for these symptoms.

71: Sometimes the presence of roofing felt will hide the underside of a natural slated roof leaving some of the surveyor's questions unanswered. In its absence much information can usually be gleaned. By the damp stains on the purlin in this case it looks as though the renewal of the roof covering could not have come too soon but the surveyor will need to enquire about the quality of the renewed slates and workmanship. The slight sagging correctly given to the felt between rafters allows drainage of wind blown rain or snow to clear from the upper edge of the battens. Tightly stretched across, wet rot can be expected in the battens in time.

72: It is still possible to view the underside of many natural slated roofs unencumbered by roofing felt and as originally laid. These slates and their fixings are still providing service after a lapse of 100 years in an area comparatively free from pollution, though it cannot be long before relaying will be necessary as the nails where seen driven through the battens are very rusty and there are indications of a lightening of colour at the edges of the slates suggesting that some, if not all, will need replacing. Could the original builder have been using up oversized bricks in the party wall left over when the Brick Tax was repealed in 1850? Front and rear walls are of normal sized bricks and there is no evidence of difficulty in bonding these to the party wall. Presumably the use of the oversize bricks is confined to the area of party walling above top floor ceiling level. On the other hand the builder could have just been saving money by using rat-trap bond where the brickwork is not seen, although this leaves a 75mm gap between the two leaves, ideal for vermin, hence the name.

73: Slipped and missing natural slates due to nail sickness and slippage of an inadequately tacked lead flashing at the change of pitch to this mansard roof. A novel way of refixing slipped slates has been used where the slate is turned head to tail and refixed through the single nail hole by a screw and washer to the batten between the two slates in the row below. Not recommended!

74: Suspicions should be aroused where, as here, exposed nailing of the slates at the tail has been carried out between the slates in the course below, sure to allow damp penetration. Have the slates been turned over and is delamination occurring at the edges? There is no upstand at the sides of the bituminous felt gutter, allowing a serious risk of flooding.

The effect on the edges of the slates is to separate the layers making up the thickness, allowing water to penetrate which, on subsequent freezing, will cause yet further disruption.

Where good quality slates have been used on a dwelling, it is much more likely that the fixings will fail before the slates themselves. This may be after a life of 40 to 60 years though it may be longer depending on the degree of pollution and exposure and the quality of nails. Nails eventually rust and a strong gust of wind lifting up the slate will cause the nail to break. If single nailed, the practice in some parts of the country, the slate will slip down the slope. If double nailed, it may hang on for some further length of time, perhaps a bit askew, until another gust breaks the second nail. The sight of many missing, slipped or broken slates is an indication of trouble of this nature. So too is the sight of many slates refixed with tingles, strips of lead nailed to a batten and turned up over the tail of the slate to hold it in position when slid back into its former place. Repairs of this type can often be continued for many

75: *Numerous slipped slates on this roof, some still hanging askew on one nail, others having slipped down to lodge in the gutter possibly impeding flow. Stripping, sorting and refixing with new nails is the least that is required, provided the slates are still sound.*

77: *Patch repairs with new natural slates, one of which is distinctly thinner, to an old slated roof with two lead tingles visible. Rather than continue with repairs, which of this nature are not really cost effective, recovering ought to be advised and seriously considered.*

76: *The sight of many lead tingles on an otherwise sound natural slate roof indicates that the fixings are corroding and that the time for entire relaying approaches, always providing that the slates are still sound. Here also the chimney stack needs repointing at least and capping with preferably a lead flashing in lieu of the shrunken and part missing cement fillet at the top of the slating.*

years, an option available and chosen by many owners, but eventually entire relaying will be necessary.

Purely temporary palliatives, which will occasionally be found applied to slate roofs include laying hessian, canvas or other mesh material and bitumen bonding it to the top surface of the slates. This spoils the appearance and is therefore usually restricted to roofs hidden from general view. Another method is to spray the whole of the underside of an unfelted roof with a bonding agent to hold the slates and nails in place. Applying a rendering or slurry to the top surface is also a traditional remedy used in some localities, but all these restrict ventilation and are likely to induce rot in the battens. In an endeavour to prevent this happening other methods have been developed and are applied to and from the underside of unfelted roofs. They include spraying a bonding agent between the battens to secure the slates and another of applying adhesive blocks to fix slates to battens and, if necessary, battens to rafters. Although offering guarantees of effectiveness for 20 years and thus overcoming the need for the fairly frequent renewal required by the other methods, such palliative treatments can only be considered appropriate for dwellings with a limited life, destined for demolition within a certain time scale and reported upon as such.

Artificial

A glimpse on the roof of the earliest type of lightweight artificial slates from the first half of the 1900s, namely those made of asbestos cement, uncoloured or in shades of red or green, need not detain the surveyor for long. As they have not been manufactured for many years and were never considered to have a very long life anyway, despite the evidence that some have lasted for well over 60 years, those seen must be extremely brittle and very much due for replacement with all the consequences which that entails. Careful removal has to be carried out in accordance with strict regulations for such work by a specialist contractor, new battens and roofing felt will be required and there is the possibility, depending on the choice of replacement covering, of the need for strengthening the roof structure.

78: Coloured asbestos cement slates shown in 2002 on a typical 1930s chalet type bungalow. Moss and lichen covered and patched with other forms of artificial slate, the roof is in serious need of recovering with a different material as has already been done on many of the other bungalows on the estate.

While there is not considered to be any danger to health from the presence of asbestos cement slates on a roof as long as they remain undisturbed and their surface condition is firm, a special warning needs to be included in any report where the slates are very old and particularly where they are showing signs of flaking, powdering or crumbling. The asbestos fibres released could pose a threat to health and therefore even more reason to recommend their total replacement as a matter of early priority.

The same care needs to be taken in regard to

those asbestos cement slates manufactured later in the 1900s, specifically as a replacement for natural slates, coloured accordingly and at roughly the same, or a little less, weight than thin Welsh slates. Those found, as above, to be flaking, becoming powdery or crumbling need to be removed and replaced with a material unlikely to become a health hazard. As to those developed later, made from Portland cement and other materials, such as crushed stone, pulverised fuel ash, glass fibre, polypropylene or from polyester resins using crushed slate as a filler, these must be dealt with as found. If comparatively new and showing no signs of premature failure, the surveyor should press the present owner for any evidence he may have as to when the slates were installed, their make and type and whether they came with a guarantee, preferably one backed by insurance in case the manufacturer goes out of business. Although there is no British Standard covering all types of fibre based artificial slates some have obtained Agrément Certificates and at least one maker guarantees its products for 30 years. Such may provide some re-assurance

79: When newish some types of artificial slate have a distinctly shiny appearance as here. There is a noticeable absence of clips to the verge which should have been provided to prevent uplift, but a roofspace ventilator is visible and there is an adequate lead flashing to the parapet wall complete with holding down lead clips. On the other hand a narrow width of slate has been left out which means that there is at one point only the thickness of two slates instead of three with a corresponding greater risk of damp penetration.

to a prospective buyer but the surveyor should point out that some of the versions did not prove to be satisfactory and that although no signs of degradation may be visible, long term satisfaction in use has yet to be demonstrated.

If the surveyor is very fortunate and is able to find out the name of the manufacturer, it may well be that he will already have, or will be able to obtain, a copy of the fixing manual supplied for use with the particular type of artificial slate. A maker's guarantee is invariably subject to the installation being carried out in accordance with the manufacturer's recommendations. The surveyor will then have the opportunity, within the limitations of his visual inspection, of verifying that having regard to the dwelling's exposure category and pitch adopted, the correct size slate and lap has been used, the correct fixings employed and any special fittings, such as ventilators, ridge, hip, valley, verge or eaves pieces are of the same make and have been installed properly. If the surveyor has real doubts in his mind regarding the installation, it may be possible to obtain advice from the manufacturer as to whether the roofing system installed could still qualify for the guarantee

Before the surveyor is able to give any form of reassurance, however, he needs if possible to make a close inspection to see whether there are any signs of undue bowing or curling in the artificial slates as laid. This is not in itself a cause of problems unless it is so severe as to straighten the clips used to hold down the tails of the slates thus rendering them liable to splitting due to wind uplift.

81: Fibre artificial slates beginning to bow on a south facing slope of a bungalow. Will it be progressive and is the single slipped slate an indication of things to come? Slates on the north facing slope are unaffected, suggesting that further hot summers could cause more damage.

In some respects later versions of artificial slates are less able to resist cracking or splitting due to wind uplift than earlier asbestos cement versions, because their bending stress is less. Careless handling before they are fixed can result in them being laid with an induced defect which only becomes apparent on weathering.

Weathering on less than entirely satisfactory artificial slates will begin to show at the edges as either a lightening or darkening of the original colour, an early give away if damaged slates have been used. In addition those slates provided with an applied surface coating will begin to lose this sooner or later depending on the quality of the manufacturing process.

80: Someone was swindled here where the artificial slates are beginning to slip down in quantity revealing roofing felt stretched tightly across the rafters but an absence of the battens to which slates are normally fixed. It is hard to see what was meant to hold them in position and clearly the whole roof needs urgent recovering.

82: These artificial slates are made of Portland Cement and a slate aggregate giving a passable imitation of thicker natural slates from a distance but from closer their uniformity and rounded top edges give the game away. Verge clips have been correctly fitted to prevent uplift. The use of red clay ridge pieces, although now probably of concrete, was very common on slate roofs in the 1800 and early 1900s.

83: A roof covered in the small random sized limestone slates commonly encountered in the Cotswolds, Northamptonshire and the Purbeck area of Dorset. Such stone slates lend themselves to steeply pitched more complicated shaped roofs with gables, valley gutters and the like. The irregularities are a characteristic of the areas where such limestone occurs and replacement by a roof of more formal appearance would be inappropriate and considered undesirable.

Stone

Of all the materials available and used for covering roofs, none is more bound up with the location of the dwelling than what are generally called stone slates, but sometimes tilestones. For example, it is almost inconceivable to find a dwelling in the Cotswolds with a roof covered with stone slabs from Horsham in Sussex. They would look quite out of place being a different much darker colour than the local stone and much larger and thicker than the stone slates common to that area. Not only that, but the practice of forming a roof of stone slates is more often than not a matter of long tradition in the area where a particular stone occurs. To pursue the example further, stone slating in the Horsham area presents a formal regular patterned appearance of slabs about the same size and thickness, whereas in the Cotswolds slates of many different sizes and thicknesses are used and the usual practice is to lay them in diminishing courses from eaves to ridge. Furthermore, the weight of the stone slabs would have made it likely that the cost of transporting such material for any great

84: The plain shallow pitched simple shaped roof of this timber framed 1400s house is here covered with sawn slabs of the Horsham stone of Sussex. The regular appearance with courses lining up and the slates of a comparatively even thickness is typical of roofs in the sandstone areas of Yorkshire, Cheshire, Northumberland, Caithness, Hereford and South Wales. The word 'tilestones' does seem more appropriate than 'stone slates' for these components. The gap left between slabs is intentional to assist run-off.

distance beyond the immediate locality would have been prohibitive for any but a very important and exceptional dwelling.

All this suggests, as do so many other aspects

of inspecting buildings, that the surveyor should take the trouble to familiarise himself with the materials typically used in his area and the craft practices adopted to turn those materials into the components which eventually form parts of a building. Certainly the practice of stone slating is one of those.

Unlike true natural slates which are always riven, that is split, from a block of metamorphic sedimentary rock compressed and fused by heat, stone slates are derived from limestone and sandstone sedimentary rocks which have not been metamorphosed. However, they are capable of being turned into comparatively thin slabs either by being split with a hammer or sawn. Also unlike natural slates which are impervious, stone slates are to an extent porous and hence have to be much thicker to take in a certain amount of rainwater and then allow it to evaporate.

Limestone slates are normally split along their bedding plane by a hammer and chisel and tend to be creamy greyish brown in colour, the brownish shades encountered in Oxfordshire, Gloucestershire, Wiltshire and Somerset, Northamptonshire and parts of Worcestershire and Lincolnshire, the greyish shades in Cornwall, Wales, Cumbria, Leicestershire and parts of Lancashire, Dorset and Scotland. They tend to be somewhat thinner than sandstone slates, usually from 12 to 20mm thick, and used in small random sizes, irregular in shape and thickness and rough in texture. These are all considered good qualities to produce an attractive roof by skilled labour at, because they are thin, a steepish pitch of about 50 to 55° and often of comparatively complicated shape.

Some sandstone can be split to form stone slates but more often it is sawn to produce regular, quite large sized, slabs sometimes up to 75 mm thick in beige to very dark brown

shades. Sandstone roofing is typically encountered in Yorkshire, Lancashire, Derbyshire, Northumberland, Herefordshire and Sussex and parts of Cheshire and Durham, South Wales and Caithness in Scotland. As the slabs are thick and large, the pitch provided is usually much lower, at around 35°, with the slates laid in even regular courses on plain simple shaped roofs.

85: Stone slates on the roof of a late 1500s stone built manor house. The need to keep the roof design simple when such thick slates are used is emphasised here by the sudden change to clay tiles to form the valley gutter.

Inevitably, roofs covered in stone are, in the main, likely to be old and on dwellings which are either listed or in a conservation area. The unevenness, mottled colour and perhaps somewhat unkempt appearance is what attracts and makes such dwellings desirable. A new owner therefore is going to be required to maintain and conserve rather than alter that appearance and a surveyor's ready condemnation of what he sees and insensitive advice to replace with something prim and proper, according to modern practice, is inappropriate. He needs to recognise that such roofs constructed and maintained over the years by craftsmen, according to the needs of local conditions of weather and exposure have often lasted for hundreds of years without the benefits of modern practice. Nevertheless items wear out and things do go wrong so that a new owner may need to dig deep into his pocket to put things right for the roof to provide trouble free protection from the elements for some years to come.

The current owner may, of course, claim that he has already done any necessary work. Accordingly, if there is no access for close inspection to either the external or internal slopes, no apparent defects visible and no signs of damp penetration to top floor ceilings on oldish decorations, the surveyor should make enquiries and if possible see any Building Regulation or listed building approval, estimates, details of the work carried out, bills and receipts, together with any guarantee which may have been provided. In the case of top floor ceilings which have been newly decorated there is all the more reason to ask questions as this will be an indication that the dwelling has perhaps been redecorated preparatory to sale.

Signs of defects on external slopes of stone slates, when viewed either from a distance or close up, include cracked, broken, loose or missing slates, delamination, an over accumulation of moss and undue settlement in the slopes or ridge or the supporting structure. The latter is more likely to be a problem with the larger and heavier sandstone slabs rather than the more accommodating smaller limestone slates. It should, however, be evident from an inspection of the roof space whether the roof structure needs strengthening or restraining if movement is progressive. One of the problems is that the battens become over-stressed if there is undue settlement, some-times causing them to split but in any event distorting and loosening the securing peg causing slates to slip down. It may well be that if the condition of the roof is so bad that the slates need relaying, some judicious packing out may be necessary to improve the future lie of the slates.

86: A roof covered with stone slates under repair showing the stout timbers required to support the weight.

As the thinner slates, in all forms of stone slating, are usually laid nearest to the ridge, these are more likely to split due to wind uplift, particularly on the side furthest from the direction of the prevailing wind. Along with as close an examination of these as possible, ridge pieces should be looked at for breakages and the pointing for shrinkage cracking, as also verge pointing on exposed gables. Similarly with abutments, as lead flashings are less likely to be encountered because of the difficulty of fixing into stone walls unless built of ashlar, cement fillets are likely to be present and need checking for shrinkage cracking.

With porous stone, there is always the danger when it is laid flat as a coping, or at an angle as in roofing, of salts migrating as rainwater evaporates towards the exposed edges. The concentration of salts then effects the binding material of the stone which becomes friable and the edges crumble away. This is a more pronounced characteristic of some stones than others.

Roofs covered with stone slates, particularly those near trees, can acquire considerable quantities of moss and lichens. While these do not cause damage in themselves, they can interfere with the rapid run-off of rainwater allowing it to accumulate and correspondingly the effects of pollution to cause more pronounced damage to the stone below. Should the growth of vegetation build up in the side joints, the edges of the slates will be more affected by being in longer contact with polluted rainwater. If the growth is pronounced, advice needs to be given for the vegetation to be scraped off rather than removed by the use of fungicide which will probably cause unsightly staining if used.

As with all roofing materials, inspection of the underside, if at all possible, could reveal whether lamination either inherent or caused by salt action, is taking place and whether this might have affected, or be about to affect, the area of slate around the peg holes. Should this area become friable and the holes enlarge in consequence, the pegs could become distorted and eventually snap allowing the slate to slip down. Nail fixing, however, indicating probably a later roof or relaying at some time in the past needs careful inspection of the nails for signs of corrosion. Torching may be found as this was a fairly common provision for stone slating. If loose and friable, but there are only a few defects in the roof, its renewal may conceivably extend the life of the roof since, apart from its original purpose of keeping out wind blown rain and snow, its presence helps to hold stone slates in place.

As mentioned previously, the surveyor should be aware that the concrete industry has been producing artificial stone slates in reconstituted stone for many years in sizes, shapes and textures to match up with the traditional varieties for the various limestone and sandstone areas. The awareness is not only necessary so that on an inspection he can correctly identify what has been used as a roof covering but in recognition of the fact that, in some instances, replacement may be necessary. If carried out in the original stone, however, the cost may be beyond the reach of either a

present or even a prospective owner. Some such artificial slates are better in appearance and weathering qualities than others, but it is to be hoped that in their own locality, surveyors will have found out, or be prepared to find out, from the local conservation officer what are considered to be most suitable for the area concerned and, from their own experience, be able to put a 'ball park' figure on work that may be required not, it may be added, for inclusion in the written report, but for discussion at any interview.

Thatched

It has been estimated that there are around 50,000 dwellings in the UK with roofs covered in thatch and, even 30 years ago, there were 800 thatchers operating. The dwellings occur mainly in country areas of East Anglia, the South and the West Country but can crop up in many other areas, even City suburbs. Thatch is one of the oldest forms of roof covering and, as it is also one of the lightest of the traditional forms, it is particularly suitable for dwellings where the walls are of unbaked earth, such as cob or clay lump, not strong enough to carry the traditional heavier forms, and where they comprise the less substantial examples of timber framing.

As no thatch seen today can be more than 80-100 years old, its presence on a roof does not invariably indicate a dwelling of any great age and other evidence needs to be sought to establish an approximate date of construction.

Nevertheless many are of some age and subject to listing or included in a conservation area where any proposed change of covering would likely be met with opposition. This would be so even though the disadvantages of thatch, the ravages of birds for nesting material, infestation by vermin, the danger of

87: A surveyor acting on an instruction to inspect a 1920s suburban semi-detached house might be surprised to find that it had a thatched roof. It looks as though the birds have been helping themselves to nesting material just above the eaves on the right.

fire and the correspondingly higher costs of insurance may prove burdensome to an owner. The availability of grants in certain circumstances, the good insulation and sound proofing qualities and the undoubted attractive appearance could well be compensatory factors influencing a decision to purchase or persevere with ownership. It does no harm to advise prospective purchasers of these pros and cons along with other necessary routine precautions that need to be taken, such as being around on bonfire night, the provision of interconnected fire alarms in the roof space and in each floor level, extra protection for wiring in the roof space, extending chimney stacks and fitting spark arrestors and carrying out work to reduce the risk of fire reaching the thatch through the top floor ceiling, any hatches and windows. The fitting of a sparge pipe along the ridge is another suggestion that could be made. This would not only be for helping to douse any possible fire but could be used to water the thatch in a dry spell.

In the past, many different types of grown material were used for covering dwellings ranging from rushes, reeds, heather, bracken and broom to the straw obtained on harvesting rye, flax, oats, barley or wheat. While examples of all these may still be found on a local or regional basis, thatching is now mainly carried out in either water reed, combed wheat reed or long straw. The 2-3m long reeds obtained from the Norfolk Broads, the marshes and the margins of rivers elsewhere in East Anglia and further south and west have long been considered the strongest and best material for thatching, but the cutting and collection has always been arduous and they are difficult to work with. This makes water reed more expensive than any other material for thatching and recently the run-off of pollutants from farmland into the water has been affecting quality, necessitating imports from Eastern Europe and the Middle East. In arable crop growing areas, where conditions for reed growing on a large scale did not exist, long straw was the most common form of thatching material in use.

There was a considerable difference in longevity, however, between water reed and long straw. A roof thatched well in water reed could last a lifetime in the drier parts of the country while a roof of long straw, even of good quality and closely laid, would only last half that time. In the warmer and wetter parts of the country if short in stem, ravaged by birds and the weather, it might only last 15 years.

However, modern growing methods and modern farming machinery have rendered straw, unless specially grown, somewhat less useful for thatching purposes in normal circumstances; it grows too quickly to the detriment of its strength, is too short and the combine harvester crushes the straws.

Consequently for thatching dwellings, it is now necessary for wheat to be grown over an extended period to produce longer and stronger stems. When cut these can then be used in the traditional way as long straw, where the stems are about 1m long and the ears, stems and butt ends can all appear in the finished surface of the thatch. However, when it is cut and specially combed it is then known as 'wheat reed', a distinct misnomer as wheat is in no way botanically related to water reed. The special combing leaves the straw clean and unbroken ready for use with all the butt ends laid in one direction. Using good quality material and with a skilled thatcher a very close similarity in appearance to water reed thatching can be achieved with a life somewhere between the longer life of water reed and the shorter life of long straw.

All thatched roofs keep the rain out because there are a great number of small components to receive and transfer it down a long series of nearer horizontal, rather than vertical, stems to the eaves. Whatever the pitch of the roof structure, the stems are always at a shallower pitch. To follow the axiom of the smaller the roofing component, the steeper the pitch; compare small tiles, steep pitch, large slates, shallow pitch, therefore the steeper the pitch for thatch the better, 50° to 55° is preferable and not less than 40°. Shallow pitches should be viewed as more prone to earlier biodegradation as the water is in longer contact with the thatching material. This is one of the reasons why long straw thatched roofs, which traditionally tend to have been provided with less steep roof structures, need more frequent replacement and this is so even when the long straw is replaced by wheat reed without, as is often the case, the pitch being altered, though admittedly the replacement material will provide a somewhat longer life. The steep pitch, coupled with the innate property of permeability helps the thatch to resist wind pressure and avoids the need to weigh it down with ropes and stones, as is sometimes necessary in very exposed positions, although

securing with steel chicken wire or plastic mesh is often desirable, particularly towards the end of its life.

Thatch roofs tend to have an overall thickness of about 300 to 600mm, rain penetrating initially to a depth of around 50 to 100mm before running off at the extremity of the eaves in a band of about the same width from where it can get blown back against the wall below. Consequently a deep overhang of 400 to 500mm is required. Gutters and rainwater pipes are neither traditional or necessary and, indeed, gutters would require extensive brackets and need to be of inordinate size to be effective. However, some provision, by way of channelling, should ideally be present for dealing with the rainwater when it hits the ground. A good check on whether the thatch drains satisfactorily is to see whether, after rainfall, the underside of the eaves is reasonably dry near the wall head on the side sheltered from the wind. Thatch performs best on simple uncomplicated roofs and any changes in pitch to sweep over top floor or dormer windows should be gentle. For this reason such windows should be spaced reasonably far apart. Chimney stacks are best sited on the ridge as back gutters are not easy to accommodate.

As with all roof coverings, the surveyor should endeavour to provide the client with a reasonable idea of how long the present covering will last before entire replacement is necessary, together with what work of maintenance ought to be carried out in the meantime.

In view of the different life spans of the materials used for thatching, identification of the type of thatch is important. In the case of long straw with the shortest life span of around 20 years in the wetter and warmer parts of the country and perhaps 40 years in the very driest, this should not be too difficult. The method of laying is different, leaving lengths of stem visible, rather like a traditional style of haircut where the hair is brushed down smooth to lie flat, strokeable in fact like a cat. This is not the case with either combed wheat reed or water reed, unless it is in very worn and battered condition when it will need renewal anyway. The identification of water reed with the longest life of 80 years in the driest parts of the country and 40 to 50 in the wettest, as distinct from combed wheat reed with a life span of 30 to 40 years according to location is more problematic, as when laid they are very similar in appearance. With water and wheat reed, only the thick ends or butts of the stems show so that roofs appear similar to a cropped hair style, close, neat and tidy, gently pattable or as one writer put it, much as a hedgehog is enveloped in its spines.

Location could, at one time, be an initial guide in that a thatched dwelling in East Anglia was more likely to be covered in water reed whereas in the West Country the covering was more likely to be of combed wheat reed, locally called Devon or Dorset reed in the appropriate areas. With the importation of water reed, this is no longer a reliable guide and neither is the presence of the traditional patterns along the ridge and down the slopes associated with water reed necessarily a distinguishing feature, since they can all be closely matched in wheat reed. What is needed to determine which of the two types is present is a close examination. If it is at all possible to pull out some stems which remain unbroken, signs of ears and a length of about 1m would indicate combed wheat reed whereas the presence of flowers, long stem lengths of around 3m and a coarser texture would be evidence that water reed had been used as the covering.

Confirmation of the surveyor's assessment of the type of material used might be gained by questioning the current owner, though with the usual reservations as to veracity. However if, for example, a receipt can be produced for examination this could reinforce the assessment and, furthermore, provide a date for when work was last carried out and of what it comprised. Contradictory evidence and any elements of doubt in the surveyor's mind should be explained to the client and advice given either that an experienced local thatcher be commissioned to inspect and report or the names and addresses of a number of thatchers be obtained from the National Council of Master Thatchers Associations, 16 Purcell Close, Broadfields, Exeter, EX2 5QS, the Rural Development Commission for England and Wales or Historic Scotland. The Commission

88: A roof thatched comparatively recently in long straw clearly showing the stems which characterise this type of thatching.

89: The stems visible on the surface indicate that this thatched roof is of long straw, perhaps half way through its life.

90: This roof was rethatched four to five years ago, the matching ridge indicating that it was carried out in long straw. Contrast the appearance of this roof with the roof in 91.

92: A roof recently recovered in a rather rough and ready manner with straw derived from barley. Straw for thatching nowadays is usually specially grown for the purpose from wheat, which produces longer straw which is also the strongest. It is to be hoped that the owner here has not been fobbed off with the weaker straw from barley in the belief that it was from wheat.

91: The velvety appearance of the main slope of this roof and the contrast with the close cropped, neat material for the ridge indicates a thatching of 'reeds' but whether of water reed or combed wheat reed would need further investigation. The roof is clearly in good condition.

93: The thickness of some thatch can be more fully appreciated on a gable such as this where it can be seen to be approximately 800mm thick on this timber framed with brick infill cottage. The thickness is due to the practice of thatching on top of old thatch, a not uncommon method.

run training courses in England and Wales for thatchers who can obtain a City and Guilds Certificate.

If he can satisfy himself on the identity of the material which has been used, the surveyor should then give some consideration as to the remaining length of life of the two principal components of a thatched roof, namely the thatch itself and the ridge feature. As with information on the type of material, this might be derived from the owner subject, as before, to assessment of its veracity and

verification by the surveyor with his own eyes and the examination of receipts and details of work carried out.

The ridge of a thatched roof should be looked at first as this is the most exposed part, receives the most battering from the weather and deteriorates faster. It is for this reason that an essential item of maintenance for thatch is the renewal of the ridge capping which may be necessary once, twice or even more times in the roof's life, depending on the type of material used on the main slopes. Where water reed is the main covering, the reeds are too stiff and brittle to be bent over the apex where the steep slopes meet and some other material, usually sedge, a coarse pliable grass-like water loving plant, is employed being the most durable or, alternatively, tough grass or straw are used, all laced normally with willow or hazel spars. Such materials are not in themselves so durable as water reed and need to be renewed every 10 to 12 years. The same form of capping is added to roofs of combed wheat reed but if long straw is the main covering the same material can be used for the capping to give greater protection since straw is sufficiently pliable to be bent over the apex. In this case the appearance of ridge and main slopes match in colour and texture.

If the ridge is showing signs of patchiness, any decorative shaping is disintegrating and generally of an uneven and unkempt appearance, particularly at the top of south facing slopes, it is undoubtedly due for renewal. Neglect to do this at the appropriate time is distinctly false economy. Deterioration in the ridge releases pressure on the thatch below which can then be washed down the slope perhaps to be blown away leaving the upper part much more exposed than before.

If the remainder of the thatch is seen to be in good condition then re-ridging together with an overall examination to tighten the fixings of any loose sections and clean off any moss may be all that is required for another 10 to 12 years when the same process may need repeating. At the penultimate time of re-ridging, or perhaps at the one before, it may be thought advisable to secure chicken wire or fine plastic mesh over the whole surface of the thatch, which helps to hold it in place when its strength is weakening

towards the end of its life.

However, it may be that renewal of the ridge capping coincides with the need to renew all of the thatch. Signs that this is due are distinct discolouration at the eaves, the pattern of the fixings showing through, considerable loss of shape to any decorative features or patterns and areas of weathering where an unduly high proportion of stems can be seen on roofs of water or combed wheat reed. With all types of thatch general untidiness, unkemptness, patterns of run-off showing unevenly and distinctly more pronounced in certain areas are all signs of degradation. Intruders in the form of rats, mice and squirrels do damage from the inside of the roof space and this can show up as many short chewed lengths of thatch lying around while, from the outside, damage by birds in spring can be extensive, in particular on long straw roofs, where they pull out whole stems to build nests. Covering with netting does not always act as a deterrent and if wildlife becomes trapped the damage can become aggravated. Patching with long straw is sometimes carried out so as to put off the day when entire recovering becomes vital but this shows up very clearly and is only an indication that the day for entire renewal is not too far away.

94: Much more than a patch is needed for this thatched roof which is badly worn and degraded. The renewal of the ridge in the not too distant past was probably a waste of money.

95: *The patching with long straw to such an extent on this roof, apart from spoiling the appearance, is only putting off the day when complete recovering will be necessary and, even so, looks as though it ought to have been extended further to the left.*

97: *It would be false economy to ignore the wear and loss of shape to the ridge at the top of this south facing slope. Renewal of the ridge to this thatched roof will give it another lease of life and as it is already covered in a fine wire mesh it should give years of further service.*

96: *The thatched roof of this late 1600s cottage built of witchert, roughcast and colour washed needs entire renewal. Neither new ridge or patching will do.*

98: *This thatched roof is badly affected by moss, being near large trees and also facing north. The overall covering of wire netting is no doubt helping to hold back areas of degraded material from dislodgement but hampers removal of the moss thereby encouraging further degradation.*

As with damage to the ridge capping, defects are likely to be far more pronounced on south facing slopes where the thatch dries out to too great an extent but also, conversely, where areas are too sheltered from drying winds. For example, where a thatched slope is sheltered by a bank of trees some of which overhang, an undue accumulation of moss can build up preventing drying out and accelerating the natural bio-degradable process through increased fungal activity.

Thatch requires a balance, hence it will usually last much longer in the dryer less humid areas of the east than the damper areas of the west.

Shingled

It is conceivable that the surveyor will at times encounter a roof covered in wood, either as boards laid horizontally or vertically or, much more likely, in smaller rectangular sections known as shingles, in effect wooden tiles. This is likely to be more rarely than would have been the case in the past.

Wood shingles are another form of roof covering of long tradition. The cedars of the Middle East provided suitable wood in early days and the oak used in Medieval times for timber framing and boat building in many parts of the UK also proved suitable, as did other hardwoods such as elm and teak. Indeed the roof of Salisbury Cathedral was at one time covered with shingles. After the many fires in towns during the 1100 and 1200s shingles were perceived, along with thatch, as a fire hazard and prohibited in many urban areas. These bans coupled with shortages of suitable timber and the ready availability of other roofing materials saw a substantial decline in their use. Elsewhere, in North America and Scandinavia their use continued in popularity because softwood in those locations had always been found to be equally suitable, larch, pine and redwood for example but, in particular, western red cedar. The importation of the latter to this country in the early 1900s saw an increase in the number of roofs covered in shingles, particularly in rural areas.

Western red cedar is particularly suitable for use in shingles because of its close straight grain. It starts off as reddish brown when first laid but turns to a darkish silver grey tone on weathering. Shingles are very light in weight, something to the order of a tenth the weight of tiles, and accordingly may be found on some of the less substantially built bungalows of the 1920s and 1930s, themselves often lightly timber framed.

Subject to the shingles being treated with a fire retardant which, unlike similar treatments applied to thatch, is not prone to being washed out, shingles are an acceptable material for covering roofs under the Building Regulations subject to certain limitations on distance from the boundary and other buildings in the same way as with thatch. They need, however, to be treated in addition with a preservative to prevent insect attack as the natural repellent inherent in western red cedar can get leached out under the heavier rain conditions likely to occur in the western parts of the UK.

As in the case with other materials, the surveyor presented with the sight of new or near new shingles on a roof needs to seek assurances as to the species of wood used, the treatments applied to improve fire resistance and for preservation purposes together with any guarantee or at least an indication of the anticipated length or life. Use of a less suitable species, such as larch, may reduce life expectancy down to 20 years or so, particularly if used in the wetter parts of the country. Western red cedar shingles, on the other hand, might be expected to last something to the order of 50 years if appropriately treated. Good quality shingles, often known as 'edge grained', are obtained from quartered logs alternately rift sawn since shingles taper down from tail to head. They are sawn outwards from the centre of the log at right angles to the annual rings so that the lines of grain are roughly equidistant apart and ensuring that the fibres run continuously for the whole length of the shingle. Cheaper shingles can be produced by sawing tangentially to the annual rings leaving the

99: Terrace houses with rooms half in and half out of the roof space showing the characteristic colouring and appearance of roofs covered in shingles. The porch roof and the house on the right have been recovered with natural slates.

annual rings far from equidistant. When laid this can lead to early shrinkage in width, and warping and splitting when exposed to the elements, as well as being more liable to decay.

Where a roof of shingles is neither new or near new and can only be viewed from a distance, signs of warping or splitting coupled with an undue gap between the shingles could indicate that the cheaper flat sawn variety have been used. If the opportunity for closer inspection exists, such defects may be more easily ascertainable and signs of decay or undue erosion on the exposed surfaces may also be seen. Warping can also occur if, on laying, the shingles are nailed too closely together with a gap of less than the 6mm normally employed to allow for swelling. If this is so, the defect will probably be apparent much sooner than, for example, decay or erosion.

Shingles are normally centre nailed but, even so, wind uplift may occur if they are not very securely fixed. Two nails are needed for each shingle and it is preferable for them to be ring shanked for greater security. Because of their lightness in weight there is a greater need than with other roof coverings for adequate tying down of wall plates and the roof structure, a point to be checked if there are any signs of disturbance.

Because the run off of rainwater from a shingled roof is acidic, signs of defects may be apparent in rainwater gutters and pipes if any type of metal fitting has been used, as can also be the case with flashings against or around any projection through the roof surface. Such projections are ideally avoided but if they are essential plastic fittings should have been used in preference.

Traditionally, shingles were nailed to battens with no underfelt to allow a free circulation of air and it was not common for felt to be provided. If it is provided nowadays, then counterbattening is essential to allow moisture to run down below and between the shingles. The sight of underfelt therefore from the roof space should lead the surveyor to enquire whether counterbattens were employed if it is not possible to check at a tear in the felt whether this was done.

Surveyors should be aware that artificial shingles can be made from wood particles bound together with Portland cement. Although thicker and heavier than wooden shingles they are similar in appearance to those of Western red cedar, but are claimed to have better resistance to the spread of fire and a longer life expectancy. They have been available abroad for some time and it may well be that opportunities for their use here in similar circumstances to wood shingles will appeal to developers and their designers. If encountered, a surveyor will need to include advice and warnings as he would in the case of other new materials brought into use without evidence of long term satisfactory performance.

Other forms of Pitched Roof Coverings

Occasionally the surveyor will encounter forms of material covering pitched roofs other than tiles, slates, thatch or shingles. The metals copper, lead and zinc can be used on the slope, the two former often used to crown features such as canopies, porches, balcony roofs and turrets, but their main purpose in domestic building is as coverings for flat roofs on the more expensive dwellings. Their characteristics are the same, as are those of zinc, and the detailing is not wildly different when used for either purpose and these will be dealt with later under the respective headings for flat roof coverings.

100: Standing seam copper roofing to a feature on an expensive dwelling after 70 years exposure. The metal in the atmosphere of this area has formed a greyish patina instead of the more common green found where there is a different form of pollution. In consequence the appearance might be mistaken for lead but the use of a standing seam joint differentiates as lead is too soft and subject to greater thermal movement to be used in this kind of detail.

101: A barrel vault roof covered in copper, showing the more characteristic green patina layer to the metal produced by the acidic pollution of the atmosphere together with standing seam joints.

102: A fairly new lead roof on the slope showing the characteristic near round batten roll side joints. Whereas for lead flat roofs the lengthwise joints need to be in the form of a drip with a change in level not less than 50mm and preferably more, on the slope they can be lapped with a clip to hold down the edge.

103: Corrugated asbestos on the roof of a two storey terrace house, probably the original from the 1950s. Clearly the house is in need of recovering with a more permanent and less health hazardous form of roofing as has already been done elsewhere on the estate, but how will the structure below need to be adapted? Costs will be increased by reason of the need to comply with the regulations for the removal of asbestos.

Other materials which may be encountered such as corrugated steel and corrugated asbestos can be discounted as being inappropriate covering materials for dwellings with any pretensions to have a life expectancy of at least 60 years and should be advised for renewal in a more permanent form. The same can be said for bituminous or other forms of felt in whatever condition. This includes felt made in the form of tiles, sometimes incorrectly and confusingly known as 'shingles' in the building trade. Failure with the early types of built up felt roofing and the workmanship involved has long been evident and while maker's assurances of longevity based on fatigue and other testing for the many different varieties of felt made available since the 1980s are often backed up by various types of certificate, it is not yet possible for them to be confirmed by evidence of long

104: Corrugated 'iron', in fact steel, to the roof of the studio section of a late 1800s listed cottage.

105: Colour coated profiled steel panel roofing to part of 1970s local authority flats. Parts of the stepped flashing at the abutment to the wall of the higher section has come away and is likely to lead to damp penetration.

106: One wonders what was said about the roofs to prospective purchasers of long lease flats in this block when it was new, perhaps 15 to 20 years ago. The reflective surfaced felt is not lasting well and is already patched above the right dormer while the flashing to the separating wall is looking suspect and the one opposite has already had to be replaced by other material. All lessees are going to have to dip deeply into their pockets for a new roof covering very soon; tiles perhaps on the sloping surfaces would have been infinitely preferable for a long term solution with asphalt or a metal on the flat sections above.

Groundwork

Those features which are included in pitched roofs below the covering itself but are not part of the supporting structure are usually referred to by the term 'groundwork'. Changes to the traditional arrangements in the last 40-50 years have had serious consequences, most not anticipated at the time they were made.

For long, groundwork was limited to either battens for securing tiles or slates for cheaper work or close boarding, either feather edged for hanging tiles with nibs or plain with battens which could also support tiles with nibs but also those without and, of course, slates, both of which would require nailing. Boarding is more expensive but while it is the traditional form of groundwork in Scotland, it is more rarely found in England and Wales; it is thought in only about 2% of dwellings south of the border.

term use in real life conditions. This is particularly so in domestic situations where the rigidity of structure more often found in commercial premises is usually not achievable.

The use of such coverings on outbuildings such as garages is a different matter, since defects do not take on such a critical importance. Clearly, however, their condition needs noting along with whether their renewal is required on the assumption that this might well be carried out in the same material in the circumstances of the comparatively small areas involved and their situation.

107: Slates nailed to battens viewed from a roof space when there is no underfelt. The condition of the nails and the edges of the slates can be examined.

109: Profiled clay tiles hung at a shallow pitch entirely by their nibs on battens viewed from below where there is an absence of roofing felt.

108: Plain tiles hung entirely by their nibs on battens viewed from the roof space of a steeply sloping roof where there is no provision of underfelt. The surveyor can check for any deposits of a whitish powder and any signs of crumbling in the nibs.

110: Torching of lime and hair plaster to the underside of clay tiles and battens which helps to keep wind blown rain and snow at bay and adds to the security of the fixing whether by pegs, nails or nibs.
The downside is a possibility of wet rot in the battens as the torching soaks up the rain and snow to remain damp for some time.

In roofs without boarding, the penetration of wind blown rain and snow was dealt with in many cases by the provision of torching with lime and hair plaster around the nibs, nails and battens. The torching had a secondary function of helping to secure the tiles and slates. While successful at keeping damp away from ceilings, not always possible if omitted, torching nevertheless tended to cause wet rot in the battens by retaining moisture. Fortunately this is an item the surveyor can check in a roof of this type along with the actual condition of the torching.

Before the introduction and general use of

bituminised roofing felt as a moisture, but not vapour, barrier over the last 50 to 100 years, wet rot could also be a problem in boarded roofs. Torching is not a possibility with boarded roofs and wind blown rain and snow could creep under both tiles and slates. Even though there was usually sufficient ventilation to prevent outbreaks of dry rot, as there was usually in the case of unboarded and unfelted batten roofs provided with torching, wet rot could affect the boarding. This was particularly a possibility if tiles were hung on the feather edged variety or slates were nailed direct to flat boarding, detailing which tends to limit ventilation in each case. The provision of battens and counter battens would improve conditions but the omission of the counter battens could substantially increase the risk of wet rot in the cross battens.

Boarding fixed to the top of the rafters allows the surveyor only a glimpse, between cracks, of what arrangement has been provided above. If the dwelling is from the earlier part of the 1900s and of good quality, as were many built at that time or even built in the latter part of the 1800s, and still appears to have its original roof covering, then it is highly probable that there is no roofing felt. Confirmation of this may well be obtained by indications of damp staining on the boards, principally on the edges, as the boards are probably of sufficient thickness and quality as to prevent moisture penetrating the total depth. It would be entirely appropriate in these circumstances, and indeed essential, in the event of the roof needing to be recovered, to provide a warning as to the likelihood of considerable renewal being required to the boarding, total renewal if the boarding was found to be feather edged. Such a warning would be even more essential if there were signs of slipped or dislodged tiles or slates indicating that battens or boards were no longer sound enough to hold nibs or nails.

As indicated, there is a danger of wet rot in both boarded and unboarded roofs of earlier dwellings but little danger of dry rot, generally, above the level of the boarding. Below boarding, however, in the roof space and in the boarding itself, there could be a danger of dry rot if ventilation is limited because of tightly closed eaves, particularly in confined parts of the structure, for example towards the foot of a valley gutter or at the corners of hipped slopes. The surveyor therefore needs to pay particular attention to these areas, even at the risk of getting a little dusty, and all the more

111: The sight of boarding below the roof covering provides the surveyor with food for thought. If the covering is of tiles or slates, has underfelt been provided and have counter battens been laid to prevent wet rot setting in to the transverse battens on which the tiles are hung or tiles and slates nailed? Damp staining to the edges of the boards usually indicates an absence of underfelt and probably counter battens but it is not possible to be certain unless there is a gap somewhere in the boarding, more likely to occur at abutments.

112: The underside of a boarded and tiled roof of a dwelling approaching 100 years in age and still with its original roof covering. The damp staining indicates the absence of roofing felt to keep wind blown rain and snow at bay. Appropriate warnings would need to be included as to what might be found to have happened to the timber when the top surface of the boarding is exposed on recovering the roof.

so if there is evidence of damp penetration nearby.

The development of felts enabled both boarded and unboarded roofs to be provided with protection against wind blown rain and snow and rendered the provision of torching

113: Underfelt to roof coverings, fixed through battens to the rafters and which prevents the ingress of wind driven rain or snow, must be fixed to sag slightly between each rafter. Not too little, not too much (and consistently) but just enough as here, to allow any water to run away below the battens into the gutter, otherwise there is a danger of wet rot in the battens eventually depriving slates or tiles of support.

114: When a section of roofing felt is allowed to sag too much and there is insufficient overlap between sections as here, there is a danger of water flowing back into the roof space.

unnecessary for unboarded roofs. Fitted across the rafters and below the battens it could, however, be the cause of wet rot in the battens unless allowed to hang down slightly between each rafter, a point always to be checked, because water could be retained along the top edge. If felt was fitted above boarding then counter battens were needed otherwise again there was a serious risk of wet rot developing in the battens for the same reason. As before the presence of boarding usually prevents the surveyor from seeing what has been provided above and he may, therefore, need to provide appropriate warnings of what might happen in the future if there are not already signs of dislodged tiles or slates providing strong evidence that the fault has already materialised.

Up to the 1960s it was rare for top storey rooms to be heated. In consequence roof spaces remained cold with the air at a temperature and water content not too dissimilar to that of the exterior. As such no problems of condensation occurred, although there were often plenty of other problems, for

example if the dwelling was provided with an unlagged cold water storage cistern. The advent of widespread central heating and the use of plasterboard and a skim coat of plaster, *in lieu* of lath and plaster, to top floor ceilings for new dwellings in the 1960s and 1970s led to roof spaces being almost as warm as the rest of the house. This could lead to a high level of condensation as the warm air, able to support a much greater level of water vapour, would condense overnight on the cold under-side of the pitched roof structure, the ground-work or covering and any metal fastenings forming part of the roof structure. Although this might cause damp staining on the timbers there would only be problems if the roof was constructed in such a way that ventilation was so very restricted as to lead to wet rot or even dry rot in the timbers, to corrosion in metal components and perhaps even damage to the electrical installation through dampness.

Considerable increases in fuel costs led to the recognition that this form of construction led to waste and to most owners adding a layer of insulating material at top floor ceiling level as an improvement and such insulation becoming a requirement under the Building Regulations for new dwellings. The degree of insulation required has been progressively

115: Early types of roofing felt often deteriorated rather rapidly and would split under the weight of any substantial amount of water. Here an old friable felt has been replaced with new and given a fairly generous sag between fixings. It would have been better to have removed the old felt.

116: Roofing felt clumsily torn away to allow a tube to ventilate the roof space to be fitted. Such tearing can be useful at times to the surveyor by permitting a sight of the underside of the roof covering but, obviously, can allow damp penetration.

increased as national commitments to reduce energy consumption and limit greenhouse gas emissions have been made. Around the same time as insulation became a requirement for new dwellings so too did a requirement for ventilation to roof spaces to counteract the risk of condensation. The attempt to provide a vapour control barrier at top floor ceiling level by the provision of foil backed plaster-board, while practical from the constructional point of view, proved totally unsuitable as water vapour condensed on the room side of the foil, ruining the plasterboard.

Current ideal practice is therefore to concentrate on providing an effective layer of insulation, now considered to be no less than 200mm thick, in both new and, if possible, existing dwellings, together with ventilation in accordance with current Building Regulations. This produces what is termed a 'cold' roof. However, even in new dwellings the principles of such a roof have not always been translated satisfactorily into sound forms of construction. Close inspection of many dwellings built since the 1970s and 1980s has brought to light examples of insulation quilts stuffed into the eaves restricting ventilation, insulation missing from areas difficult of access, introducing cold

bridging, and heavy condensation in small areas causing damp stains on ceilings. Further problems are access traps not insulated and inadequately sealed, dormer windows not ventilated, insulation carried under cold water storage cisterns, a lack of cross ventilation where there are separating walls and poor ventilation to monopitch roofs. All these points need checking by the surveyor if inspecting a new or near new dwelling.

117: When insulation at top floor ceiling level is encountered it is usually around 100mm thick, roughly to the top of the ceiling joists as here in fibreglass. This amount is not now considered adequate, but does at least enable the surveyor to step gingerly, if safe to do so, on to joists and move around to inspect storage cisterns, wiring etc., as well as the underside of the roof covering or boarding.

For existing dwellings, the provision of what is considered the effective 200mm of insulation will have made for a colder roof than previously so that if there was no evidence of previous condensation problems it is probable that no further work might have been considered necessary when the insulation level was increased. Such a thick quilt of insulation, however, covering the ceiling joists, would be a nightmare for any inspecting surveyor, owner or contractor's workmen attending to an emergency in the roof space. Boarded walkways would be an essential provision in the circumstances, otherwise a foot through the ceiling would be the likely outcome.

Evidence of previous condensation problems, however, would necessitate investigation into ventilation aspects. To achieve Building Regulation standards, many existing roofs would need modification at both eaves and ridge level. For projecting eaves, forming ventilation apertures in the soffit may not present too much of a problem but with flush closed eaves where the guttering is virtually at the wall head, complications would arise. Ridge ventilation should not, however, be too difficult, as would the insertion of ventilators in the roof slopes to provide cross ventilation air currents to dead areas around roof lights and dormers. It is also possible to insert these in to the lower ends of roof slopes when ventilation through closed eaves is not a feasible proposition.

Although not so attractive from the constructional point of view as in the case of flat roofs, a 'warm' roof, where the insulation panels are placed between and a vapour control layer placed below the rafters, is still a very suitable form of construction for new dwellings where rooms are to be provided in the roof space, an option increasingly exercised by developers over the last 20 years or so. It could also be a suitable solution for

118: Surveyors, of course, are not the only persons needing to be able to move around in the roof space. Plumbers, electricians and not least the occupier may need access, sometimes in a hurry. Partly floored or with defined boarded pathways provide a useful feature. These are not really adequate for the purpose in this roof space with its somewhat less than 100mm of mineral granule insulation.

upgrading an old house if the roof covering has to be entirely renewed. However, if the roof space is not going to be used for any purpose at all most owners would consider it too wasteful to allow it to be heated through a lack of insulation at ceiling level. Upgrading, nevertheless, often includes the formation of an attic bedroom, study and bathroom in at least part of the roof space, in which case the provision of a warm roof overall, omitting all insulation at top floor ceiling level, is by far the most straightforward solution and obviates the need even to insulate a cold water storage cistern or pipes positioned in space otherwise unusable. Hopefully, any access to such area for maintenance purposes will provide the surveyor with a safe and feasible opportunity to enable a check to be made on the construction.

Only a few dwellings at any point in time comply with the very latest amendments to the Building Regulations. An endeavour to bring existing dwellings up to current standards would not, in most instances, be considered worthwhile and is certainly not a requirement of the law. A lot of money could be expended without necessarily any great advantage being obtained or any real savings in running costs being achieved. This is not however the case with the services providing heating and hot water and the Government's committed aims can usefully be said to coincide with the aims of most householders to minimise their costs. Even apart from any requirement to provide an Energy Efficiency Report following the inspection of a dwelling the surveyor's report should indicate the shortcomings of the roof in regard to its ability to reduce heat loss, since for a comparatively limited expenditure, reasonably good annual savings in energy costs can be made and a pay back period of under ten years often achieved on the expenditure.

119: By adding another 150mm of fibreglass insulation on top of the 100mm existing, starting at the extremities no doubt and skilfully withdrawing towards the trapdoor, in itself however left uninsulated, the contractors have left nowhere to step, not even gingerly. The surveyor should not disturb in any way, but report the limitation on his inspection and advise that something needs to be done to improve the situation.

Flat Roof Structures

The Inspection

Although the surveyor's ladder should enable him to gain access to the upper surface of most single storey structures with flat roofs, it is best for safety reasons to commence the inspection internally to endeavour to establish

whether the roof is of hollow or solid construction and whether subject to leakage. A sharp tap on the ceiling with the surveyor's measuring rod, or a handy broom handle, will indicate by a resonant sound that it is hollow or a dull thud that it is solid. There might be pattern staining on a ceiling not recently redecorated and, if the roof is poorly insulated, an indication of the spacing of joists. In the majority of cases for domestic premises, the construction will be found to be hollow and if the structure lower down matches the rest of the dwelling, then it will probably bear a distinct relation to the floors elsewhere. This is likely to be true even if the structure is found to be solid, as it might well be in the case of a flat roof to a block of flats.

A consideration of the general condition both internally of the ceiling and externally of the roof should be made next to help towards a decision as to whether it would be wise to walk on the roof. Undue neglect, damp stains on the ceiling and a peep from the ladder in the case of a single storey structure revealing an aged covering with visible defects, would indicate that it would be unwise to do so. Hollow timber constructed flat roofs have been known to collapse when warmth, damp and a lack of ventilation provide ideal conditions for outbreaks of dry rot. The failure to warn a prospective purchaser of an upper maisonette in a converted house of the lack of ventilation to a hollow flat roof and the possible consequences of damp penetration through badly constructed upstands without angle fillets, or cover flashings, to the felt roof, led to a surveyor being found negligent in the case of *Hooberman* v *Salter Rex* 1985. It transpired that chipboard had been used as a decking with the vapour control layer placed on top instead of underneath, resulting in its disintegration through condensation and extensive dry rot in the timber structure. It will be noted that the surveyor was found negligent for failing to warn the purchaser of possible consequences derived from what he could see, not for failing to report on unsound construction that he could not see.

Being unable to see the construction is, of course, the bugbear on providing suitable advice on the condition of flat roofs. The surveyor will no doubt be familiar with the differences between cold deck, warm deck sandwich and inverted warm deck flat roofs, the relevant constructional details of each and their relevant advantages and disadvantages. Such knowledge, while necessary for a surveyor to have, will be of no use to him when presented with a waterproof covering on one side and a ceiling on the other and no opportunity to see what lies between.

Considering the approximate age of the property might be thought to be a help, but it does not provide any degree of certainty. Cold deck flat roofs likely to be found on dwellings built before the 1960s are not thought of highly nowadays, because of their need for ventilation and the difficulty of its provision. If the covering has been renewed and there is no ventilation, there is no telling for certain whether the roof has been upgraded to either a warm deck or inverted warm deck or left as it was, with possible dangerous consequences. The presence of ventilators, on the other hand, whether old or new does provide the assurance that this aspect of the construction, essential if a cold deck roof has been retained, has been provided.

It is best for the surveyor not to speculate on what might be and to confine himself to stating whether construction is hollow or solid and whether ventilation is provided or not, with appropriate warnings of the possible consequences if not.

Nevertheless, a close visual inspection of the ceiling can provide crucial information. Damp stains have already been mentioned but cracks in the plaster can be caused by deflection in the structure. This could be due to inadequacy in its strength through, for example, the use of joists of insufficient depth, the formation of a trapdoor without proper trimming or the placing of heavy items on the roof not envisaged at the time of construction, perhaps a cold water storage cistern or a tank, or the formation of a terrace. Many flat roofs of timber construction are supported on wood plates set half way into one brick walls very much exposed to driving rain which penetrates the brickwork and which causes either wet or dry rot in the plates with consequent deflection in the structure. Compounding the defect, deflection is a frequent cause of failure in the covering itself.

The deflection might be a permanent condition and the surveyor's ladder will enable him to scan across at ceiling level to see whether it is bowed as well as cracked or it might even be just bowed. If the deflection is intermittent from foot traffic or the occasional heavy fall of snow, the ceiling is more likely to be merely cracked and not bowed as the structure will recover.

If the key to a lath and plaster ceiling has failed because of deflection it could be bowed to a greater or lesser extent, depending on the circumstances, and possibly in danger of collapse. A gentle push with the palm of the hand should indicate whether the key has failed or not. Whether to do this must be at the discretion of the surveyor who ought, first, to obtain the vendor's permission having already explained the possible consequences. It should not be too difficult to envisage what the surveyor should include in his report if the various scenarios above are presented to him. Warnings need to appear as appropriate and these need to be coupled with the conclusions derived from the visual examination of the covering.

Coverings for flat roofs ie normally considered to be those below 10° in pitch, differ substantially from those, in the main, used to cover pitched roofs. While those used to cover flat roofs can be used satisfactorily to cover pitched roofs the reverse is by no means the case. Because water is in contact with the flat roof surface for much longer periods, conditions are much more onerous and coverings have to be impervious whereas those for pitched roofs can afford to absorb a proportion of water for it to evaporate when the rain ceases. The components which make up most pitched roof coverings are small and capable of accommodating, and adjusting, to a limited amount of movement in the structure below, far more so than those used normally to cover flat roofs. Ideally flat roof coverings are seamless and have no joints. Sprayed on coverings are a current development and conceivably could have a future but the only material in use now which fulfils that ideal is asphalt. All others are delivered to the site in sheets and have to be jointed appropriately as they are laid according to the type of material, be it lead, copper, zinc or felt.

Deflection in the structure has already been mentioned where it is evident from an examination of the ceiling below, but its effect above will often be the cause of ponding, because the fall, which should have been provided on construction to counteract those permissible amounts of deflection allowable in all forms of design and construction, will be compromised. Even on seamless asphalt, ponding can cause problems by encouraging a build up of moss or other organisms, particularly near trees, and obstructing outlets. In extreme cases, it can allow a build up of water to such an extent that flashings to pipes or stacks are covered, resulting in damp penetration. With sheet materials the build up of water can breach the joints themselves, particularly if blown by the wind. For felt

sheeted roofs which have stuck or welded joints, the persistent presence of water will soon reveal any faults in the workmanship.

There was a time when, for seamless asphalt roofs, it was thought better to construct without a fall and to form a reservoir with a weir, to control the level of water so as to obviate the affect of the sun on the material. Certainly for domestic purposes, excessive deflection, obstruction of the weir and drying up in periods of drought have tended to provide greater problems and provision of a proper fall and a pale light coloured reflective surface to the material are now considered preferable.

The effect of relatively intermittent ponding will show, when dry, on the surface of the roof covering by the whitish colour of the deposits left behind by the slow evaporation of a quantity of water. More persistent ponding will reveal itself by the vegetation which has taken root and which will be encouraged and more firmly embedded if the covering be rough rather than smooth, for example mineral surfaced felt or the more corroded metal surfaces.

Flat Roof Coverings

Mastic Asphalt

Although naturally occurring bitumen mixed with sand had long been used for waterproofing purposes in the Middle East, mastic asphalt did not come into use as a covering for flat roofs here until the latter part of the 1800s. Initially, the material would have been entirely the imported natural mineral rock found in parts of Europe containing a percentage of bitumen. Another bitumastic product laced with mineral material was found in a lake in Trinidad and refined and imported. Later, the distillation of petroleum was also found to produce bitumen and this derived product when mixed with a limestone aggregate produced a synthetic mastic asphalt. Cast into blocks it could be delivered to the site in the same way as the natural product. The synthetic product can be modified with the addition of polymers to avoid cracking in colder climates and strengthened against indentation with further mineral additives to produce a surface suitable for foot traffic.

120: Surface crazing to a mastic asphalt roof, made more pronounced here by the application of a lighter coloured surface treatment, is usually brought about by over trowelling bringing the mastic content to the surface. It is generally of no consequence.

A surveyor inspecting a flat roof covered in asphalt will not be able to tell whether the material is natural or synthetic or how it may have been modified, since there is no observable difference when laid. On the inspection of a new or near new dwelling he might, however, have the opportunity of examining the specification and be able to pass on its contents with comments to the client and with the appropriate advice that it has to be taken on trust, even if substantiated by receipts, since there is no possibility of checking whether it has been carried out in every detail. There are British Standards for mastic asphalt, one for material containing natural rock aggregate and another containing limestone aggregate, a further Standard for sampling and testing and a Code of Practice for laying to which,

hopefully, any specification for a new or near new dwelling will refer so as to enable the surveyor to comment. The Mastic Asphalt Council Ltd, Claridge House, 5 Elwick Road, Ashford, Kent, TN23 1PD, Tel 01233 634411 produce useful publications, which can also be used for the purpose of checking details of work carried out.

Although it might be thought and indeed it is believed to be so by BRE, that the natural might be more long lasting than the synthetic, any difference there may be is materially insignificant since it is usually the detailing rather than the material itself that lets the covering down. Well laid to a sound specification a life of 50 to 60 years may be expected from asphalt, sometimes even longer.

On delivery to the site, the blocks of asphalt are heated in a cauldron to become a thick paste which is spread, usually in two coats about 19mm thick, on an isolating sheathing membrane to separate it from the structure below, either a timber decking or a cement screed on concrete. Both decking and structure should be as firm and stable as possible although no movement joints are required as the asphalt will take up any slight imperfections and accommodate any slight movement.

The fact that the blocks of asphalt have to be melted down to produce a spreadable material provides a clue to its main weakness. Laid on surfaces no more than 10° from the horizontal, it will perform well, only softening under the effects of hot sunshine not expanding as do the metals used for flat roof coverings. Laid at an angle, particularly the 90° required at abutments to walls, chimney stacks, roof lights and pipes, in the full glare of the sun, it will sag and creep downwards by gravity and its own weight, a fault commonly found. Angle fillets of 45°, with lathing to provide a key at the junction of horizontal and vertical

121: Straight cracks in the asphalt surface are usually the cause of damp penetration, as here. Movement in the supporting structure is often the reason and needs investigating. The fact that the asphalt skirting to the chimney stack has had to be replaced in felt, nevertheless suggests further trouble and that total renewal is required. Such advice needs to be coupled with warnings as to what might be found underneath.

122: All the evidence here suggests an asphalt roof at the end of its useful life and in need of renewal, together with any work necessary to the underlying support. The renewal should include the provision of a more adequate skirting detail.

surfaces, are therefore essential with, where there is brickwork, well raked out joints to a depth of 19 to 20mm as well. Furthermore, a reflective white or silver coloured paint should be applied to the asphalt before a flashing at least 75mm deep, preferably of lead, is well tucked into the brickwork and dressed down over the top of the asphalt. In this respect, an inverted warm roof where insulation is placed above the asphalt, also included at the abutments, is a preferable type of roof construction for asphalt along with a hefty type of ballast to hold the insulation down against wind suction. These have, however, only come into use since the 1980s and the surveyor is only going to be able to report either on specification details, if available, or merely on what he can see and this is likely to be fairly limited should such a roof be encountered.

Blistering of the asphalt is another defect fairly frequently encountered, though much more so if the structure below is of concrete rather than timber. It is due to entrapped moisture expanding on warming by the sun and pushing up the material. The fitting of small dome

123: Asphalt skirting to a flat roof beginning to sag downwards through the action of the sun and pushing the flashing outwards due to a lack of clips and which will eventually cause damp penetration. A case where the upstands could be renewed where the near horizontal asphalt surface is sound and intact.

hooded ventilators at the time of construction is the way now adopted to anticipate this problem, but if not incorporated at that time it is possible to have them inserted at a later date if blistering appears. The normal method of poulticing with hot asphalt to carry out repairs can be used to remove a small section of roof covering for their insertion. If nothing is done, there is a possibility that the blister will expand further until it bursts, with almost certainly consequential damp penetration.

Deflection in flat roof construction has already

been mentioned and this might be the cause of any crazing seen on the top surface of the asphalt where excessive trowelling on laying has brought a more fluid and less dense coating of bitumen to the top. This is usually of no consequence and can be distinguished by its patterned affect from cracking through the total thickness brought about by movement in the decking as distinct from the structure. If expansion occurs in the decking this could produce an upward pressure and induce longer straight cracks in the asphalt.

Cracks in the asphalt, other than crazing, together with indentations caused by foot traffic perhaps in excess of that originally envisaged, can be repaired. Also repairable are skirtings which have lost their adhesion to any projections through the surface of the asphalt. There will come a time, however, when the extent of the visible defects, coupled with a worn degraded appearance of the general surface, will suggest to the surveyor that he must advise his client that the covering should be renewed with all that entails. Warnings of what might be found if there are already signs of damp penetration need to be given since, in his view, further repair would prove uneconomic. It would then be up to the client to decide whether to continue with further uneconomic repairs or to follow the surveyor's advice to renew.

Lead

Along with zinc, lead is the most likely metal to be encountered on flat roofs over dwellings. It is heavier, softer and more malleable than zinc and therefore more likely to be found as the material chosen for covering porches and canopies of complicated shape as well as balconies and dormer windows. The difference in cost between lead and zinc usually finds the former provided on the better quality of

dwellings but less so nowadays than in the past. The difference in cost is justified by the difference in the length of life, something like 100 years for the currently available milled lead covered by BS1178:1982, perhaps double that of zinc. In earlier times lead was cast on sand in sheets which came out much thicker and when reversed produced a rougher texture for exposure to the weather. This could last, astoundingly, to the order of 400 years depending on the amount of pollution in the atmosphere. The use of cast lead, nowadays, is generally confined to historic buildings.

Although a shiny silver when new, milled lead soon develops a patina of lead carbonate when exposed to the atmosphere which turns it to a medium grey shade but provides the metal with a strong degree of protection against pollution.

Lead has a relatively high coefficient of expansion which has to be allowed for in the design of joints. Side joints to sheets are usually accommodated in near round wood batten rolls in the direction of the fall. Standing seams, although possible, are too easily damaged on flat roofs and, if encountered, will usually be found to be so. Lengthwise joints, so as not to interrupt the fall, are welted drips and involve a slight change in level, not less than 50mm and preferably more. Any wind blown build up of leaves, moss or grit accordingly can cause flow problems towards outlets and perhaps compromise the joint.

The need for a change of level at lengthwise welted joints can cause design problems in long flat roofs and even more so in parapet gutters where there is limited height. This sometimes results in detailing which is less than ideal and intermittent problems of damp penetration, particularly after heavy falls of snow. A critical examination of the detailing is therefore essential in such circumstances so as to warn the client of possible problems. The details as found can be checked against those shown in BS6915:1988, which covers design and construction for the fully supported metal, and which along with BS1178:1982 for the production of milled lead can be used when a new or near new dwelling is being inspected and there is an opportunity to examine the specification. Also available for this purpose are the publications of the Lead Sheet Association details of which can be obtained from www.leadsheetassociation.org.uk or by telephoning 01892 822733. Checking details is facilitated by the malleability of the metal which permits it to be lifted at rolls, drips and flashings to see whether the detailing is sound. While checking the detailing is relatively simple, checking that lead of the specified Code has been used is less straightforward and delivery notes and invoices need to be seen where a new or near new dwelling is being inspected.

Over the years, however, on older roofs, the continual intermittent movements under the effects of hot sunshine induce metal fatigue and the development of ripples, buckling and eventually of splits and cracks. Although these can be repaired by burning in new lead, eventually the frequency of their occurrence will be such as to render repair uneconomic and renewal will be required, which may also involve renewal of the decking. This would be the case even though the roof was in such an area as to be comparatively free from the acid rainwater, resulting from atmospheric pollution, or attack from the acids in the run off from copper, perhaps in a lightning conductor, oak details in sills for example, cedar shingles or from other timbers treated with preservatives or from moss or algae. The alkali in mortar also attacks lead unless it is protected with a coating of bitumen. This protection was often not provided in the past, leaving the tucked in section at the top of flashings vulnerable to

attack, often resulting in a split at the right angle bend. This can be compounded by the weight of the metal preventing its recovery after sagging downwards in hot weather.

124: An old flat roof over the projecting ground floor part of a late 1800s dwelling. Two batten rolls are shown but the lead towards the bottom of the photograph has buckled and split. To compound matters, the rainwater gutter has been badly aligned and holds water.

Damage from foot traffic, much more likely from flat roof use in summer when the metal is hot from the sun but also from maintenance operations at other times, can cause indentations and eventually holes to form unless the roof is provided with duckboards or some form of raised platform or terrace, but these are seldom found.

Zinc

Zinc came into use as a covering for flat roofs in the latter part of the 1800s as a cheaper substitute for lead. It will frequently be found covering the roofs of the back additions, offshots, of dwellings constructed in that period or the early part of the 1900s.

Like lead, zinc is a shiny silver when new but is far less malleable. When laid, it soon develops a carbonate, grey in colour, which gives it some protection against pollution but by no means effectively as the carbonate of lead does for that metal since it is not as dense and adherent. If thick, say 14 gauge, and laid at a steep slope exceeding 45°, it might last 50 years but life expectancy when laid flat is more likely to be 30 to 40 years, at best, and perhaps much less, since it is much affected by both acids and alkalis. Acids in particular in the form of acid rain from polluted atmospheres causes much damage as do the run offs from moss, lichens and algae and cedar, oak and other species of timber treated with preservatives. Any run off from copper is also very damaging as well as causing staining.

126: A zinc covered flat roof probably half way or so through its life. The defects in the decking allowing ponding of rainwater are the places where pollution affects the metal to the greatest extent and where holes and splits will appear first of all. It may be that the essential clearing of the build up of snow, always a bugbear with flat roofs of zinc and to a lesser extent lead, has not been carried out as diligently as it might, allowing damp penetration under the cappings at the side joints and rot to develop in the decking. In the background a flashing is visible against a chimney stack but the covering over the top of the adjacent parapet wall leaves much to be desired.

125: A fairly new zinc flat roof showing the characteristic near rectangular batten side joints and cappings and in the background a generously sized flashing.

Many zinc covered flat roofs on fairly old dwellings have had to be renewed, probably quite a few times, since first laid and the surveyor may well encounter a new or near new zinc covering. If there is an opportunity to consult the specification the material can be compared with British Standard BS6561: 1985 and other details compared with the Code of Practice 143 Part 5 1964.

The jointing of sheets follows somewhat similar principles as those for lead since the coeffiicient of expansion is about the same but the battens for the side joints are more rectangular than the near round shape used for lead, as the material is that much less malleable and they

are provided with a capping over the turned up sides of the sheets. These tend to be displaced rather than scored or damaged by foot traffic. When this happens, as it often does, there is a serious danger of water penetration. Not only that, even with the caps in position, there can be damp penetration where a heavy fall of snow reaches the top of the capping, starts to melt and penetrates the roll joint since it is formed in a way less effective than it is with lead. Against the fact that the metal is not so soft as lead and therefore more difficult to work, is the advantage that it does not creep or sag in the same way. Nevertheless where it is turned up the sharper angle so formed tends to hold polluted rainwater for longer and it is often at these points that the material eventually splits. In reverse so to speak, at the top of flashings, the sharp turn for tucking in to the brick or stonework is also prone to splitting due to alkali attack from the mortar used for pointing unless the metal is coated with unsightly bitumen to protect it, a precaution not often taken because of the affect on appearance. Often if there are design difficulties through lack of height, both in flat roofs and in gutters, long lengths of zinc will be soldered together. These form a weak spot and will frequently be the first place where a split will appear.

127: Zinc on a flat roof fast approaching time for complete renewal with the pattern of the decking showing through and not helped by the presence of rusting metal objects. The zinc capping to the parapet is a crude attempt to overcome the problem of cracked cement rendering. A joint effort by both owners to do the job properly would have been better.

Over time, the patina becomes whiter, brittle and more crusty, less able to accommodate movement, either in the structure or in the metal from temperature changes, and what is left of the metal begins to corrode and split. While in the relatively early stages soldered patch repairs can be carried out this soon becomes uneconomic and total renewal will be required. The sight of attempts at such repairs or patching with strips of felt and bitumen or even covering the whole roof with canvas and bitumen are sure signs that the time for this is fast approaching.

Copper

The least likely covering to be encountered for flat roofs over dwellings nowadays is copper. This is because its initial cost is high and,

accordingly, it does not feature at the top of the list of choice for developers, even for houses on up market estates. Its use is more likely to be found on individually designed houses for the wealthy or as part of, or as a renewal on, a listed dwelling, with historical associations, anticipated to be a feature of the landscape for at least another 100 years or so, the probable life of the metal itself.

Although more expensive than lead, copper does have certain physical advantages. For one it is very light in weight, about one fifth the weight of lead when used for the comparable area of roof. For another it expands under hot sunshine at only about half the rate of lead, or of zinc for that matter. Both of these factors tend to favour its use on steeply sloping surfaces, such as domes and cupolas and the roofs of balconies, since the structure can be lighter and the metal does not sag or creep to the same extent. These advantages are also useful when it is used on a flat roof. When new, copper is a reddish brown colour but what clinches it for designers of visible features, however, is the attractive shades of pale green, see 37, 101 and 129, brown or even greyish patina, see 100, depending on the type and degree of pollution in the atmosphere, which develop over the first 10 to 15 years of exposure.

The patina comprises a layer of copper carbonate which protects the metal from attack by the acids in the atmosphere. The development of the patina is a slow process but which can be accelerated by the presence of salt in the air of coastal regions. While alkalis do not affect the metal, it can be subject, however, to severe deterioration through electrolytic action if, for example, nails or screws of any material other than brass are used for fixing or other metals are used in conjunction in any way.

128: New copper, on an indoor demonstration panel, showing its characteristic reddish brown colour, a standing seam joint used on sloping roofs and features and a square batten roll joint used on flat roofs.

129: The surveyor will need to establish what material has been used here for roofing, cladding and as a capping to the brick on end parapet wall of this recently completed block. The makers probably thought it might pass off as copper at a quick glance, but it is far more likely to be of coated steel or aluminium with a much more limited life span.

Although not so subject to expansion and contraction as lead or zinc, allowance nevertheless has to be made for what is likely to take place under the hottest sun and, accordingly, side joints on flat roofs are in the form of batten rolls. As with zinc, they cannot be near round in shape since both metals are not so malleable as lead. The rolls are often made triangular in shape although round top, undercut or even more decorative shapes are also used to improve appearance where a flat

roof can be seen from upper floor windows. While standing seams, or stand-up welts, can be used for inclined roofs not subject to any form of traffic, staggered double lock cross welts are usually adopted for joining sheets end to end and a form of drip is only required in parapet gutters.

Flat roofs covered in copper on the more expensive houses built in the latter part of the 1800s and the early part of the 1900s could well have been recently renewed or there might conceivably be a new copper roof on a recently built 'executive' mansion. Should the opportunity arise to see the specification, British Standard BS2870 covers the material and a Code or Practice BS CP 143 Part 2 1970 sets out the details with which comparisons can be made. The publications of the Copper Development Association are an additional invaluable source of details which should be in the possession of every surveyor. They can be obtained from its address at Verulam Industrial Estate, 224 London Road, St Albans, Hertfordshire, AL1 1AQ, Tel 01727 731216 or viewed and downloaded from its website www.cda.org.uk. As for the material, 24 SWG (Standard Wire Gauge) is normally used for

130: A flat roof covered in copper about 50 years old showing batten roll side joints and the characteristic green patina. It is adjacent to a flat roof covered in mastic asphalt but arranged so that there is no run off to give rise to unsightly stains. The asphalt roof has the typical dome hooded ventilators to the underside referred to in that section.

flat roofs although 23 SWG might be used for superior work. Because the metal is so light, bays should be strictly limited to the recommended sizes, otherwise there is a danger of wind uplift splitting the sheets.

The surveyor should be aware that the run off from copper roofing can damage zinc and steel, so that gutters at the edge of flat or sloping roofs will be affected if in either of these materials. Undesirable replacements of original coated cast iron gutters are a possible source of trouble in this respect and may need replacing, perhaps in good quality plastic if the original is thought too expensive. Such run off can also severely disfigure stone and brickwork, the staining being almost impossible to remove. The aggressive qualities of the metal do, however, have one upside in that it does not sustain organic growth so that moss, lichens or algae will not be found adhering to the surface. It is still very necessary, however, to keep copper flat roofs clear of leaves and any other form of debris or deposits which accumulate.

As with all metal roofs, signs of undue wear, buckling, splits or patch repairs are indications that the time for total renewal is fast approaching.

Aluminium

Although there was a vogue in the 1960s and the 1970s for the use of aluminium, despite its high cost, as a covering for both shallow pitched and flat roofs, it is unlikely that the surveyor will now encounter many dwellings with roofs covered with this material. Most will already have been replaced.

Although bare aluminium has a comparatively slow rate of corrosion it rapidly becomes whitish and rough on exposure resulting in

poor appearance made worse by a propensity to pick up dirt and collect pollutants which, of course, accelerate its corrosion. Anodising the surface in colour suited well the styles of the time and which in theory was to lengthen the life by producing a layer of oxide on the surface of the metal. However, the anodised coating is brittle and easily damaged on installation and fixing so that length of life was not significantly extended beyond the 20 to 25 year span of the unprotected metal, attacked as it can be by both acids and alkalis and in even greater danger of corrosion if in contact with other metals. Frequent washing or periodic painting have been put forward as ways to extend the life of the metal but it is not considered that many owners would wish to incur this liability either physically, by way of payment to contractors or through a service charge. Other problems also materialised through condensation below the metal affecting the decking and the need to very securely tie down the covering because of its lightness was not always appreciated, with the result that quite a few roofs got stripped off by wind suction.

It is, accordingly, unlikely that aluminium will be encountered as a roofing material on new or near new dwellings. If it is, the surveyor needs to be very prudent in the advice he gives since experience indicates that the longevity found in most other metals used for roofing does not extend to aluminium. This is so despite the material being covered by a British Standard and the detailing provided in 1973 as Part 15 of the Code of Practice 143 on sheet roof and wall coverings. No doubt developers will, however, endeavour to reassure purchasers and their surveyors with elaborate Agrément Certificates to cover any further developments in ways of protecting the metal.

Built-up Felt and Other Membranes

Although it is, of course, true to say that as far as roof coverings are concerned, it is only a matter of time before renewal is required, for someone purchasing a dwelling it is better to be reasonably assured that the need is not likely to arise within, say, the next 10 to 15 years. This at least gives that person time to settle in, make those changes to fittings and decorations that most aspire to and have a good few years, hopefully, to enjoy them.

Unfortunately, where built-up felt or other membrane roofs are concerned that reassurance is never really possible. Most of such roofs, usually on extensions or back additions, have been provided in the first place on the grounds of cheapness. Where covering whole dwellings they were often provided in the 1960s and 1970s at a time when it was fashionable to embrace the modern movement. Many of these failed very quickly and have since been removed and replaced with pitched roofs, but quite a few still remain provided now with renewed coverings. Others on new or near new dwellings could be a case of a developer chancing something new and innovative out of sight to keep down costs. Where encountered, more often than not, the poor materials and even poorer workmanship will inspire no confidence. The sort of defects which the surveyor is likely to encounter are often all, generally some but seldom none, of the following:

1) Blistering of the layers, often due to moisture entrapment.
2) Embrittlement and cracking of the covering due to the exposure of the oxidised bitumen content of the felt to solar radiation.
3) Splitting and cracking due to laying in cold weather which can split the sheets as well as causing a lack of adhesion.
4) Rucking, buckling and splitting due to differential movement with the decking or insulation material or poor bonding on laying, perhaps due to a lack of, or inadequate cleaning of surfaces before laying.
5) Ponding, either due to inadequate falls, sagging of decking between joists or any of the above faults interrupting the flow. This can rapidly seek out weaknesses of adhesion at joints, particularly if manufacturers recommendations for overlaps have not been followed.
6) Inadequate solar protection to the surface, either by omission at the design stage of reflective materials or the washing away of materials such as stone chippings applied for the purpose.
7) Mechanical damage from maintenance traffic, particularly where surfaces are uneven.
8) Missing or inadequate flashings, for example a lack of angle fillets to upstands and abutments, particularly around flues.
9) Poor detailing.

131: There is no mistaking the appearance of a bituminous felt covered flat roof even, as here, in good condition. Sure indications are, the colour, the surface texture and the stuck down side joints and where the lengths cut off from the rolls overlap. Apart from the quality of the material itself, it is on the integrity of the joints that weather tightness depends.

While some of the above faults can be found, at times, in relation to other types of roof covering, most are unique to built-up felt roofing. The main problem with such covering is that while the material itself may be reasonably satisfactory with a life of maybe 15 to 20 years or so, perhaps even longer for some of the newer materials, the fact that the material is in sheets and laid to overlap means that it relies on the workmanship of bonding for its durability. It is difficult to ensure that this work is done correctly or satisfactorily and flaws will not necessarily become apparent immediately. When they do, maybe in under ten years, it may be too late to prevent widespread damage and the flaws may prove to be so extensive as to make repair uneconomic.

Even when presented with a new or near new built-up felt covering to a flat roof the surveyor should, as advised by BRE, be wary of giving any view as to future performance. A current specification for built-up felt roofing may comprise one of a bewildering variety of possibilities for both materials and installation methods. There are those based on the felts covered by BS747:1977 Specification for Roofing Felts. This Standard has been extended down the years so that while Class 1 organic fibre based and fibre and hessian based felts and Class 3 glass fibre based felts are still included, it is to be hoped that for any part of a dwelling intended to be permanently occupied, Class 5 polyester or polyester polymide felts would be specified nowadays. These can be with the oxidised bitumen content modified either with atactic-polypropylene (APP) or styrene butadiene-styrene (SBS) to improve the performance when exposed to solar radiation and thereby reduce embrittlement. Felts containing the former are normally torched or hot air welded whereas those with SBS are still stuck down with hot bitumen. Installation of these felts would no doubt be covered by reference to BS8217:1994 Code of Practice for Built-Up Felt Roofing which does, however, allow for a combination of glass fibre and polyester based felts to be used.

Surfaces of the top layer of felt would either be sanded ready to receive a coating of bitumen and reflective chippings applied on site or the top layer be pre-bonded with mineral granules.

Then there are those newer systems based on felts not covered by the British Standard, for example those comprising a core of continuous fibre, spun bonded polyester fleece or polyester sheathed in polymide with a coating of bitumen and thermoplastic rubber. Some of these will be covered by a British Board of Agrément Certificate including the Board's opinion that, if properly installed in accordance with the BS Code of Practice and the manufacturer's instructions, accelerated tests had confirmed the likelihood of a life of at least 20 years. One firm will put this into the form of a guarantee for 10 years, subject to the use of its specification, materials and nominated contractor and with the availability of an extension to 15 and 20 years, subject to an annual inspection and the entering into of a maintenance agreement. Such a guarantee could clearly be of advantage to a developer issuing the likes of a 10 year NHBC warranty, but would be considered of doubtful value to a purchaser for the 15 to 20 year period, since it would involve expenditure added to the prospect of probable renewal at the end of that period.

132: The dip in this bituminous felt covered flat roof, discernable from the line of the brick courses above the lead flashing to the parapet, and the lightening of colour are indications that ponding occurs and where the first defects will probably appear.

133: One of the problems, apart from their short life span, with bituminous felt roofs is that their rough mineralised surface is an ideal base for the growth of moss and algae. Sometimes the accumulation is so great that the drainage route to the outlet becomes blocked and causes a flood.

To add yet more variety to the possibilities of what might be encountered there are also those single sheet bituminous felts which can be torched down to the surface, plastic sheets which can be softened and welded together and synthetic rubber sheets joined with solvents, all forming a single sheet covering and proving an irresistible attraction to some developers and their designers. Some of these, also, may well be offered with the backing of British Board of Agrément Certificates and there is even now European Union of Agrément guidance on the assessment of APP polymer bitumen felts, reinforced with sheets of polythene film. All should be rejected when reporting to a buyer as being inappropriate as roof coverings for dwellings other than those of a temporary nature.

Paving, Decking and Lantern Lights

Flat roofs, particularly in closely packed urban areas, encourage use for outdoor gatherings and some, in their normal finished state, are more suitable for this purpose than others. The metalled, lead, zinc and copper should

not be used because of the risk of damage to their projecting side and lengthwise joints. Asphalt because of its clear surface is more encouraging to use but at the time of year when such outdoor gatherings are likely to take place it is at its softest and can be damaged by pointed heels, chair and table legs. The provision of suitable lightweight paving tiles is sensible to reduce the incidence of such damage and if in a light colour, do much to combat the effect of hot sunshine. By taking the paving as near as possible to any abutment and dressing a lead flashing as far down as possible good protection can be provided.

134: The provision of paving to asphalt flat roofs, particularly in an even lighter colour than here, does much to combat the effects of hot sunshine.

To an extent, the addition of paving to a roof of bituminised felt which unpaved, should not be used for gatherings because of the risk of tearing, does much the same. Nevertheless, the drawbacks of felt and its susceptibility to bad workmanship still applies round the edges and, in particular, at the turn up and flashings to abutments to limit its life span.

Roof decking, as distinct from paving, on the other hand presents the surveyor with the difficulty that he cannot see either the roof covering or the way the decking is supported. The decking can be raised up to override the projections from the surface of the metal roofs or to bring level the surface of asphalt

135: *Decking on a flat roof prevents the surveyor from seeing what is below and therefore in addition to describing the construction round the edges in the report, warnings must be included on what might be underneath and the possible consequences.*

136: *A traditional type of lantern light on an asphalt covered flat roof. The overhang of the patent glazing obviates the need for a flashing but as a result there is no tight seal between the glazing, which should be wired cast not plain clear as here, and the curb resulting in massive heat loss. The lead wrapped over the top suggests past trouble.*

137: *A proprietary type of domed roof light which should have been fitted in accordance with the manufacturers instructions. Is it secure from possible intruders?*

or bituminous felt originally provided perhaps with an over generous fall. Bearing in mind the case of *Hooberman* v *Salter Rex*, 1985, see p117, warnings must be included in the report based on what can be seen at the edges and the likely consequences of such decking being placed on top of old and worn out or badly detailed construction below.

Flat roofs are very often provided with lantern lights in different forms to light a part of the interior below. Badly designed and or badly installed they can be an endless source of trouble. Traditional types with properly detailed upstands and patent glazing generally perform well, within their limitations, if maintained satisfactorily but the surveyor should seek out any deficiencies and any evidence from below of rain penetration. Their limitations are heat loss and condensation problems. Glazing should be wired cast for safety but this is not always found to be the case.

Small much cheaper alternatives to the traditional type are available in proprietary form with impact resisting polycarbonate, one piece, domed lights. They come with instructions for fitting, particularly with regard to ensuring that they cannot be lifted off by

potential intruders from the outside. These are not always followed so that as with traditional lantern lights the surveyor needs to examine them carefully for flaws on installation giving rise to possible damp penetration.

Rigid Sheet Over-Roofing

Surveyors who carry out inspections naturally do quite a bit of wandering around and will become familiar with the street scene of their area. This may include quite a number of low and medium rise blocks of flats, built either privately or by the local authority. If they have not been seen for a while, it may come as a surprise when they are observed to have roofs nothing like what they had before and which alter their appearance quite substantially. What has happened is that their owners have indulged in the increasingly popular sub-terfuge of over-roofing former troublesome flat roofed blocks of flats with new pitched roofs of profiled steel or aluminium sheet. Appearances are changed considerably, no doubt to the dismay of the original designers, if still alive, or to cause those who have passed on to turn in their graves. Those listed icons of the modern movement with leaky flat roofs cannot, of course, be subjected to such outrage, possibly more to the dismay, this time, of the owners and occupants. Sometimes such work is accompanied by over-cladding the elevations and when the two are combined in this way, the components making up the new appearance fit together rather better than when the over-roofing is carried out on its own.

138: These three storey blocks of local authority flats from the 1970s were originally provided with flat roofs. They were overclad in the early 1990s with profiled steel sheeting which at first glance gives the impression of pantiles.

We are not concerned here with the huge bills presented to those tenants who have exercised their right to buy or to subsequent purchasers who are in occupation. What is of concern is what the surveyor, asked to inspect a flat in the block for a new purchaser of a

long leasehold interest, is to make of the work if newly completed or completed within the last few years. How, for example, will it affect the purchaser's liability for his share of maintenance costs?

Hopefully, the seller of the flat will have details of what work was carried out, when it was done and what it cost and access will be available or will be made available through the freeholder for the surveyor to inspect at least part of the work, if not all. That would probably be the best the surveyor could hope for but far more likely he will have to make the best he can from a lot less, thereby limiting the advice he can give.

139: The vertical upstand necessary for the overclad roofing system shown on 138 is clad in the same profiled steel sheeting as the monopitch roof where it is easier to see the size of the panels. One panel here appears to have partly lost it's fixing.

As might be expected, there are many different ways in which over-roofing can be carried out and much will depend on the initial assessment which will have been carried out to determine how much extra load the structure will carry and what effect wind loading will have on any pitched roof structure if added. The name of a well known and reputable firm of structural engineers associated with the scheme will provide some re-assurance to the surveyor. In particular he will be seeking to advise the purchaser that the lessons learnt from the failures due to wind suction of shallow metal sheathed pitched roofs built in the 1960s and 1970s have been put into effect. This will be because, almost invariably, the material deemed suitable for over-roofing will be some form of metal sheeting, due to its lightness in weight, supported on either a light framing of timber or steel. There is a necessity for such covering to be very securely tied down to the framing and for the whole of the new roof structure framing in turn to be tied to sound construction below, points which it might be possible for the surveyor to check.

140: Not all profiled steel sheet panels used for overcladding set out to imitate other materials in appearance. These are more distinctive in shape and colour and are provided with a trim in a contrasting shade. There are signs of former dampness problems immediately below what might well be the underside of the original edge detail.

It will be recalled that some metal sheeting is just the type of covering that has been mentioned earlier in relation to pitched roof coverings that should be considered inappropriate for any dwelling with aspirations to a long life. Indeed some of the types of roof covering used for over-roofing, such as steel panels made to give the impression of concrete tiles, are to be found on 'mobile' homes set out in parks to give an impression of permanence, see 39 and 40. Others of plastic coated profiled steel or aluminium sheets in bright colours with contrasting coloured edge trims are to be seen and might

be thought more appropriate to warehouse or distribution centres.

In the case of examples of over-roofing, however, the decision to go forward with such materials has probably been taken on the basis that there is no practical alternative and that over-roofing is considered the best solution to long standing problems. The outcome, nevertheless, is a roof covering, but not necessarily the roof structure which supports it, which may have a limited life of around 20-25 years at most. The coated metal coverings generally used are usually subjected to cutting and much other mechanical damage often occurs on installation. Though the manufacturers normally specify what is to be done to damaged edges in such circumstances, workmanship may be such that the procedure is not followed. Consequently corrosion can set in at an early stage. The surveyor, given the opportunity of a close inspection, needs to look carefully for signs of corrosion, not only on the outer surfaces but also, if possible, on the underside where undue levels of condensation have been known to add to the problem. Corrosion can also be accelerated by incorrect laying of the sheets. Sheets should

141: It would have seemed incongruous a little while ago to see a 12 storey block of local authority flats with a hipped pantile roof, but perhaps not so much now that there is no overriding fashion. Sensibly as far as appearance is concerned panels between the windows were provided to match the profiled steel sheeting on the roof.

be laid with the exposed edge of side laps facing away from the direction of the weather, but this precaution is often overlooked.

The noise of rain, but even more so of hail, falling on a metal covered over-roof is bound to be very much greater than when it fell on a felt covered flat roof. The problem might be somewhat less if it is necessary to use an insulated double skin form of sheet metal covering to improve thermal values but this is not always needed. This is because the existing flat roof is usually left in place and should already have the appropriate level of insulation within its thickness as well as a vapour control layer. If, however, a single sheet covering has been used an extra layer of plasterboard should have been fixed to the top floor ceilings to assist in keeping noise at bay, but it is not always possible to check whether this has been done.

There is no way either that a seller or a prospective purchaser can alter the situation as found and all that the surveyor can do is to describe what he sees, warn about what is likely to happen in the future, and when, and to explain how the service charge operates, hopefully in the form of an annual sinking fund to cushion the impact of the cost of major repairs, otherwise there could be nasty surprises in store.

Abutments to and Projections Through Roofs

Among the least troublesome of pitched roofs are those presenting plain rectangular sloping surfaces to the sky from where rainfall can be discharged into gutters clear of the walls of the property below. Apart from those roofs provided around the period 1960 to 1980, similar to those illustrated on p88, few such exist. Most have projections such as chimney

stacks, ventilating pipes and roof lights, as, of course, do flat roofs, and sometimes dormer windows. Some pitched roofs change direction necessitating a valley gutter between slopes and all types of roof may abut walls rising higher. Such features introduce potential trouble spots. While there are long standing ways and materials for overcoming the problems posed, there are also many less than ideal ways available both for new construction and for use when repairs are needed.

Abutments to Roofs

Probably the simplest junction to make is that between a flat roof and a wall rising higher, either as part of a building, a chimney stack or a parapet wall. Usually this is formed by turning up whatever material has been used to cover the flat roof for at least 150mm, considered the limit of splashing, against the wall. Preferably the material so turned up is tucked into either a horizontal joint, if the wall is of coursed brick or stonework, or into a groove cut into uncoursed stonework. However, this is not always done and the material is merely left turned up. Either way the tucking in, or the top, needs protecting with a flashing. It is, moreover, essential that the top of the flashing is tucked into the wall, ideally for 25mm for a secure fixing, so as to make it watertight. To avoid any chemical reaction between different metals, it is important that the material for the flashing is the same as that for the roof covering, 125 on p124 is a good example, zinc for a zinc flat roof. If it is not, the surveyor should look as closely as possible for any signs of degradation from chemical reaction but if none is seen, include a warning in the report of its possibility in the future. Degradation can also occur if lead and zinc are not coated with bitumen before being tucked in because jointing or subsequent pointing in Portland cement or lime mortar will result in alkali attack on the metal.

Given reasonable construction, the above type of junction is basically foolproof. However, the mere fact that it is necessary to bend material through 90°, or even 45° if a tilting fillet is used, introduces a weakness and it is often at this point that a split or tear will occur allowing dampness to penetrate. Such splits or tears are often difficult to see but a defect at this point is often the earliest sign that the material of the roof covering is approaching the end of its life. Earlier patching with canvas or felt and blobs of bitumen will provide a clue that this is beginning to happen.

In all cases with flashings, inadequate tucking in, a lack of wedging, shrinkage over time of the mortar used for pointing or physical damage can cause them to become loose and require refixing. This can also happen if they are not held down in place by tacks at regular intervals and are lifted up by the force of the wind.

142: Physical damage to a zinc flashing, here to a gutter also of zinc which is turned up against the brick parapet wall. The damage was probably caused when the old slates, bits of which remain in the gutter as potential obstructions, were removed and replaced with artificial slates. Damaging may have been made easier by inadequate tucking in to the brick work of the parapet and subsequent pointing but it leaves the top edge of the gutter turn up exposed and thereby a possible path for water penetration.

More complicated joints are necessary where the material for pitched roof coverings, such as tiles or slates, is such that it cannot be turned up against the wall. A traditional and

cheap way of forming this joint is by the provision of a triangular fillet of cement and sand. The life of such fillets is very much determined by the quality of the mortar used for the purpose. Builders have a tendency to think that the more cement in the mix the stronger the mortar. However, such strong mortars shrink and crack far more than weaker mortars containing coarser aggregates and lime and eventually part company with both the vertical surfaces and the roofing material itself. In an attempt to slow down shrinkage in the mortar, resort may be had to embedding pieces or tile or slate in the top of the fillet. Conceivably this could lengthen the life of the fillet but only by a comparatively short period. If cracked, shrunk or part missing fillets are encountered on a tiled or slated roof the surveyor's advice should be to replace them with something better if that is a possibility, but inevitably this would increase the cost and the advice usually falls on deaf ears.

Something much better could consist of the full array involved in providing the ideal way of forming this joint, that of lead soakers and stepped flashings. The soakers are short strips of lead, the length of each tile or slate turned up against the wall. They are either hooked

143: Shrinkage of a cement fillet to a chimney stack. Although not causing visible damp penetration at present, timbers are probably being affected and this along with the condition of the slating, will cause more problems to develop in the future. Ideally it should be replaced with a lead apron.

144: Often the best the surveyor can hope to see is a neat cement fillet in reasonable condition to the joint between a chimney stack as here between natural slates, clay ridge tile and brickwork.

145: The incredibly crude pointing to the tucking in makes the presence of this lead flashing very obvious. It may be possible to check whether soakers have been provided from the roof space. The crack and displacement of the brickwork in the parapet requires further investigation in case there is a danger of collapse.

over the top end of each tile or slate or nailed to boarding and laid in an overlapping manner so that if a defect occurs in the flashing, or it is dislodged, any penetrating rainwater will follow a path down to discharge into the gutter. Soakers cannot be seen from above being covered by the flashings but might be seen from the roof space in the absence of boarding or roofing felt. The risk of damage from run-off or staining is too great for this form of detailing to be an appropriate use for copper, but using zinc instead of the more expensive lead is a cheaper, but not so long lasting, alternative. A combination of lead and zinc has to be avoided, however, in view of the possibility of electrolytic action between the two in damp conditions.

146: Stepped lead flashings to a tiled roof on a demonstration panel. At the top, a soaker is shown before being covered by the flashing and the top tile.

A sort of half-way house, to save on cost, is to provide soakers but instead of a flashing to use a triangular cement fillet to cover the joint. This is better than the same fillet without soakers, provided the soakers are well coated with bitumen, before installation, to protect them against alkali attack from the chemicals in the fillet or in the brickwork and pointing of the wall above. It is a method adopted in areas where the local stone is so hard that cutting a groove for tucking in the top edge

of the flashing presents undue difficulty. However, if the top edge of the fillet shrinks away from the upstand there is no protection against rainwater getting behind the soakers and perhaps causing stains internally as well as dampness in the timbers with all the problems that could entail.

Although leadwork is to be preferred, where it is possible for it to be used on account of its greater durability, there are quite a few situations where there is no alternative or

147: Junction of a pantiled roof, parapet wall and chimney stack formed with a cement fillet in which pieces of slate have been embedded to reduce shrinkage.

where the cement fillet might be preferred. On the brighter coloured profiled tiles for example, the sight of strips of lead dressed over the tiles as flashings is not considered attractive, both at side abutments and even dressed down on to the tiles as an apron flashing. A coloured mortar in combination with matching tile insets, galleting, looks neater and less obtrusive.

Against uncoursed stonework, where stepping a flashing is not possible, the alternative of cutting a groove in the stone can be difficult depending on the type of stone and could cause unnecessary damage so that even with ordinary tiles and slates a neat cement fillet provides a better solution.

Occasionally on older stone constructed dwellings, the joint between the pitched roof covering and an abutment, usually a low parapet wall but sometimes a chimney stack, will be provided for when the stonework is laid by extending the coping or a section of stone outwards so that it overhangs the roof covering. This leaves the gap between the underside of the overhang and the top surface of the roofing material to be filled with mortar and pointed to provide the joint. This can be quite effective over a period of time.

The method clearly required considerable forethought and is usually only found on the best quality of dwelling.

Projections through Roofs

Ventilating Pipes

Although it is best to bring ventilating pipes on a swan-neck out and away past eaves gutters, sometimes there is no alternative but for them to be taken through the roof slope. Providing a suitable flashing is obviously essential but is not always achieved.
The preferred method is by way of a lead 'slate' dressed around the pipe. Securing the joint of slate to pipe presents difficulty unless a special section of pipe with a shaped collar to receive the lead is used. If this is not provided there can be problems of damp penetration around the pipe and the surveyor should always try to see, from the inside, the ceiling area around the point where the pipe

148: Innocent looking ventilating pipes projecting through pitched and flat roofs can be a source of damp problems if not properly detailed. On pitched roofs the nearer they are to the eaves as here on this 1930s semi-detached the more likely they are to cause trouble, though in this case any signs would first appear on the underside of the projecting eaves. The tall boiler flue on the flank wall probably developed its not inconsiderable lean before the advent of metal liners. If the boiler was replaced by a type utilising a balanced flue the stack could be reduced in height and capped off.

penetrates the covering, in a cupboard perhaps but very often within a casing which, unfortunately, prevents the closest of examinations. Nevertheless, if the problem has persisted it may be that staining will have extended beyond the casing. This is more likely to happen where ventilating pipes are taken through flat roofs when there is a shorter distance between outside and inside.

Chimney Stacks

Chimney stacks can present the surveyor with some difficulty on account of their inaccessibility. It is generally impossible to look closely at back gutters on pitched roofs and often only at side and apron flashings with the aid of binoculars. Back gutters are prone to acquiring debris and an accumulation of silt which can interrupt or even dam the flow, even assuming there is a good fall provided to either side in the hope of preventing rainwater from accumulating, not always the case by any means. If poorly designed and installed rainwater can build up to the extent of over-flowing. To this extent chimney stacks spanning the ridge are to be preferred and cause far less trouble.

Apart from back gutters and flashings, attention should be paid to the structure itself. Stacks are sometimes built unduly tall, particularly when rising from flank walls, so as to avoid the possibility of down draught. In consequence, they can develop a lean which can be aggravated by sulphate attack on the bricks, mortar or any rendering present. In the past chimney stacks were seldom provided with the damp proof courses which would be or ought to be incorporated today. Because of their position they are subjected to the severest of weather conditions and saturated brick-work can soon take on a sorry appearance if the frost gets to it. High winds can subsequently bring about a collapse so that the surveyor needs to note all cases of cracked or leaning brick or stone chimney stacks and all cases of defective pointing or rendering and any defective flaunching, if seen, to the tops of stacks. With the advent of central heating many chimney flues are now redundant and many will be found to be capped. It is essential however to maintain a through current of air to all old flues to avoid excessive condensation and if ventilation at the top has not been provided, the point needs noting and possible consequences explained in the report.

149: *The same remarks about the flue on the flank wall in 148 can be made about the flue on this late 1800s semi-detached instead of endeavouring to restrain movement with an adjustable tie bar.*

150: *The chimney stack on this late 1700s semi-detached house is of an inordinate height for no apparent reason, leans badlyand ought to be reduced in height. These houses both have distinctly unusual angled bay windows extending through all four floors.*

151: Where there are clear signs lower down that disused fireplaces have been blocked up, it is important to check that through ventilation is still provided to the flue. This is usually done by fixing an unobtrusive ventilator grille in the room itself. At roof level there are various ways it can be done. Here the two far end pots have been fitted with clay ware plugs in which there are ventilator holes on the underside of the projecting caps. These are less likely to be dislodged than the half round ridge tiles fitted to the others. In other instances the pots are removed and such tiles are bedded in the top of the flaunching.

152: Dormer windows can take on many different shapes. In the north of England and in Scotland they are often in the form of bay windows which can make an examination of the sides and the junction to main roof a little easier. Old age at this point can cause many problems. These seemingly overlarge versions from the early 1800s have roofs at the ridge level of the slated pitched slopes. The dormers retain the original small squares of glazing which unfortunately have been substituted by large squares elsewhere in the dwelling.

Dormer Windows

Within the limitations of a visual inspection and the availability only of a 4m ladder, dormer windows which are in effect mini-structures in their own right with roofs, side walls and windows, present similar difficulties for inspection as chimney stacks. However, as with other projections through the roof the consequences of any defect may be apparent by way of damp staining on adjacent surfaces.

The presence of stains on new decorations suggests that perhaps a temporary repair has not been successful and obviously something more needs to be done. On old decorations, there may be an assurance on offer that the problem has been solved and that a new owner can redecorate to his choice. Like all information supplied of such a nature, it either needs checking or provided in the report with a warning that it has not been

153: The two large dormer windows on the left have angled sides, slate hung, and flat roofs while the smaller dormer in the roof of the adjacent house on the right is in the form of a bay window with a pitched roof, hipped at the front. Here at least at first floor level, the dwelling on the left has retained the glazing in small squares with the very slender bars appropriate to the dwelling's early 1800s period.

verified. Whether it is possible to check the information will depend on the circumstances, in particular the position of the damp staining. If it is on the ceiling, it is unlikely to be possible, but if at the sides or below the sill, opening the window and peering around could well provide useful information on construction and condition for inclusion in the report.

154: The danger with dormer windows of this shape is that the shallower pitch of the dormer roof compared with the main slopes may not have been appropriate had the designer inadvertently chosen plain tiles for the covering. Here with profiled tiles on this 1980s dwelling there ought to be no problem. However, if there are damp stains from the entry of wind blown rain a check on this aspect should be made. The use of plain tiles in any case would have looked better and used on the sides of the dormers would have avoided the need to show such a large expanse of lead. Being the 1980s, the open fire is back with chimney stack, though probably not originally with the rather quaint baffle.

155: Much expenditure is required to the roof covering of this 1700s dwelling along with substantial work on the dormer windows. Even the replacement of the zinc sides and the substitution of a zinc flat roof for the existing felt would be expensive but much better would be a lead roof and tiled sides which along with good detailing for side gutters and aprons could mean long life with little need for attention.

156: Skylights in pitched roof slopes were almost always a problem, whether of wood or metal, until the advent of the Velux window. Well detailed, well made, relatively easy to install and simple to operate they amply justify their good reputation. This one fitted in a roof covering of double roll profiled tiles should give no problems. If Velux windows do produce problems it is invariably due to the installer not following the maker's instructions and this does sometimes happen when fitting to existing roofs.

Parapet Walls

As with chimney stacks, parapet walls are subjected to the severest of weather conditions and, if old, are seldom provided with features such as adequate copings or damp proof courses of an appropriate sheet material type, all of which would help to keep them in repair. If they are allowed to fall too far into disrepair, parapet walls can be in danger of collapse which may not only have consequences on an owner or his family, but may also injure a passer-by and perhaps result in a claim. As much as is possible should be seen of all parapets and all defects noted in the report, along with any warnings it may be necessary to provide. It is useful to bear in mind the Building Regulation requirement that parapet walls should not exceed four times their thickness in height and the recommendation that they should not be less than five courses of bricks high.

157: Parapets receive much battering from the weather and need to be examined carefully for any signs of sections becoming dangerous. These appear sturdy enough from a distance but it is not unknown for neglected sections to become dislodged and fall to the possible danger of passers by or occupants.

158: Many parapets will be found to be short of the ideal in regard to pointing of brickwork for weather protection, defective rendering, lack of copings with an overhang and few will have any damp proof courses at all, let alone the two recommended. All these flaws can cause troublesome damp stains in the room below.

159: The sight of bituminous felt wrapped around a parapet in this fashion should immediately put a surveyor on guard. What is the felt hiding? A suitable instance of the necessity for the insertion of appropriate warning where needed in the report.

Rainwater Disposal

A quick and efficient system for transferring rainwater from the roof to underground drains or, at second best, surfaces at ground level is an essential feature for any dwelling.

The system should be able to cope with the heaviest falls of rain even when accompanied by high winds without leaking or undue splashing of adjacent areas. The normal method is by external gutters and rainwater pipes, the safest where gutters are fixed on a generously overhanging eaves to a pitched slope. Not so safe when they are fixed to a fascia at the top of the wall and least of all safe, but vulnerable to leakage, when they are provided behind a parapet wall or as a valley gutter or in the form of a secret gutter immediately above internal accommodation.

The inspection of a dwelling during a period of high winds and heavy rainfall, although unpleasant for the surveyor, at least enables the system employed to provide a demonstration of its effectiveness. Water dripping down from holes in gutters or defective joints, water cascading over the gutters or trickling down between the back of the gutter and the roof covering, or fascia, due to too great or an inadequate fall are all indications that repairs or renewals are required. Split hopper heads or rainwater pipes, particularly at their backs, along with blocked and overflowing pipes are a frequent cause of other serious defects when water soaks into adjacent areas. It may be that a parapet or secret gutter is not actually leaking at the time of the inspection, although such cases have been encountered but since, obviously, internal leakage is even more undesirable than leakage to the exterior, special pains must be taken on the examination of disposal facilities from the roof or in roof space areas referred to later.

160: A matter for urgent investigation. The extent and the degree of damp staining on the brickwork adjacent to this lead hopper head and rainwater pipe on a 1700s mansion suggests that it has been going on for a long time. It might only be that the hopper head is blocked by leaves from the tree nearby but it could be more serious. Such defects are a frequent cause of dry rot in timbers adjacent to the damp and in this case floors will need investigating and window frames and shutter boxes.

161: It would seem that the provision of guttering where none existed before on this late 1600s house has been a recent phenomena. However, it is a pity that plastic was used and the arrangement not better or the workmanship, as the appearance is spoiled and a gap has been left where the alignment is astray. The defects need correcting urgently while at the same time the stone slating needs some attraction.

When conditions are dry, there will be no demonstration of effectiveness but there may well be evidence of flaws in the system, particularly if their consequences have been continuing over a period. The sky can sometimes be seen through holes or gaps in gutters when looking up from ground level and they can be viewed from a distance to check their alignment, which should be gentle rather than steep. Viewing from a distance also provides an opportunity to check whether there are any damp streaks on external walls below gutter level, a check which can be extended to the areas traversed by rainwater pipes. Continuing damp will be indicated by darker areas behind rainwater pipes while white streaks will indicate where previous areas of damp have dried out. This can give rise to the question was the repair carried out in a temporary or permanent manner, needing investigation accordingly.

Gutters and pipes come in various sizes and should have been provided with a capability related to the area to be drained. This is not always the case and reliance often seems to have been placed on rules of thumb or merely on what had been provided before in similar circumstances.

Attention has been given to this aspect in recent years and BS6367:1983 Code of Practice for Drainage of Roofs and Paved Areas provides methods of calculation for all shapes and situations of roofs. However, a comparatively simple method of checking whether gutters and pipes are adequate is available for use on dwellings which is sufficiently accurate for the purpose.

In order to assess the amount of rainwater running off a roof to each gutter it is, of course, necessary to calculate the roof area concerned in square metres. For a flat roof the area on plan is taken, for a pitched roof the area on the slope, plus 25% for roofs with a pitch of 50° or less and 75% for steeply sloping roofs above 50°, to allow for the greater speed of run-off. For a roof which abuts a wall, half the area of the wall should be added.

The catchment area in square metres for the length of gutter so found should then be divided by 48 which will provide the run-off in litres per second, assuming the heaviest of thunderstorms which are estimated in the Standard to unleash 75mm of rain per hour wherever the dwelling is situated. Depending on the size and shape employed, a gutter with a flow capacity equal to, or better still greater than, the run-off should have been provided. For example, true half round gutters have a better performance than ogee, or other shaped, gutters even though they may be sold as of the same size. The flow capacity of both true half round and ogee gutters of various sizes are shown below:

Gutter Size (in mm)	Half Round (litres per second)	Ogee (litres per second)
75	0.38	0.27
100	0.78	0.55
115	1.11	0.78
125	1.37	0.96
150	2.16	1.52

A length of gutter will perform better if it has two outlets a quarter of its length in from each end; twice that size of gutter will be needed if the only outlet is in the centre and four times the size if the outlet is positioned at one end ie the whole of the run-off will be carried by the section of gutter nearest the outlet. Bends in gutters, often adopted in better quality dwellings to avoid rainwater pipes appearing on the front elevation, affect flow capacity which needs to be reduced by 20% if the bend is within 1.8m of the outlet but only half this amount if the bend is further away.

As to the size of rainwater pipes, provision is generally far in excess of strict requirement, for example a 75mm downpipe, the most commonly found size on dwellings, has a flow capacity three times that of a 100mm gutter.

It is not usually size that is the problem with rainwater pipes but splits and blockages. It is best that joints are left open so that a blockage at a bend or a shoe at the base of a pipe will rapidly become apparent at the first joint above.

Gutters and pipes come in various materials. The earliest were lead which gave long service provided they were not damaged. Often when early dwellings were provided with a wooden carved cornice, the top would be shaped to form a gutter and lined with lead but when these cornices were found to contribute to the spread of fire, gutters in lead were formed behind parapets.

162: A square shaped lead rainwater pipe in good condition after nearly 80 years service on a good quality house.

Later, cast iron was common for both gutters and rainwater pipes. They too give long service but do need regular painting to keep rust at bay. Downpipes, for this reason, should have been provided with distance pieces on the ears to enable them to be fixed at least 30mm from the wall, but this is seldom done. The failure to paint the insides of gutters and the backs of rainwater pipes leads to earlier failure than might otherwise be the case. The small hand held mirror comes into its own here for examining the backs of rainwater pipes if cracks in the metal are suspected. Steel employed to the same profiles as cast iron makes the difference difficult, to detect but steel is reckoned to be less long lasting as it is thinner, although still needing regular painting in the same way as cast iron. These rigid types are less vulnerable to damage by ladders leant against them than gutters and pipes made of zinc which can easily be distorted in shape. Gutters made of zinc also suffer from the disadvantage that in manufacture they are provided with horizontal bars at the top for strength which tend to provide an obstruction to the free flow of rainwater should there be an accumulation of leaves or silt.

Plastic is now very widely used both for new and replacement work. It has its advantages in cheapness, ease of availability, simplicity for installation and lack of the need for maintenance. Its disadvantages are its appearance and its degradation under the effect of UV light causing it to become brittle and easily split. It can also be melted by undue heat and therefore needs to be kept clear of boiler flues. Plastic gutters and down-pipes, available readily in all shapes, sizes and colours, are attractive to the handyman. The light weight of the components, generally seen as beneficial since assembly seems relatively easy, can actually be a disadvantage when the fixings are given inadequate attention, leading to the sagging gutters and the unstable downpipes often encountered.

At one time sectional concrete gutters were available which combined the lintels for top floor windows and a rebate for a wall plate. These were lined with either asphalt or felt and bitumen, but differential movement tended to open the joints and cause damp penetration. They used to be condemned on sight but they can in fact now be lined satisfactorily with lengths of rubberised plastic sheet, hot air welded on site to form one long length of sufficient weight merely to need spot bonding to the concrete which avoids the effects of differential thermal movement.

Regular clearance once a year of sludge, grit and leaves from gutters pays dividends if a dwelling is near trees of a deciduous type and more frequently if the trees are coniferous, whose needles can form a dense mass of obstruction. Such clearance provides the opportunity to check and, if necessary, coat the insides of metal gutters with bitumastic paint, if such painting has been neglected in the past. The fitting of wire balloons helps but, if not cleared regularly, sludge gets washed down and soon causes blockages in rainwater pipes at bends or a shoe at the base which can be difficult to clear.

Mention was made at the beginning of this section of the least safest type of gutter at the foot of a pitched roof slope, namely the gutter behind a parapet wall. In the absence of rooms in the roof space, trapdoors to the roof space and the exterior or a long ladder, little can be said other than whether there is evidence, old or new, of leakage on the ceiling below. Evidence of leakage necessitates the surveyor warning of the consequences to the timber supporting structure from rot and advice to treat the matter of opening up and repair as urgent.

The absence of evidence of leakage should not be left on the assumption that all is well. Evidence of new decoration to the ceiling below should arouse suspicion. A blob of bitumen could hold back rainwater temporarily until the dwelling is sold and the burden of repair transferred. Even the absence of staining on old decorations should not be passed over and the client advised that if he wishes to have information on the condition of the gutter (and this could probably be extended to the condition of the front roof slope, since it may well be impossible to see much of this from the front of the dwelling at ground level) then he needs to commission a further investigation necessitating the use of a long contractor's ladder.

However, where there are rooms in the roof space or trap doors every opportunity should be taken to examine parapet gutters and to note the material used, whether the design is appropriate to that material and the condition. As mentioned previously under the section dealing with chimney stacks and parapet walls the inspection should be extended to include the parapet wall itself and the coping.

Intruders

Dwellings provide an appropriate habitat not only for human beings but also for other creatures from the natural world. Those that are welcome become pets and usually, apart from the odd cat, move when the owner moves, although leaving their aroma behind for a while. The unwelcome ones remain and since many new owners have individual dislikes, sometimes amounting to a 'horror of ..... and I wouldn't have bought the place if I had been told' the surveyor needs to keep a wary eye open for any visible evidence of intruders so that a purchaser's or seller's attention can be drawn to any implications. The roof space, being an area of the dwelling infrequently visited by the human occupants, offers the ideal area for roosting and nesting along with ready access to other parts of the dwelling so that it is appropriate to deal with the unwelcome intruders here even though they may also be encountered elsewhere. The surveyor, accordingly, needs to spend a reasonable amount of time looking at the timbers in the roof, at the upper surfaces of ceilings and around, even into the farthest corners, for any visible evidence. In many cases he will not be disappointed. Intruders are not, of course, restricted to the interior of a dwelling. The garden, a joy to many, can be disrupted by various creatures, those that burrow, such as moles, those that dig and

knock things over such as foxes. Inspecting and reporting on the surroundings features as Section 7.

Birds

Birds of many different species will take the opportunity of nesting in roof spaces if access is available. Evidence of their presence are the nests and the birds themselves, fouling by droppings, feathers and the remains of their food. They can cause considerable disturbance to occupants by their movements and should cold water storage cisterns be left uncovered, contamination to the water. The sight of a dead bird floating on the water is not unknown.

It can be a constant battle to keep birds at bay but necessary as they carry parasitic mites which can cause dermatological problems on the skin of human occupants. Seeking out and closing up all openings likely to provide access is the solution, taking care not to eliminate ventilation to the roof space in the same process. Even when this is done and the space cleaned and fumigated, it may be found that others, particularly those that visit for the summer months, will nest below the eaves or in hopper heads or the tops of ventilators or flues. These must be looked out for and reported on as at one time there were a number of deaths caused by asphyxiation through birds nests blocking the flue outlets from instantaneous gas water heaters. In and near coastal areas, gulls have been known to peck at and damage the gaskets to plastic pipes and have a habit of congregating tiresomely, as do pigeons, which are a particular problem where they tend to cluster as a group and are hard to dislodge. Specialist methods are needed both in terms of dispersal and prevention by means of small strips of spikes on ledges and by scaring devices. A surveyor was called back to an 1890s house and was at a loss for words when the owner pointed out a small innocuous hole in an outer wall which provided access for pigeons to an enclosed roof void. The purchaser had bought not only a house but an aviary!

Rats and Mice

Along with other parts of a dwelling, rats and mice can use the roof space for nesting purposes and as runs to give access to other areas. The rat most commonly found is, fortunately, not the Black, the Plague carrier, but the prolific grey common Norway rat which seems to have arrived in the early 1700s and which produces about 40 young a year. Even so all rats and mice are vermin and present an infestation hazard causing a nuisance by fouling of surfaces which has health and hygiene implications, being responsible for heptospirosis (Weil's disease, which can be fatal), salmonella and tape-worms. They have also been known to gnaw at the insulation to wiring for electricity and telephones although this is comparatively rare.

Droppings provide evidence of their presence, comprising smooth cylindrical black pellets about 20mm long for rats and about 5-8mm long for mice. Eradication can prove difficult, requiring the blocking of access and runs, baiting and poisoning. It can be expensive depending on the number of visits required and on the attitude of the local authority. Some accept responsibility to deal with infestations, some charging for the service and some not, others merely list contractors. The British Pest Control Association also list members on www.bpca.org.uk.

Ultrasonic devices which emit a high frequency sound beyond the range of the human ear but sufficiently distressful to encourage mice and other vermin to pack their bags and

move elsewhere have proved to be successful in some circumstances.

BRE has reported problems from infestations by edible dormice in parts of the counties of Berkshire, Oxfordshire, Buckinghamshire, Bedfordshire and Hertfordshire. These dormice are a protected species and a licence is required under the Countryside and Rights of Way Act 2000 before steps are taken for their removal. Surveyors in these counties should check with English Nature and the local Wildlife Trust for sightings in the area and whether they would wish to inspect the dwelling under consideration. Dormice are much smaller than ordinary mice and the sight of clusters of very small droppings to the order of 2-3mm long should put the surveyor on his guard.

Bats

The Countryside and Rights of Way Act 2000, English Nature (www.english-nature.org.uk) and licences loom again in connection with another intruder to roof spaces. Bats, an endangered and protected species, colonise and roost in roof spaces but producing only one baby a year, their presence is not always noticed by occupants. They tend to be quiet, emerging only at dusk to fly about and seek food in the form of bugs, insects and moths. Even although the fouling of areas below their roosting places could be considered to have hygiene implications, until recently they have not been thought of as a health hazard. However, the death in Scotland of a wild life handler in 2002 of rabies, after being bitten by a bat, may lead to a modification of that view. It is certainly best not to assume that owners are or will be happy at the thought of bats in their roof space.

The presence of bats, even apart from any

possible health hazard is, however, tiresome with cost implications in that it is an offence wilfully ie if their presence is known about, to harm or disturb bats or their roost, incurring a fine of up to £5000 if the law is broken. Any proposed works in the roof space, such as forming a room, moving or installing cisterns or tanks or carrying out preservative treatments against wood boring beetle infestations need a licence from English Nature, who it is said receive 5000 calls a year in relation to bats. English Nature will consider whether the roost can be moved, whether it can remain subject to modifications to the proposed works or the formation of a 'bat box' enclosing the roost but leaving access for the bats, or in the case of beetle infestation, specify the form of pesticide to be used. In particular, a request may be made for works to be delayed until young bats are able to fly. Where bats are found in residential property English Nature will arrange free of charge for a representative or the local Wildlife Trust or a Bat officer to inspect.

Bats may, of course, be seen hanging from timbers in the roof space and other evidence of their presence may include the remains of feeding, moths wings and the like, and droppings, which can be distinguished from those of rats and mice by their irregular appearance of a size 7-12mm and sometimes brown in colour as well as black. It is perhaps unwise to assume that bats only favour very old buildings with large roof spaces full of cobwebs. Many do, of course, but the pipistrelle species, the most common, will happily take up residence in modern dwellings, given the opportunity.

Squirrels

The grey squirrel population increases apace and, as intruders to roof spaces, they are probably the most troublesome and difficult to remove. Classed as vermin, they are energetic and resourceful and are able to shin up rainwater pipes and jump from nearby trees to gain access to vents in facias and soffits and gaps in tiling or slates. Sealing up the gaps does not always keep them at bay as they are quite capable of lifting flashings or removing grilles and the odd tile to get back inside.

Once inside, they can cause extensive damage because like all rodents they have front teeth which grow continually and need to be constantly trimmed and sharpened. They do this on the woodwork and metalwork in the roof space and on the insulation to the electric wiring. A group of them larking around at night is enough to keep the soundest of sleepers awake and during the day the odd mysterious clatter will cause occupants to jump. It is debatable who is the more surprised, the squirrel or the surveyor when, on opening the hatchway, their eyes meet. Apart from the animals themselves, often no doubt outside in the immediate locality if they have chosen to go out for the day, remains of food and excrement to say nothing of the visible damage should leave the surveyor in no doubt of their intrusion.

Extermination can be carried out by trapping and poisoning but only without causing unnecessary suffering and only between 13 March and 13 August. Accordingly, specialist contractors need to be brought in to use designated hopper traps and a Warfarin, blood thinning, bait. This needs to be coupled with effective ways of preventing access. One successful method at the eaves is to remove the lower two or three courses of tiles or slates and fix, and fold over the fascia and soffit, lengths of expanded metal lathing, subsequently refixing the tiles or slates. It will be noted that it ill behoves a surveyor to miss

163: *Gaps in the detailing around roofs provide opportunities for access by unwelcome guests. This one would be a bit difficult for squirrels to reach at the top of a three storey terrace house of the 1960s but there is always a possibility that birds or bats might gain entry. Better to advise that it be repaired and sealed up and provide a warning of the possible consequences if not done.*

the intrusion of squirrels. It has been known to cause owners to move without necessarily giving their presence as one of the reasons.

Insects

Older dwellings with large roof spaces may, of course, provide space for insects such as cockroaches and pharaoh ants, which may also be found elsewhere. Any sightings by the surveyor should be mentioned in the report though the presence of such will probably be merely further evidence that a major infestation, perhaps extending to lice and bed bugs, exists which will need drastic action. The surveyor may, with good reason, think twice about continuing his inspection beyond an initial reconnaissance until such work is carried out.

The presence of other insects, such as wasps which can form substantial nests in the cleanest of dwellings, need commenting on in the report, but obviously on no account should be

disturbed for safety reasons. Wasp nests usually take the form of a white paper-like enclosure down towards the eaves and although the surveyor is not likely to trip over one, prodding with a folding measuring rod is not to be recommended.

164: *Not to be disturbed. The white paper-like enclosure of a wasps nest in the eaves of a slated roof.*

Although wasps can be a nuisance in the garden to owners, and their families, by their coming and going during the day and returning in numbers, probably tired and irritable on a summer's evening towards dusk, even more so when drunk in the autumn, they tend to occupy a nest for one year only. As they do not usually return to the same nest the following year, it can be more sensible to put up with the nuisance for one season rather than hazard an attempt to exterminate them one evening when they are all, hopefully, back in the nest. If the problem persists the next season with another nest, then consideration could be given to fixing a 3-4mm wire mesh over ventilation slots or other access points during the following winter.

Such a mesh will not, however, keep out those much smaller beetle insects which lay eggs and whose larvae tunnel and consume the structural timbers, emerging later through flight holes to continue the life cycle. Commonly known as woodworm, probably because the larvae might seem like a worm burrowing through the ground, they are capable of flight and can cause an infestation with more, less, or even no real damage, according to their type. The surveyor needs to ascertain whether flight holes are present and, if so, their size, shape and positions along with the extent of any damage already caused.

It is unlikely though still a possibility that the surveyor will see any beetles flying around in the roof space. Apart from the House Longhorn beetle which is about 12mm long but which won't be found outside an area comprising the northern parts of Surrey and parts of Hampshire and Berkshire, areas where the Building Regulations require pre-treatment of new roof timbers, most are small extending only to about 3 to 6mm in length. Furthermore their flight is quiet and not rapid. Unless he peers intently and has good reason to look at a particular spot, in an area of a damp patch perhaps, he is unlikely to notice any that have alighted on timbers, since their appearance and colour tends to blend in with the rougher and dustier surfaces of the unwrot timber unusually found in roof spaces. It is also a fact that most look very similar, small beetles with heads and wing cases, so it is not very likely that the surveyor would be able to distinguish one type from another if he did see one, unless he was an expert. For practical purposes the appearance of the adult beetle can be discounted as a means of identification. Having said this, however, the sight of a number of little brown beetles in the Spring and Summer months, clean fresh dust and new looking holes in the timber should set alarm bells ringing.

What helps to identify the type of beetle is the flight hole the adult makes when biting its way out of the wood. Although the holes left by wood boring insects which can cause structural damage can look very similar in size and shape to others which cause little or no

damage, it is from the type and condition of the wood from which they emerge and possibly the bore dust from within the hole that pinpoints the different species. The table on p150 should clarify the deduction process.

As will be seen from the table (overleaf), consideration of the size and shape of the exit hole, from what wood and from where the adult beetle has emerged should point towards a possible identification. Knowing, or assessing, the age of the dwelling could help to eliminate six of the possibilities where beetles, some who come or are brought in from the forest, die out within a year or so of construction or another which only infests wood which is already damp or decayed, probably due to other defects in the structure.

That leaves four which if active can cause structural damage if an infestation is left untreated. Surveyors practising to the South West of London will no doubt be well aware of possible infestation by the House Longhorn beetle and indeed the sight of its exit holes is enough to bring anyone up sharply. However, its activities are confined to date to a comparatively small area. In practice infestations by the Powder Post beetle are also comparatively rare since attacks are confined to hardwood timbers, more likely to be a feature of the more expensive dwellings. Since it is unusual, and indeed wasteful, to hide attractive and expensive hardwood, the effect of any infestation is often readily apparent.

In practical terms, damage by wood boring beetles generally narrows down to infestations by two types of beetle, both fairly common and both of whom can cause structural damage if left to their own devices but which are easy to differentiate. In regard to one type, the common furniture beetle, it is believed that around a third of all dwellings with softwood roof structures, floors and fittings, which

Table 1: Identification of Wood Boring Insects

Exit holes in ascending order of size		Wood affected	Condition of wood and whether infestation likely to be active or inactive	Bore dust resulting from active infestation	Likely type of insect and whether liable to cause structural damage	If active whether treatment required
Actual size	Shape					
1.5mm	Circular	Softwoods and hardwoods but mainly sapwood	For active infestation cool and damp conditions preferred. Tends to die out with central heating	Cream coloured pellets. Fine granular gritty texture	Common furniture beetle. Yes, liable because of successive infestations to structural timbers	Yes
1.5mm	Circular	Hardwoods only, particularly sapwood of those with open pores such as oak, ash, walnut and elm	Dry and warm, generally employed for framing, flooring and decorative features	Flour like texture, of colour according to wood	Powder Post beetle. Yes, liable in flooring and framing where such woods used	Yes
1.5mm	Circular but ragged	Softwoods and hardwoods in damp areas	Only infested when damp and decaying	Fine, granular pellets	Wood boring weevil. Not liable, but earlier damage from other causes would probably require renewal	No
2mm	Circular	Softwoods but only in the bark or sapwood	Only found adjacent to bark and dies out naturally due to limited amount	Brown or cream, bun shaped pellets	Waney edge borer beetle. Not Not liable as does not affect heartwood where strength lies	No
3mm	Circular	Hardwoods only, usually oak or elm when old. Both sapwood and heartwood	Damp and poorly ventilated wood preferred but attacks extend into relatively dry areas. If active may be heard tunnelling	Oval or bun shaped pellets in dust, visible to the naked eye	Death watch beetle. Scourge of old timber framed dwellings and and oak framed roofs. Very liable to cause damage	Yes
3mm	Circular	All woods. Sapwood and heartwood but only when growing in forest	Infestation dies out on conversion and seasoning into usable building timber. Blue or black stain may be left around flight holes	None	Pinhole or Shothole Ambrosia beetle. Not liable	No
6mm	Circular	Imported Hardwoods but only in the sapwood	Dies out within a year of importation	Flour like texture	Bostrychidae (species of Powder Post beetle). Not liable	No
6mm	Circular	Sapwood and heartwood of softwoods only	Dies out within a year of of construction as normally lives only in trees and logs	Chips of wood tightly packed	Woodwasp. Not liable	No
6-10mm	Oval	Sapwoods only of of softwoods and hardwoods	Originates and normally lives only in trees and logs. Dies out	Mixture of pellets and wood splinters	Forest Longhorn beetle. Not liable	No
6-10mm	Oval may be ragged	Generally the sapwood of softwoods only is attacked first	Dry warm timbers favoured but attacks will spread. Larvae may tunnel for years without much external sign except possible blistering but leave a mere shell. Attacks limited to date to areas mentioned in Building Regulations	Mixture of fine wood dust and short compact cylinders of excrement fill the galleries formed by tunnelling	House Longhorn beetle. Very liable to cause structural damage necessitating replacement and treatment	Yes

comprise the vast majority, are or have been infested, rather more than a third if built before 1900, rather less if built after that date. The other type, the Death Watch beetle, favours old hardwood timbers, particularly oak and elm. These are usually found in dwellings going back in time before the 1700s, either wholly timber framed or with masonry walls but with an oak framed roof and floors. There are very few such buildings which have not been infested by Death Watch beetle at some time or other since construction and many that still are. However, the exit holes of this beetle are about twice the size of those made by the common furniture beetle making recognition easier than it might otherwise be.

The importance of the surveyor pointing out the difference in the report between the two types of beetle if one or the other is found to be present, or even both as that is by no means uncommon, was brought out in the case of Oswald v Countrywide Surveyors Ltd. in 1994. The claimants had bought a timber framed Essex farmhouse for £225,000 although the survey report had indicated that the old timbers were much attacked by woodworm. However, the surveyors did not state the type of insect which had caused the attack and which the Judge considered vital in the circumstances. It transpired that one of the claimants had a horror of Death Watch beetle. The plaintiffs suffered a plague of the insects after moving in, some of whom got into their bedding. The judge awarded around £50,000 in damages and a subsequent appeal by the surveyors in 1996 was dismissed.

Having regard to the information about the extent of wood boring beetle activity, the surveyor might feel a little uneasy at not finding any exit holes thus far on his inspection. Although he might find some elsewhere in the dwelling, he needs to be able to leave the roof space reasonably satisfied that he has looked as fully as is possible and has still not been able to find any. After all if the estimates are correct, around 70% of dwellings containing structural softwood should show evidence of beetle infestation on surfaces which are visible. Having found exit holes further diligence is needed to establish whether there is evidence that the infestation is active. Activity can be deduced, and accordingly reported, if the exit holes are clean and freshly made, by the presence of bore dust in the holes, on surfaces underneath and perhaps by the beetles themselves. Advice to have the infestation treated must therefore follow.

Unfortunately, the sight of old exit holes, the absence of bore dust or beetles is no indication that activity has ceased and that no infestation is present. Once the beetles have emerged, they will be looking around for somewhere to lay their eggs. The nearest and most convenient timber for this purpose is right to hand and, accordingly, re-infestation is a very common phenomena. For this reason it is never possible for a surveyor to say that activity has ceased. It could be going on under his, and everybody else's, nose without anyone knowing. An infestation by woodworm is a natural hazard for dwellings and if there is visible evidence of exit holes the surveyor has done his duty by reporting on their presence. It is up to a new owner to decide whether he is prepared to wait for further evidence of activity later on, deduced from regular inspections, before ordering treatment, recognising that this could be more disruptive once a dwelling is occupied.

Of course, regular inspections for the above purpose should be part of the maintenance procedure carried out by all owners, not just the new, and for a seller, information from his surveyor about the evidence of woodworm will come as a surprise, if not a shock, if he has been neglecting such a task or missed the evidence in the process. The surveyor has to provide the same advice as he would to a buyer. This would be, order treatment if the infestation is visibly active but if merely the existence of flight holes is being reported then it must be for the owner to decide whether to have treatment carried out or not. Clearly, however, with this option a seller needs to take into account the effect on any prospective buyer, or lender, of information in the Home Condition Report of the presence of exit holes and no treatment having been carried out. It is thought that a seller with the resources to do so would have treatment carried out before the Report was completed so that it could be reported as having been carried out in the Report and that a guarantee was available for inspection. Alternatively if the resources were not available, the seller would need to provide an estimate for treatment and an adjustment to allow for its amount would have to be made to the asking price.

Ancillary Matters

Having inspected the roof space for evidence of intruders, there are other matters to which the surveyor should give his attention before moving downstairs.

Insulation

There are comparatively few dwellings, even very old ones, encountered nowadays which are not provided with some form of insulation to reduce heat loss through the roof. In order for the Energy Report to be prepared it is necessary for the surveyor to take note of the construction in successive layers from the ceiling to the exterior. At some point the insulation, if present, will be apparent and its type and

thickness needs to be noted. If there is none, that too needs to be recorded. The Energy Report will be the place where recommendations will be made for improvements to insulation levels.

Top floor ceiling construction will generally be either plasterboard or lath and plaster. For a dwelling built before 1940 the latter would be expected. If the former was found that would indicate a change perhaps brought about by a burst water cistern or war damage. If the latter there may be evidence of other damage from this cause in the floors below.

165: Insulation in the roof space should of course be extended to any item which could be vulnerable to a drop in temperature. Cold water storage cisterns, a requirement by some water companies, if in metal are at risk but even more so are the pipes and stop cocks carrying and controlling the flow of water. Clearly insulation is required to the services here.

Electrical Wiring

The roof space provides one of the best opportunities to examine some of the wiring of the electrical installation. The dwelling without any insulation at ceiling level provides the greatest opportunity and may well reveal a view of one of the oldest types of wiring, if nothing has been done to the dwelling for years. A sight of anything other than PVC or copper insulated wiring, for example lead or rubber cable, eases the surveyor's task by indicating the need for entire renewal to modern standards. Where insulation is present at ceiling level, it may be possible to lift a section to secure a sight of the wiring, subject to the insulation being carefully replaced. However, on no account should old wiring be disturbed, bent or tampered with, otherwise there is a danger of short circuiting and a possible fire hazard.

If PVC insulated cable is present, an idea of its age might be gauged from any discolouration or the amount of dust present, but an examination of the switchgear and fuses in the floors below will provide a sounder guide.

Cisterns

It is not always the case that part of the dwelling's plumbing system can be seen in the roof space, since not all dwellings have a cold water storage cistern or central heating feed and expansion cisterns. Where they are present, however, covers, if fitted, should be removed and the fittings and interiors examined and the type and condition of any insulation noted.

Galvanised cast iron cisterns eventually rust, which is a prelude to failure and can cause flooding in consequence. Signs of rust should

166: A badly rusted cold water storage cistern in need of urgent renewal before it bursts.

lead to the conclusion that renewal is becoming urgent and should be carried out before disaster occurs. Fortunately these are becoming rare nowadays with replacements now normally in plastic.

There are a number of different materials in use for cisterns and a careful note needs to be taken of the material used and the size for use when the report on the remainder of the plumbing installation is prepared. A listed dwelling might contain a lead lined cistern, the renewal of which in a different material might not be approved by English Heritage or Historic Scotland unless it was actually leaking. A new owner, however, might argue strongly that its retention along with any lead pipework constitutes a health hazard, particularly if the dwelling was near or in an area renowned for its soft rather than its hard water.

Apart from noting details of pipework, stop-cocks and materials, an important aspect is to check the run of overflow warning pipes. These should be well insulated and take a reasonably straight run with an even fall to discharge in highly visible positions. Failure to note and warn about any omissions in this regard could lead to serious consequences by way of flooding should a ball valve become defective. On no account, however, should the surveyor attempt to test the overflow provisions himself by holding down the float and allowing the water to reach the overflow position. The consequences of flooding could be disastrous should the provisions be inadequate, the ball valve stick and the main's water stopcock be broken, all attributable to the surveyor' actions. If the surveyor has doubts about the provisions, however, he should recommend such a test being carried with a plumber to hand, he having checked the stopcock and ball valve beforehand and ready to take action at the first sign of a problem.

Solar Installations

The idea of deriving energy from the sun for domestic use has been around for some time but it is believed that so far only one in about 40,000 homes has any form of installation. The comparatively low take up since the 1980s when the idea began to catch on is probably due to the fact that at the time equipment and installation costs were high and as fuel costs were relatively low, the pay back period was lengthy, something to the order of 25 years or so. Since then, technology has improved, installation costs have come down and as fuel prices are continually rising estimated pay back periods are reducing, some sources indicating a possible 7-10 years, others 10-15 years.

In addition, there is now considerable encouragement for owners to do more in the interest of lowering the consumption of fossil fuels as much as possible and reducing emissions of carbon dioxide. Since the beginning of 2003 a grant of £500 has been available from the government under its Clear Skies policy to any homeowner who installs solar panels to produce hot water. In some areas local authority grants of a matching amount have also been available, reducing the total cost of installation by £1000. This can amount to roughly a quarter to a half of the cost and it seems likely that many more systems will be installed.

Ideally, what is required is a south facing, unshaded, pitched roof slope for the installation of the panels, which are usually between three and four square metres in total size. Also required is a dual coil cylinder in the airing cupboard, one coil to operate with the panels, the other with a conventional hot water system arranged so that it cuts in when the performance of the panels fails to raise the water temperature to whatever is required.

This is usually in the morning even in sunny periods when most hot water is normally used and often leaving a surplus available in the evening. The problem, of course, is the somewhat indifferent levels of sunshine available in the United Kingdom and while it is possible at times to achieve a 90% saving in water heating costs in summer, that level drops to 50% in spring and autumn and only 25% in winter, annual savings being reckoned about a quarter.

167: *Not to be confused with an array of Velux windows, these are solar panels in the south facing slope of the pitched roof of a 1950s semi-detached house, covered in double roll profiled tiles. What can the surveyor say about them? He will not be able to check the tale of savings with which he will no doubt be regaled.*

The surveyor on encountering a solar installation will not be able to do more than describe what he can see from the outside, in the roof space and around the hot water cylinder. He should note, and may quote, the savings he is told about but he will not be able to verify them and should recommend an inspection and the obtaining of a report from an independent professional heating engineer on the type and condition of the installation and its likely capabilities.

Separating Walls

Where the inspection is of a terraced or semi-detached dwelling, a sight should be obtained in the roof space of the walls, or wall, separating the dwelling under consideration from its neighbours.

Occasionally, such separating walls, or party walls as they are also called, will not be found as there were times and circumstances when they were not thought necessary. The danger to a dwelling without separating walls is, of course, from fire but there is also a possibility of burglary. It is essential to include comments in the report to this effect if no separating walls are found. The principal danger is, of course, from fire. A fire next door will soon vent itself by breaching through the flimsy nature of most top floor ceilings. It will take longer to penetrate the roof structure and covering but, in this time, with doors and windows being opened below will be fiercely drafted into the roof space of the adjoining dwelling immediately above top floor bedrooms. Accordingly, for safety's sake,

169: *Flues within a party, separating or flank wall soundly rendered and complementing the good standard of brickwork and pointing elsewhere in the roof space. There should be no fume hazard should the flue be utilised for a space heating appliance but, nevertheless, the surveyor should advise a smoke test in the interest of safety.*

surveyors should recommend the building up of a barrier between roof spaces equal in fire resistance to that pertaining in the party walls lower down.

Where walls to adjoining property are in evidence in the roof space, their condition should be noted as they often provide an

168: *Defective brickwork with loose and missing pointing to a flue incorporated in a separating wall and constituting a hazard from fume leakage should any form of space heating appliance, open fire, stove or gas fire, utilise it. The remaining brickwork and pointing to the separating wall is also poor suggesting a fairly low standard of construction for the dwelling originally.*

170: *Pipes such as this running through the roof space in a comparatively new dwelling should be checked out for their purpose, support and the soundness of any jointing and taping. Manufacturers instructions are not always followed. Do they comply with any relevant Building Regulations or by-laws?*

171: Even more so than in 170 these pipes in an older dwelling could be the work of a DIY enthusiast and need checking out in the same way.

indication of the quality of construction to be found elsewhere in the dwelling. Although not seen from outside, they should be of reasonable construction without gaps or broken bricks and be well jointed. Chimney stacks should have been rendered originally as a precaution against any untoward leakage of gases from a flue, but this was not always done. With the increasing popularity of open fires and stoves this aspect is assuming more importance and flues running through the roof space of older houses and flats, whether individual or as part of a separating wall should not be considered redundant. If unrendered and in poor condition with defective brickwork and pointing then the hazard of fume leakage should be indicated in the report. What is also important is to note their position so that they can be checked against the presence of chimney breasts in the rooms below. Sometimes to increase the floor area, chimney breasts are removed from rooms without providing support to the remaining brickwork above, a failure which can have serious consequences should the brickwork collapse. The presence of support must be checked in these circumstances.

Access

The types of access to roof surfaces, both flat and pitched vary considerably and very often there is none at all, reliance necessarily having to be dependent on contractors long ladders. Most two storey and many three storey dwellings fall into this category, particularly those built in the last 100 years or so. Flat roofs very often provide the easiest of access, up a narrow flight of stairs, through a doorway and you are there. Even trapdoors to flat roofs are often hinged and provided with a ladder. The danger with flat roofs lies where there are no parapets or balustrades and even when the former are in place, represent a hazard if too low.

Taller dwellings of the 1700 and 1800s with their flat fronts topped by a parapet with a gutter behind and rooms in the roof space can provide relatively easy access through the dormer windows to view at least part of the roof surface close up. Best of all perhaps for access to see all the pitched roof surfaces are those dwellings where butterfly roofs were provided in the latter part of the 1800s. Given access to the roof for clearance of snow from

172: An access as here to the slopes of a butterfly roof with doorway and internal steps up to it is ideal.

173: More often found than the ideal access to the slopes of a butterfly roof shown in 172 is the trapdoor, only openable by a hefty shove leading to a loud clattering as it slides down into the gutter. Sometimes when lead covered such trapdoors can be very heavy and require some ingenuity on the surveyor's part who not only has to get back inside after his inspection but re-arrange the trapdoor up towards the separating wall so that it can be pulled and drop down after him. Some might not think it worth the candle and too risky but the roof slopes are in fact accessible.

the central valley gutter and for necessary repairs and maintenance, they might have a doorway approach via a short flight of steps rather similar to the better type of access to flat roofs. On the other hand they might just have two trap doors, one in the ceiling of the top floor and one in one of the sloping sections of the roof giving access to the outer surface via a roof space of restricted height.

As in all matters of access, the surveyor needs to use common sense as to whether it is safe and sensible to use the available means in conjunction with his own 4m ladder. There are times when it will be and times when it won't. However, he may not get the chance to complete his inspection if he takes unnecessary risks!

Section 4 External Walls

The Inspection

Beginning the detailed inspection of the dwelling with the roof corresponds with the required order of presentation set out in the majority of pro-forma for reports. Descending from the roof, it might be thought that the inspection should automatically follow the pro-forma in the same way and continue with an external examination of the walls. While, of course this is always possible it might not, however, be the best course to follow.

The purpose of inspecting the external walls is to ascertain their construction, so far as is possible, and their suitability for fulfilling the basic functions of supporting loads without deforming, keeping the elements at bay and insulating the dwelling's occupants against undue cold, heat and noise, all at a minimum cost as far as maintenance is concerned. Faults in design and defects in performance can best be described in a report if the construction is known or an assumption can be made to a reasonable degree of certainty. Consequently the inspection of the external walls needs to be devoted initially to ascertaining construction.

Construction is likely to be governed to a very large extent by the age of the dwelling which will either be known from information given or have been assessed on the initial reconnaissance. For example, there is generally a very fundamental difference in constructional form between the nearly six million dwellings built before 1920 and the 18 million built since. The former will almost exclusively have external walls of solid load bearing construction even though the components within the wall's thickness may differ, as in the case of expensive stone backed by cheaper brickwork.

By way of contrast, from 1920 onwards the external walls of dwellings will be found to be of either solid or, much more likely, cavity load bearing construction. Although some external walls were built before 1920 with cavities, they were few in number. Load bearing cavity construction vied with solid construction in the first 20 years from that date up to about 1940 but it became the predominate form thereafter and continues to be so. However, BRE has identified that over this same period since 1920 about half a million low rise dwellings were built of framed construction. The frames were of steel, timber or prefabricated reinforced concrete in about equal proportions, the first and last dropping out of the running to a large extent by about the late1970s but the timber continuing strongly in favour. Many of the framed dwellings can be considered to be of cavity construction since the outer leaf is tied to the frame across a gap, even although the constructional principle is entirely different from that of load bearing construction, both in its solid form and when a cavity is included. If about another half a million dwellings identified by BRE as being constructed over this same period of no fines in situ concrete are included with those of solid load bearing wall construction, then the 18 million dwellings built since 1920 can be split into 14 million of load bearing cavity construction, nearly 60% of the total stock of dwellings in the UK and four million of solid load bearing construction.

The above figures ignore the relatively small number of dwellings provided in high rise blocks. Although a number of high rise blocks in recent years have been demolished quite a few still remain constructed of reinforced concrete frames or an assembly of storey height prefabricated panels bolted together. While commenting on the structure of such blocks is beyond the scope of reports envisaged in this book, the walls enclosing the flat being reported upon need to receive the same consideration by the inspector as the walls of houses and those of flats in low rise blocks as to construction and performance.

Most load bearing walls of solid construction in the domestic field are made up of small units through their total thickness, even when comprising a mixture of materials. Such small units are able to adjust their position through their jointing material and thus accommodate movement. This is particularly so if the construction is in the softer lime mortar used in former times rather than the harder cement mortar used today. Even when such movement is of some consequence, the external evidence may be difficult to see. On the other hand, the internal face of solid external walls when covered with an expanse of homogenous plaster will crack relatively easily as there are no joints to take up movement. In many instances these cracks will be strikingly visible.

On an internal inspection, plasterwork around windows, in corners, at the junction of ceiling and external walls as well as general wall surfaces are all possible locations where cracks might indicate movement of some sort, and, accordingly, need to be noted in each room when they are found to be present. The note should extend to taking down details of the size of the crack using the crack gauge at both ends if tapering and its direction, horizontal, vertical or diagonal since it will be necessary at a later stage to decide whether it can be attributed with reasonable certainty to problems above ground level, settlement, or to foundation and ground movement, subsidence. Sketches may be considered advisable at times to form part of this notation. It is best to avoid trying to standardise subjective descriptions of cracks, such as fine, moderate or serious, as was attempted at one time by BRE, since perceptions vary as to the implications of such words.

At the same time, other defects, such as damp penetration, perhaps due to inadequate thickness or defective pointing, rising damp perhaps because of the bridging of a damp proof course, defects in horizontal features such as sills producing damp staining below windows, can be seen more readily and noted. In many places, when it is safe to do so, the opportunity can be taken to open windows to look out at the sides and head to see if there are any gaps or disturbance to the frame and to see at close hand what might be wrong.

Quite often solidly constructed load bearing walls will be rendered externally as well as being plastered internally; sometimes the rendering will extend to the whole of the elevations and sometimes only to part. In the latter case, the material used for constructing the wall can usually be ascertained but if the elevations are totally covered, then means of identification other than by appearance need to be used. Whatever the situation, the pattern of defects seen from the interior may be repeated on the exterior rendering whereas they might not have been so apparent on an unrendered surface.

The reasons given above suggest that if the age of the dwelling is known, or it can be assessed with reasonable confidence, as one of the six million built before 1920, an inspection first of the internal face of external walls will reveal more of the defects than would have been the case if prior attention had been given to the exterior. Clearly this has to be recommended course to follow for pre 1920 dwellings and it would be so whatever material is presented to the outside world.

Those dwellings built after 1920 which will have walls of either load bearing solid or cavity construction or those in the form of a frame with a covering attached, need ascertaining as to type before a decision is taken as to where to inspect next. If the dwelling is one of the four million of solid construction then it would be appropriate to start on the interior first. If of a cavity construction it can be a matter of choice, since defects can affect both leaves in different ways and some may affect one leaf only. This can be so irrespective of whether built in small or comparatively large units.

Initially it is important to ascertain the type of construction used for the external walls of the dwelling, so far as it is possible to do so on a visual inspection, together with the materials adopted for the purpose. For this to be achieved both internal and external faces need a thorough inspection and this should extend from top to bottom externally and through all rooms internally at each floor level. Walls should be related to each other and to other structural features such as the roof, floors and partitions which could be providing essential restraint to a length and height of wall which would be far too weak to stand on its own. All this needs to be done before any descriptions or comments are compiled about the walls for inclusion in the report.

Ascertaining Construction

Ascertaining the construction of the external walls of a dwelling involves collecting a number of items of information. There is the appearance, which might already have contributed towards an assessment of the dwelling's age, in itself an important item of information, there is measurement of thickness, there is the sound made when tapped sharply with knuckles and evidence which may have been derived from elsewhere. There is also the question of quality which may need to be taken into account on an overall view.

As to appearance, the initial reconnaissance will have revealed, among other information, what materials the dwelling presents to the outside world. Brickwork will be the most frequently encountered, but there will also be stonework, occasionally concrete block, sometimes in imitation of stonework, but also applied surfaces such as rendering, slate and tile hanging or weatherboarding tending to hide the true construction underneath. The elevations may present the same material on all faces or they may be different according to aspect. Different materials may be used on parts of the same elevation. All need to be noted.

Mention has already been made of opening windows if possible and looking out and around to see if there is any evidence of factors contributing to defects noted internally. This should be combined with measurement of wall thickness and ideally should be carried out at each floor level and through a window or door opening on each elevation. Measurement of the thickness will throw up a number of possibilities as to construction which when combined with appearance, tapping both faces and knowledge of or an assessment of the age of the dwelling should narrow down the likely form.

Tapping both sides of external walls is an essential requirement on all inspections. The sound made will help to determine whether rendering or plasterwork is on a solid background of bricks, blocks or concrete or on a hollow background of wood or metal lath or internally comprising a layer of plaster-board on studding.

Evidence from elsewhere in the dwelling could help to narrow down the possibilities even further. In the section dealing with roofs, mention was made of the fact that the only way to tell that some apparent brick dwellings were steel framed was by the presence of steel sections in the roof space. Correspondingly, the bare appearance of the inside face of gable or separating walls may provide a clue to construction lower down, where the bare appearance will normally be covered by plaster and decorations and even the exterior may be totally covered by rendering. The presence of a cellar might also be very useful in providing information from surfaces not normally decorated and even if they were covered, the cellar might still provide a glimpse of the inner face of the outside walls from beneath the ground floor.

As to quality, it is to be expected that dwellings for the top end of the market will be built better than those built at the lower end and this is as true today as in the past. While the law requires a certain minimum standard from plans submitted for new work, the implementation is not always as thorough when it comes to the construction on site, particularly now that house builders through their own trade organisation pass the plans and approve the constructional work. Horror stories abound. Prior to the introduction of national applied building regulations in 1965, some of the model bylaws were adopted by some local authorities and some were not, so that less than satisfactory structures were sometimes built. While most of the really awful poor quality housing of the past was demolished in the slum clearances of the 1950s and 1960s the refurbishment of the remaining older dwellings is not always direct- ed to matters which an inspector acting for either buyer or seller would consider a priority. At times it can be more cosmetic, leaving basic design faults as they were.

Brick Faced

For ascertaining construction it is probably

fortunate from the inspector's point of view that the majority of dwellings present a face of brickwork to the outside world. Coupled with other information, an elevation of brickwork can reveal factors about construction in a more immediately straight- forward manner than many other materials. For example to achieve effective strength to carry loads, a solidly constructed brick wall needs to be built so that the bricks bond together. This bonding can be seen on the outside face as a pattern of headers and stretchers. If there are no headers to be seen but only stretchers, then the wall is only a half brick's length in thickness. Such a wall has its uses in light structures such as garages and bin enclosures and also to define boundaries, particularly when strengthened along its length by one brick thick piers. As part of a dwelling, however, the sight of stretchers only will be an almost sure indication of load bearing cavity construction. The half brick outer leaf will be tied to an inner leaf carrying most of the load but some of that load will inevitably be transferred to the outer leaf by the wall ties.

Conformation of load bearing cavity wall construction can be obtained by measurement and soundings. Both leaves consisting of brickwork at half brick thickness with the usual 50mm cavity in between and plaster internally will measure between 290–300mm compared with only 240–250mm thickness of a solid one brick wall with plaster. Both walls will sound solid when tapped and both are perfectly satisfactory for a purely load bearing point of view for the average two or three storey building. Larger dwellings can, of course be constructed in both load bearing cavity and solid brickwork and measurement in excess of each of the above figures by about 115mm would indicate a more substantial wall of another half brick thickness appropriate for dwellings taller or of larger area.

In the case of cavity construction a measurement of some 10–15mm less with a house built since the 1950s could indicate an inner leaf of concrete block rather than brick and this would be a reasonable assumption for a dwelling built in the last 50 years, provided that tapping the inner leaf still produced a solid sound.

174: Many dwellings built since the 1920s will be of cavity construction, load bearing over the thickness of two leaves or a non load bearing leaf attached to a frame across the cavity. On a brick faced elevation, stretcher bond as here is the first indication of such, but what lies behind requires further investigation.

If the sound is hollow when the inner face of the wall is tapped with the knuckles, then with any measurement of thickness around the 240-300mm mark the assumption must be of framed construction. For a dwelling built since the 1980s the framing would almost certainly be of timber but from before that date it could be of either timber, steel, or, less likely, prefabricated reinforced concrete in the absence of any clear positive indications in the roof space of steel or timber framing. Evidence in this instance of which type would have to be sought elsewhere internally. One method of seeking confirmation of whether the dwelling is timber framed, is to examine externally around window and door openings where either a 6-12mm gap or a compressible filling of about that thickness should have been left to allow for differential movement between a casing of brick, or whatever

175: Dwellings built with solid load bearing brick walls require the bricks to be properly bonded for strength showing a pattern of both headers and stretchers on the elevations, as here, known as Flemish bond. There are other types of bond. Dwellings with bonding such as this can be from all periods. Both this brickwork and that shown on 174 were built in the late 1990s and there are indications on this photograph around the arch that the bricklayer might not have been too happy with what might have seemed to him something like an innovation these days. All the more reason to check. For example, how were the Building Regulation requirements for insulation met?

176: Stretcher bond on this 1970s brick clad house suggests load bearing cavity construction but tapping the inside face with the knuckles produces a hollow sound indicating an outer leaf of brick on framed construction. Further evidence would need to be sought to determine what type of frame. The age of the dwelling points towards timber framing confirmed, perhaps by the allowance left for differential movement around openings between the timber frame and the brick outer leaf and a sight of the inner face of the gable wall and perhaps the separating wall in the roof space.

material has been used for the outer leaf, and the timber frame.

Render Faced

Presented with the sight of rendering to all of the elevations from ground to eaves, or parapet on a dwelling built before 1940, the initial assumption will be that the wall is of solid load bearing construction. Rendering was usually necessary in many parts of the UK for weather proofing purposes, particularly on one brick thick walls. Cavity walling had been specifically developed to avoid that need and therefore rendering should not have been required. There is a little more to it than that, however, but combined with knowledge of or an assessment of age, measurement and tapping could provide the inspector with a much more reasoned assumption of what lies below the rendering. His knowledge of the locality could be of help too.

Rendering of any type will usually add to the order of 15-25mm to any of the previously mentioned figures for both brick faced solid and cavity construction. A solid sound when tapped on the inside face, will merely confirm that the wall comprises masonry of some sort but a hollow sound could indicate a framed structure, or in comparatively rare instances a dry lining on internal studding. Bringing age into the equation should narrow the assumptions down. Totally rendered dwellings from before 1940 are far more likely to have been provided with solid load bearing walls, the rendering to keep the weather at bay or to cover a material not particularly attractive in appearance, cheap bricks or ugly stone perhaps. In the latter case the wall would, of course, be a great deal thicker.

If on a dwelling built since 1940, measurement indicated load bearing cavity construction and there was a solid sound when tapped, obscuring a material of unprepossessing character would probably be about the only reason why rendering would be employed. If

177: The totally rendered dwelling, as here, presents difficulty for ascertaining construction. Measurement of wall thickness, assessing age and even type of rendering can be pointers. This early 1900s house with its roughcast rendering is almost certainly of solid load bearing construction, probably of cheapish brick throughout. It should be noted throughout that chimney stacks are no guide to the basic construction of a dwelling as even an old timber framed wattle and daubed cottage with a thatch roof usually has a brick or stone stack in an attempt to keep fire at bay.

178: A typical 1930s private sector semi-detached house totally rendered. Normally of solid load bearing construction only the first floor would be rendered leaving no doubt as to the basic construction. Here it would not be wise to make that assumption without measuring the thickness of the walls and sounding the inner face. Although unlikely, cheap bricks and cavity construction could have been used.

measurement and a hollow sound when tapped suggested a cavity and framed construction, then again the rendering is probably presenting a face to the outside world more attractive than would have been the case if the rendering had been left off.

Insulating concrete blocks provide an example of this when they are provided as the outer leaf for some of the framing systems in use about this time and these certainly needed to be covered over.

Of course, where rendering has parted company with its substrate, tapping will produce a hollow sound but this should only occur in localised positions in most cases and checking elsewhere should prevent any confusion. However, the 1992 Building Regulations specifically permit insulation to be placed on the outside, as well as on the inside, of solid external walls and this could produce a hollow sound when tapped. 179 is a case in point, where concrete block construction of the 1990s for a small block of flats is covered externally with insulation and a thin rendering coat producing an overall thickness of 275mm. This has its advantages and disadvantages. Whereas the grumbles by occupiers that providing secure fixings for wall cupboards and shelves when insulation is placed on the inside are obviated, such construction performs poorly particularly under impact. Some systems have already had to be withdrawn.

179: Since 1992 the Building Regulations permit insulation to be placed on the outside, as well as the inside of solid external walls, protected by a coat of rendering as here on a 1990s small block of flats. The construction is of concrete blocks with external mineral wool insulation protected by a thin coat of rendering sounding hollow when tapped. The blistered appearance suggests that all is not well.

As to the assumptions about the construction of totally rendered external walls and in the absence of any firm evidence from elsewhere in the dwelling they must, however, remain as assumptions and should be reported as such. It may be that the inspector will know from the location of the dwelling that it is one of a particular privately developed estate known for its shortcomings or that it is one of a local authority's built using a particular system. 180 is a case in point. The inspector might know that an estate had been developed by the local authority using a no fines *in situ* mass concrete system. Such systems account for about 2%, roughly half a million dwellings, of the total stock but as examples are unlikely to be encountered very often they could confuse if the inspector lacks the benefit of such local knowledge. Walls of such construction tended to get thinner as time went on. Those built before 1951 were commonly 300mm thick, those built in the 1950s and early 1960s about 250mm with those built since down to 200mm thick.

180: Totally rendered walls of a 1970s three storey block of flats. Measurement indicates a wall thickness much less than cavity construction and even slightly less than solid load bearing construction. The elevation is very plain, untypical of private sector flats at this time but the windows are sheltered slightly by the bell mouth formed in the rendering. All this evidence points to no fines in situ mass concrete construction much favoured by local authorities. The rendering has probably been renewed. Tapping the inner face would, of course, have elicited a solid sound.

181: This dwelling from the 1930s, totally rendered externally but sounding solid when tapped from the inside is in fact a steel framed Dorlonco system house. Although all Dorlonco houses are characterised by being double fronted and have steel roof trusses which can simplify recognition, enquiries would need to be made about its history and how it has performed over the years. The conclusion by BRE is that most steel framed houses even of this age, while subject to some corrosion are still likely to provide long service if well maintained. Nevertheless, a prospective buyer would need to be advised to have the structure opened up at an exposed point to examine the steel for corrosion and it may well be that the metal lathing as a base for the rendering has almost totally corroded.

Another case in point helping to ascertain construction might well be where the surveyor is practising in an area known for building in the local unbaked earth. The chocolate box white cottage with its thatched roof, the retirement dream of many, will have the appearance of a dwelling with walls rendered in lime plaster and lime washed, but the rendering may cover load bearing structural walls of cob, clay mixed with straw and water. These cottages are common in the West Country and South Wales but also exist elsewhere. For example, in areas where chalk occurs in southern England and East Anglia this too becomes part of the mix, the material known as witchert in the former area and clay lump in the latter. Dwellings constructed of witchert are found in a band across Buckinghamshire and Oxfordshire and south into Hampshire and Dorset. The construction

is similar to cob but because of the chalk content of the subsoil, the wall thickness need only be to the order of 350-450mm instead of anything from 600-900mm for cob walling. In East Anglia and up into Lincolnshire the clay lump is produced from the chalky subsoil mixed with straw and water in moulds and air dried to produce mud bricks, again to be built off a stone plinth as in the case of walls of witchert and clay lump.

182: The base of the walls of this lime washed and roughcast rendered cottage are of rubble stonework. To an inspector from the area where it is situated this usually indicates construction to the upper part of chalky earth and straw mixed with water and left to dry before being rendered to protect it from the weather. Solid sounding on both exterior and interior faces, overall thickness will be somewhere in the region of 350-450mm and the construction is known locally as witchert.

Another form of building with unbaked soil is *pisé-de-terre* or *pisé*, where the mix minus the straw is rammed between temporary shuttering which is struck later to allow for drying. This produces walls, again raised off a stone plinth, about 350-450mm thick but much quicker than by the other methods. The method found much favour in Scotland, particularly along the south coast of the Moray Firth east of Inverness and in pockets of coastal areas in the east between Aberdeen and Dundee, having been introduced from France in the 1790s.

183: In the same area where the cottage at 182 is situated, portions of the rendering to the upper parts of other cottages are left off where owners seem to be saying 'look my house is not built of earth and straw' and providing visible proof. Overall wall thickness would probably have helped to differentiate, anyway, for rubble walling is usually 600mm or more in thickness.

Knowing or assuming the age of totally rendered dwellings is a help towards making a reasonable assumption of what lies below since some types of construction are less likely to be found and can therefore be eliminated as possibilities. For example, the 1900s are unlikely to have produced more than a few examples of construction in unbaked earth. This is because the more widespread adoption and enforcement of by-laws in the late 1800s and early 1900s tended to count against the use of these methods since they were not covered. Most of the examples which exist are thought to have been built in the 1700 and 1800s, spurred on by the introduction of the brick tax in 1794, though some may have been built much earlier and some are actually said to have been built in the 1920s.
The increasing emphasis on conservation towards the end of the 1900s has led to a re-assessment of the use of unbaked earth for repairs and extensions to existing dwellings and to the construction of new dwellings.

Invoking the functional requirements first introduced in the 1985 Building Regulations for England and Wales and which are much more flexible than the specific requirements of the old Model Bylaws and the earlier Building Regulations, walls constructed of unbaked earth can, under Regulation 7, be shown, by the experience of buildings in use over many years, to be capable of performing the functions for which they are intended. Consequently approvals under all parts of the Building Regulations are readily obtainable from some local authorities, subject only to the provision of additional insulation.

Accordingly, it is conceivable that individual dwellings recognisably built in the late 1900s or early 2000s will be found to be constructed of unbaked earth since both this method, and the thatch with which it is usually accompanied, employ 'green' materials, much in favour at the current time.

As well as the possibility of encountering dwellings of unbaked earth, surveyors in the far South West of England need also to appreciate the likelihood of encountering dwellings built with a local unsatisfactory material that was all too readily available at one time. Cornwall, since Greek and Roman times, was for long the world's major producer of tin and in the late 1700 and early 1800s, underground mining also produced half of the world's copper supply. Production only ceased at the end of the 1900s after 300 years, but the waste material left behind from under-ground mining was very considerable and was there for the taking by small local plant operators and local contractors. Along with the waste material from china clay working, that other long standing but still operating Cornish industry, the waste material from mining was initially thought to be highly suitable as an aggregate in the making of concrete blocks, which were intended to replace the use of expensive natural stone for building or the cost of transporting bricks for the purpose. This followed on from the general use of

Portland Cement in the early 1900s, but particularly between 1920 and 1950.

While the waste from china clay working and some of the waste from underground mining proved suitable as aggregate for concrete block making, others proved to be highly unsuitable. Substantial deterioration was found in the 1960s and 1970s in a number of dwellings identifiable as being constructed with blocks made with aggregates specifically from some of the underground copper and tin mines. The deterioration was so severe that a number of dwellings had to be demolished. The problem was diagnosed as being attributable to the presence of mundic, the old Cornish Celtic word for pyrites, iron sulphide, in the aggregate and it was thought that many dwellings built between 1920 and the mid 1950s would be affected. Sulphate attack is the cause of the degradation when the sulphides oxidise but this was on a scale far in excess of what happens to brickwork. The blocks, being made of Portland cement and an aggregate containing iron sulphide, it only needs the presence of dampness to initiate a sulphate attack. The resulting large expansion and loss of particle adhesion affects the whole of the structural walls and is made much worse if the dampness is trapped behind cracked rendering.

The Council of Mortgage Lenders became alarmed and banned lending on suspect dwellings in 1992 until a reliable screening test was established. An acceptable test, based on petrographic study of concrete specimens was introduced by the RICS Mundic Group, with BRE input, in 1994 restoring 80% of the suspect dwellings to the market.

It had been found possible to identify the sources of many of the aggregates, to link them with consequential degradation and establish a database. Perhaps unsurprisingly, it was found that the highest proportion of dwellings with defective concrete corresponded with the districts which had had the greatest mining activity, Cambourne, Redruth, Perranporth, areas in East Cornwall and the Tavistock area of south Devon. BRE publishes an atlas of mine wastes to assist those unfamiliar with the specific areas of mining activity and this needs to be in the possession of all surveyors practising in the south west of England. Dwellings known or assessed to have been built between 1900 and the mid 1950s, when local production of the concrete blocks ceased, in the known areas where degradation has occurred must be tested if a mortgage is required and, obviously must be if a purchaser is to be assured of future performance even when a mortgage is not needed. The testing is invasive and beyond the scope of a visual inspection which has to be limited in the first instance to identifying the need for testing. If a prospective seller in one of the known areas has not already commissioned a test, then he needs to be advised to do so. A test will determine whether the aggregate used was either Class A, when a dwelling would be acceptable to a lender, or Class B, the latter containing the potentially deleterious material for which BRE was hopeful of being able to predict the rate of degradation when a further test was developed.

Stone Faced

An elevation of stone, real or artificial, presents the inspector with more difficulties for ascertaining construction than does brick-work but, on the other hand, less than is the case where walls are totally rendered. Once again, as with both brick and rendered surfaces, there is no reason why the external walls should not be of either load bearing solid or cavity construction with a further alternative of a casing over a frame as yet another possibility and much the same procedure should be followed for ascertaining the construction. Unlike an elevation of brickwork, however, appearance is not much help in ascertaining the type of construction. The shapes visible on the face of the stonework representing, as they usually do, one or other of the traditional types of walling are no guide as to how the wall was put together. Age is an important factor on the other hand, since there is also the matter of distinguishing between the real thing and the artificial variety, which is in effect a concrete block, albeit the aggregate in manufacture includes stone particles.

Both the natural and the artificial can be produced in a thickness of around 115-120mm appropriate for inclusion in the outer leaf of a cavity wall and in varying heights and lengths so as to enable that variety to be maintained on the elevations which is so much a feature of most types of stone walling. For low rise dwellings built since the 1960s, both solid and cavity load bearing construction are available to meet the requirements of the conservation movement which, since that time, has come to the fore. Since the 1960s, building in the areas where stone was the traditional material has been required, or at least strongly encouraged, to match in with its surroundings. For a special one off privately built dwelling, it may be that the owner could rise to the expense of real stone. For estate development, however, even relatively top of the market, the expense of the natural product would be such as to turn the developer towards the artificial. The products of the reconstructed prefabricated stone industry have improved immeasurably over the years and are acceptable to most planning authorities. The inspector is therefore more likely to encounter the artificial than the real on dwellings built since the 1960s.

Measurement of thickness to the same order

as with a facing of brick, ie around 300mm and a solid sound when the inside face is tapped will suggest load bearing cavity construction. Hollow sounding from the inside face on the other hand would suggest framed construction which is, of course, the other alternative to load bearing construction. For a dwelling built since the 1960s, the framing is far more likely to be of timber since by this time the use of prefabricated reinforced concrete and steel for domestic low rise construction had fallen out of favour.

Having ascertained the type of construction of the post 1960s dwelling with an apparent stone facing, a close examination of the facing should reveal whether the stone is real or not, although this information is not vital from a structural point of view. Reconstructed artificial stone can be made to look remarkably

184: The 1960s detached house with partial stone facing. Although lightly loaded, as roof and floor loads are carried on the brick flank walls, the gable wall could be either load bearing cavity construction or a frame with a skin of blocks. The former would be more probable as the gable wall does not just consist of glass and panels, which would be so typical of the time and there should be a fire separation between the garage and the rest of the house. Tapping the inside face will determine. The artificial stone concrete blocks are not a very convincing representation of squared rubble built to courses and a careful inspection should reveal the presence of repeated patterns and possibly joints cast in the mould in contrast to the real joints between blocks.

real but although the panels may be made from a range of masters to avoid undue repetition, a thorough examination may reveal some panels which are identical along perhaps with joints moulded within the blocks as distinct from genuine joints between blocks, duly pointed.

For dwellings built in the periods 1920 to 1940 and 1940 to 1960 the use of artificial stone can be discounted since in the first period the conservation movement was comparatively weak and certainly in the latter period, the concentration was on producing dwellings in large numbers as quickly as possible to make up for wartime losses and a growing population. The appearance of stone on a dwelling would have seemed an unnecessary luxury and there would hardly have been any demand for the real let alone the artificial,

185: This pair of 1930s semi-detached houses present a distinct contrast, perhaps through neighbourly rivalry. The one on the left has had unnecessary rendering applied to the load bearing brick cavity wall construction. The one on the right hand has had its entire front elevation covered with an imitation stone veneer to represent squared rubble built to courses. Because of its location on an estate, the 'stonework' cannot be mistaken for either real or artificial stone even though natural stone-blocks 115-120mm thick were available at the time this house was constructed and used in the building of cavity walls for similar type houses. Applied to all elevations on a detached house, however, an inspector might be confused. The waney edge boarding on the gables gives a rustic feel to the suburban setting.

leading to a near fatal marked decline within the industry. However, in the 1920s and 1930s, particularly the latter, the typical suburban house of the time could be produced in a range of finishes according to location and what was perceived as likely to sell. The bay windowed three up, two down brick or rendered semi-detached or its detached equivalent could therefore appear at the top end of the market in natural stone in load bearing cavity construction. The pre-cast concrete industry at this time was not geared to producing artificial stone in the form of realistic walling panels, although it could produce smooth well formed sections for sills, copings and door surrounds.

Going back in time, stone faced dwellings where squared blocks were required for better appearance, built between 1920, around the time when cavity construction began to be used more often, and the time around about 1840 to 1860 when the railways began to transport mass produced common bricks to more or less all parts of the country, walls could in many cases be of composite construction. This consisted of stone with a backing of brick, which not only saves on the cost of the stone but also produces a better surface for internal plastering. The stone would be reduced to blocks in sizes of multiples of a brick to enable a satisfactory bond to be maintained without showing brickwork on the face.

Wall thicknesses would depend on the suitability of the local stone for reduction to blocks of small size with the equipment available at the time. With some types of stone, blocks with the thickness of a brick around 115-120mm were possible and accordingly walls could be built of a comparable size, the thinness of the blocks being balanced by lengths and heights greater than a brick providing a variety of shapes on the face of the stonework. However, a British Standard of the 1950s should be borne in mind which points out that solid stone walls, even with a backing of brick, should generally not be less than 400mm thick since the width on bed of most building stones varies from 150-300mm and difficulties with bonding could be encountered with any lesser thickness, a point to be remembered if a stone faced wall of unsteady characteristics is encountered.

186: This small 1900s terrace house is wider at first floor level than its neighbours as it extends over the 'gennel' giving access to the gardens at the rear. Nevertheless, in an area where much of the housing of the same age and of similar size is built of brick, this house would be considered a bit up market with its genuine stone regular coursed rubble facing backed by brick, which is in evidence on the rear elevation and on the flank walls at the end of the terrace.

Where a rustic or rural finish was acceptable, for example uncoursed random rubble or flint, the backing might still be of brick but the overall thickness of the wall is more likely to be to the order of 400-500mm with a fair amount of small stones and mortar making up space around the through stones which are providing the transverse bond, though these, of course, will not be seen.

Where stone faced dwellings were built before around 1840, local custom will have prevailed so that in a brick producing area the aspirational owner could have what appeared to be stone house so as to be thought superior

187: At the end of the market somewhat lower than the dwelling shown in 186, the narrow fronted brick built and faced terraced houses of the late 1800s and early 1900s can be 'improved' and set apart from their neighbours by the application to the brickwork of an imitation stone veneer as here. Strangely the representation of squared rubble built to courses is more convincing than on the artificial stone concrete blocks shown on 184.

188: Stone faced dwellings before the early 1800s, as here, in areas where bricks were not produced will traditionally be constructed in two leaves with through stones for transverse bonding across a hearting of rubble and mortar. The thickness will seldom be less than 600mm and often more, particularly in dwellings of more than two storeys.

to his neighbours but where the stone would be backed by brick. In areas where bricks were not produced locally, the long standing traditional method of constructing a stone wall would be employed, two stone leaves with a hearting of rubble and mortar in between. Wall thicknesses in these cases tend to be at least twice those of the comparable requirements for brick, though will often be found to be far more, depending to a large extent on local custom and the materials available.

Information on age, the appearance, thickness, the sound on tapping and, perhaps, evidence from elsewhere should provide the inspector with what he needs to ascertain or make a reasonable assumption of the construction of the external walls of dwellings with elevations mainly of brickwork, rendering or stone, the latter real or artificial.

Tile, Slate, Shingle, Timber, Steel or Other Material Faced

Other materials which dwellings present to the outside world are somewhat less common and are generally, apart perhaps from weather boarding, on part only of the elevation. Slate or tile hanging for example, is usually confined to upper floors in view of the danger of accidental damage or damage by vandals at low level. While both can be applied to solid walls they are more often found applied to timber framing so that tapping the inside face is always a necessity, even if the dwelling's basic construction has been established from elsewhere. There might have been a change in the form of construction from one floor level to another or between parts of the dwelling.

Weather boarding, whether of timber or plastic, can be considered in much the same way as slate and tile hanging. Although not so subject to low level damage and therefore occasionally found as a material covering the whole of a dwelling, generally it appears on part only so that usually the basic construction can be ascertained from elsewhere. However, it too can be applied to both solid and cavity load bearing walls as well as framed walls so that tapping the inside face is always necessary.

189: Tile hanging at first floor level on a late 1800s dwelling. The inspector would need to ascertain that there is not a change in the form of construction at first floor level from load bearing rendered solid construction to timber framing. Could the tile hanging and the rendering have been a subsequent application to bare faced brick walls to cure damp penetration and if so why and are they working?

190: With not a brick in sight, there is little doubt that this 1960s two storey tile hung and weatherboarded house is if timber framed construction. Tapping inner surfaces will confirm or disprove and an examination of the inside face of the gable wall from the roof space will provide the answer to the type of framing, or the material used for the walls in the unlikely event that they are load bearing.

Knowing or assessing the age of a dwelling is still a useful guide to ascertaining what might lie below slate, shingle or tile hanging or weatherboarding even though these finishes have been applied to all types of construction in all periods. Where it can be a really significant help is in identifying newer materials

191: Knowing the age of a dwelling is a help towards ascertaining what lies beneath weatherboarding or other forms of timber or plastic cladding. On this two storey 1980s house, the inspector would need to ascertain whether load bearing cavity brickwork is the form of construction, as might appear, or whether both brickwork and the timber are carried on a frame. Tapping and an examination of the inside face of the gable wall could provide the answer.

192: The wholly weather or clapboarded house on a timber frame has been a cheap and wholly suitable dwelling for many hundreds of years. This two storey example from the mid 1700s is being refurbished but the inspector will need to check that the proper balance between insulation and ventilation is being maintained for the avoidance of rot in the future.

193: Part of the front and side elevations of a 1950s two storey semi-detached Unity system built dwelling showing the distinctive pattern of the external leaf concrete blocks with their continuous vertical joints and the typical tile hung gable wall. Built for both the public and private sectors, the system used was one of the 27 designated under the Housing Act 1985 because of possible limited life due to carbonation of the prefabricated reinforced concrete frame and inadequate cover to the reinforcement.

cover the frame, prefabricated reinforced concrete panels probably comprised the majority, arranged in a variety of ways with different finishes, but there were also steel and asbestos sheets, timber boarding and proprietary panels of insulating material with factory produced external finishes. Some framing systems offered alternative types of finish and all told there were up to two hundred different types of frame and finish available.

There are also dwellings where the walls are covered with coloured vitreous enamelled steel panels, profiled aluminium sheets or glazed tiles. It is more likely to be a matter of speculation as to what lies below these finishes but the same procedures should be followed to arrive, hopefully, at a reasonable assumption having regard to all the possibilities which were available at the time of the original construction and also, perhaps, at the time of any refurbishment or upgrading that may have since been carried out. Needless to say, during the course of the inspection there may

or materials and systems which were used for particular reasons at specific times. As already stated the period 1940 to 1960 saw the use of various systems of framing to speed the process of house building in a time of shortage of both materials and skills. Different systems employed different external treatments to

194: The consequences of designation of prefabricated reinforced concrete dwellings under the Housing Act 1985. The unsuspecting purchaser under the right to buy legislation would have got a grant from the government to the order of £20,000 (at the 1986 level) to remove the frame and replace it with load bearing brickwork for his Unity house on the right and make it thereby mortgageable. The local authority owner of the house on the left and purchasers of private sector built Unity dwellings were entitled to nothing on the principle of caveat emptor, let the buyer beware. The presence of the adjoining dwelling and probably others nearby, will mean that the inspector will have no problem ascertaining the original 1950s construction of the dwelling on the right, but he will need to see the Certificate and Warranty issued by PRC Homes Ltd., the organisation set up by the NHBC to oversee the work.

196: The undulating appearance of the cladding to these 1970s three storey terrace houses with integral garages may cause puzzlement. Tapping will reveal that the panels are of off white vitreous enamelled steel to the exterior and there is solid sounding construction internally. Construction could be of brickwork but the wall thickness is less than that required for load bearing cavity work. Enquiry would elicit that it is in fact a load bearing no fines mass concrete in situ form. The replacement white painted metal windows spoil the symmetry but may be more practical with their top hung opening vents.

195: This two storey semi-detached house of the 1950s is a British Iron and Steel Federation system steel framed dwelling with a rendered ground floor, profiled steel sheeting to the first floor and either steel sheeting or corrugated asbestos to the roof. It has the distinctive large ground floor windows which are usually retained in shape if the dwellings are upgraded, which is a help towards recognition when the overall appearance may well be substantially altered.

be evidence available from the performance of materials that might alter a preconceived view or help to reinforce it, deformation or delamination are such. The ingenuity of the suppliers to the building industry are limitless and new 'wonder' finishes are continuously being developed. All the more reason for the surveyor to keep abreast of current developments.

A word about fakes. These not only lull an owner into thinking he has something that he has not but can also mislead the surveyor carrying out his inspection if he is not careful. Some fakes are fairly easy to detect even from the exterior. Mock Tudor framing, for example, consisting as it does of planks stuck on the outside, with football studs added to simulate dowels, lacks the drunken air brought about by the shrinkage of the green oak timber of which the genuine was constructed. Those early timber frames which have been entirely

197: *Brown vitreous enamelled steel panels, glazed tiles and brickwork cover the exterior walls of this 1970s designed block of medium rise flats on a cast in situ reinforced concrete frame. To know what lies behind the panels needs enquiry since tapping will reveal only whether the construction is solid or hollow.*

198: *Although there are dwellings of much the same height and appearance from the 1500s, the inspector ought not to be mistaken into believing that this is a genuine timber framed example. Even without knowledge of its age, its too precise appearance, measurement and sounding will reveal that it is a block of three storey load bearing brick constructed flats with applied decoration, the 'Mock Tudor' for which there was a vogue in the period 1920 to 1940.*

rendered are likely not be fakes in the true sense but treated thus to keep out the draughts. On the other hand, there were many timber framed dwellings which were given a make over in the 1700s to become, from the exterior at least, up to the minute examples of fashionable building. The interior will probably tell a different story but the surveyor needs to be careful all the same.

To be wary indeed must be the advice for any surveyor inspecting dwellings prior to reporting to buyers or sellers. It is best to anticipate the likelihood of meeting many different types of construction, some more frequently than others. With this attitude, and the avoidance of hasty assumptions, meeting the unexpected will not come as too much of a shock.

If the comparatively rare is encountered then the surveyor must be prepared to make enquiries from the current owner, or the local authority if necessary and to use his resourcefulness to track down information if it should be available. The surveyor's own library should be of sufficient scope to help, websites can be searched and enquiries made of BRE or

English Heritage, local historical societies and perhaps, in appropriate areas, organisations of local enthusiasts dealing with an unusual type of dwelling common to a particular area.

There can be no argument with a designated order of presentation which requires the external walls to be dealt with after roofs. Most clients would consider the two elements to be of the utmost importance and one Building Society not so long ago reported that, between them, roofs and walls accounted for around 50% of all defects found by its surveyors. Accordingly that order is being followed for the sections of this book even though in the remainder of this particular section there will be references to how deficiencies elsewhere in the dwelling affect the condition of the external walls and, vice versa, how their defects affect other elements of the interior. All these aspects need to be considered by the inspector before the section on the walls is written.

Of the roughly 24 million dwellings in England,

199: A certain impreciseness in appearance coupled with the jettied construction at first floor level, measurement and tapping would suggest that these are genuine timber framed structures perhaps of mixed dates, the dwelling on the left earlier than that on the right, certainly above first floor level. A check on the listing on the English Heritage website, www.imagesofengland.com, would confirm a mixture of 1500s and 1600s construction.

200: The true construction of this totally rendered two storey dwelling is given away by the projecting jetty supporting the front wall above first floor level. Such construction was a feature of the 1400 and 1500s. The rendering was probably applied for no other reason than to keep out the draughts.

201: The less than wary surveyor on arrival might think here that he was about to inspect a fine example of an early 1700s dwelling. Once inside he would find that, instead, this is a timber framed structure from the 1500s given a makeover so that its owner could keep up with the latest fashion.

Scotland and Wales it is believed that about nine million are enclosed by solid mass load bearing walling, brick, stone or concrete or a combination of these materials with a relatively few built of unbaked clay or chalk. The majority of the dwelling stock, however, something like 15 million units or about 60% is enclosed by walls of cavity construction, mostly load bearing but with perhaps around a million or so hiding a frame.

The external walls of all dwellings have a number of basic functions, one of the most important of which is making sure that rain driven against the outside face does not penetrate to the interior. Having ascertained the construction it is necessary in the first instance to consider whether that construction is adequate to cope with the vagaries of the local weather.

Assessing Exposure

The climate is a never ending presence and the United Kingdom's dwellings should be, but are not always, built to withstand that climate. The question arises, were the external walls of the dwelling under consideration designed and built to withstand the local climate and, in particular, is the wall facing the prevailing wind suitably built for the purpose? It is the prevailing wind which at its strongest can drive rainfall at its heaviest through the external walls to the interior, a fairly common defect encountered in dwellings.

Local custom and tradition often used to govern the form of construction of the external walls. What was successful in keeping out the rain in a particular area would be repeated and what was not would be abandoned. Accordingly, in the wetter and windier parts of the country, thick walls of harled stone are seen as successfully keeping the rain at bay but in the drier and calmer parts, unrendered one brick walls perform the same function equally well. Such custom and tradition would be ignored, however, by landowners and builders for example, putting up dwellings at the least cost possible for renting out during the Industrial Revolution. Such practices continued even though by the 1900s local laws often imposed some minimum standards. From the 1950s onwards, cost limits on public housing have not always meant that the best decision is taken as far as construction to cope with exposure is concerned. In the private sector the volume builders are governed by the bottom line and on their estates the best case scenarios rather the worst case, may govern the type of external wall construction adopted throughout.

Codes of Practice in the 1950s for both brick and stone walls broadly divided exposure conditions into three groups.

Sheltered: conditions pertaining in districts of moderately low rainfall where a wall is protected from the weather by the proximity of buildings of similar or greater height. The lower two storeys of dwellings in the interior of towns in such districts are cited as usually within this group.

Severe: conditions pertaining where a wall is liable to exposure to a moderate gale of wind accompanied by persistent rain. Walls which project well above surrounding buildings may be severely exposed even if not on a hill site or near the coast.

Moderate: these conditions exist where the exposure is neither sheltered or severe.

Such description did little more than sum up in words the common sense which ought to have been applied to traditional and customary construction in the various parts of the UK. There was at that time no measure of rainfall or wind speed and direction to indicate in which exposure group a particular location should be placed. This was left entirely to the designer or builder and a surveyor inspecting the dwelling at a later date would need to exercise his own judgment as to whether the construction, as found, was reasonably adequate to cope with local conditions, aided by the evidence from its performance. The Codes did, however, proffer recommendations for suitable construction for the three described categories of exposure, pointing out that a soundly constructed hollow wall did provide an absolute barrier to rain penetration. These recommendations were of assistance always provided the decision taken on the exposure group was correct, based as it had to be at that time on local custom and observation, rather than quantified evidence of wind and rain.

To provide a sounder basis for relating construction to weather patterns, BRE and the Meteorological Office from the 1960s onwards maintained a study of rainfall amounts and the speed and direction of the wind at various locations throughout the UK. As far back as 1976 BRE produced a report, Driving Rain Index, which included indices based on the product of the average annual rainfall on the horizontal and the average airfield wind speed. The maps for this annual wind-driven rain index appear in BS5618:1978

Code of Practice for Thermal Insulation of Cavity Walls (with masonry inner and outer leaves) by filling with Urea-formaldehyde (UF) Systems, as amended in 1982, which will be referred to again later when cavity walls are considered in greater detail.

Although rainfall varies considerably across the UK it is largely unaffected by local features unlike the general wind speed, which does not change very considerably but is much affected by local features, such as trees and buildings, and whether the ground is flat or steeply rising. There was a need for correction factors to take these into account.

Further work, and the advent of computer analysis, enabled the Meteorological Office to produce more realistic values over longer periods and to allow for the fact that heavy prolonged rain was usually associated with stronger than average winds. Maps were produced showing the quantity of wind driven rain falling on vertical surfaces during the worst likely spell of bad weather in any three year period. The data formed the basis of a draft British Standard DD93:1984 Methods for Assessing Exposure to Wind Driven Rain, which enabled local spell indices to be calculated in terms of litres per square metre per spell of rainwater driven against a wall for 12 different orientations at each location.

Following shortly after DD93:1984, came a new Code of Practice for the Use of Masonry (brick and blockwork) BS5628 and Part 3, 1985, Materials and Components Design and Workmanship, specifically referred to the Draft for Assessing Exposure. Because of the improved Meteorological data it contained and the problems which the filling of the cavity in existing hollow walls had produced in the 1970s, it increased the exposure categories from three, which had been used since the 1950s, even when it was possible to

quantify them, to six, by subdividing the amount defining each of the three earlier categories. In this Standard, each of the six new categories was provided with recommendations for appropriate construction which equated in respect of the new category of 'Very Severe' with that subsequently provided for in Approved Document C4, Resistance to Weather and Ground Moisture, Section 4, Walls, paragraph 4.8 of the Building Regulations 1991.

The Technical Committee which produced the 1985 edition of Part 3 of BS5628 recognised the need to further develop methods of assessing exposure to wind driven rain which could be applied to many aspects of the design of masonry. There was interest not only in the aspect of rain penetration for which the local spell index is the most significant factor but also the average moisture content of masonry, for which the local annual index is relevant. More research was needed into the influence of local environments on the incidence of rain penetration and the effect a building's size, its shape and its design features had on the amount of rain actually falling on a wall. When BS8104:1992 Code of Practice for Assessing Exposure of Walls to Wind Driven Rain appeared, to replace the Draft of 1984, it contained new maps from Meteorological Data collected over 33 years from 1959 to 1991, based on improved methods of interpretation and which showed substantial increases in some spell indices over figures previously produced.

Following the publication of BS 8104:1992 a full revision of BS5628 Code of Practice for the Use of Masonry Part 3:1985, Materials and Components. Design and Workmanship was undertaken and it is to BS8104:1992 that the 2001 edition refers for an assessment of local wall spell indices to determine the risk of rain penetration. The six exposure categories of the 1985 edition are reduced to four as set out in the following table, No. 11 which appears in the Standard.

Table 11 from BS5628:Part 3:2001 Categories of Exposure to Local Wind Driven Rain

Category of Exposure	Calculated quantity of wind driven rain (a) litres per m² per spell
1 Sheltered	Less than 33
2 Moderate	33 to less than 56.5
3 Severe	56.5 to less than 100
4 Very Severe	Not less than 100

(a) Maximum wall spell index calculated by the method in BS8104:1992

Guidance on the recommended thickness of single leaf masonry for brick and concrete construction for the four categories of exposure is given and these will be discussed in the detailed consideration of wall types, materials and defects which follows. It is, however, the assessment of exposure conditions which is of immediate concern and it is BS8104:1992 which provides the current guidance. It may be, however, that construction following the recommendations of the earlier 1985 edition of Part 3 of BS5628, to cope with the spell indices produced at that time, may be less than satisfactory and this may be the reason if problems of rain penetration are encountered in dwellings built in the 1980s and 1990s.

It should be mentioned that in a note below Table 11, the Standard indicates that there is a simplified procedure for walls up to 12m high, intended for use with low rise domestic construction. This is available in BRE Report No. 262 of 1994, Thermal Insulation: Avoiding Risks, and is based on a single small scale map showing broadly the areas where the same four categories of exposure apply as in Table 11. It assumes worst case conditions

and the note points out that it could restrict choice to the extent of recommending a higher standard of construction than a specific assessment using BS8104 would produce. This is the map which is reproduced in the National House Building Council (NHBC) Standards and which registered builders are expected to use for assessing exposure conditions and determining the form of construction to be adopted, unless a full assessment under BS8104:1992 is carried out.

However, it is considered that on large estate development it could be more profitable for the developer to produce a full assessment by the BS method if that would recommend a lower standard of construction for the most exposed wall. This would permit all other external walls on the estate to be built to that lower standard, since it seems unlikely that a developer would wish to go to the extra expense of having different forms of construction for different dwellings on an estate. A surveyor, using the simplified method, could accordingly provide incorrect advice in the case of a new or near new dwelling, perhaps maligning the developer. Accordingly the simplified method will not be considered here. A further reason for not considering it, of course, is that there may well be walls over 12m in height requiring consideration.

BS8104:1992 the Code of Practice for Assessing Exposure of Walls to Wind Driven Rain provides every surveyor with vital information. In the first place, without inspecting anything, it can provide him with the exposure category for the area in which he practices, not just on a generalised basis but in a quantified form which he can then use, if necessary, to assess the exposure category of the wall most exposed to the prevailing wind for the dwelling which he is inspecting. For this reason every inspector needs either to possess a copy of the Standard or at least be able to consult it readily, so that he can assess the Local Spell Index for the area in which he practices and find out the precise area to which that Spell Index applies before it needs recalculation.

For the above purpose the Standard provides 19 wind driven rain maps at a scale of 1:1,000,000, i.e. 1mm on the maps represents 1 km on the ground, in regions and sub regions, the latter identified by a symbol with the geographical increment shown between the contour lines in steps of one up to six and in steps of two above six. For each sub region spell rose values are shown giving the figures to use for each of the 12 possible orientations from each location in the sub region concerned. The highest figure, or figures, on the spell rose, or 'clock' represent the strongest wind speed and the direction, or directions, from whence it comes. Annual rose values are also shown and these will become useful when the further research on the relation between exposure and the properties of materials, such as the absorptive capacity of bricks and rendering etc and the effects of evaporation and moisture content are investigated as mentioned in the Standard as part of future work.

Using the maps, which are divided into 10km squares on the national grid, and the grid reference of the location being considered, it is possible to determine the 'airfield spell index' which, by combining two definitions given in the standard, provides the 'quantity of driving rain that would occur 10m above ground level in the middle of an airfield from a given direction during the worst spell likely to occur in any three year period, in litres per m² per spell'. The height of 10m above ground level is included in the definitions because, of course, (most) meteorological data on wind speeds and direction is taken at that level on airfields for passing to pilots.

Taking for example Brighton as the location and being positioned 10m above ground, assuming the location was a wide open space, and facing due South from whence the strongest prevailing wind comes, according to the spell rose, the airfield spell index would be 143 litres per m² per spell. Being exposed to that amount of wind driven rain would be categorised as 'Very Severe' from the categories of exposure given in Table 11 of BS5628:Part 3:2001 which is probably about what one would expect by the coast in such a situation facing into the teeth of a howling gale.

While the above figure and its exposure category is what a surveyor practising in Brighton (corresponding information can easily be determined for every location in the UK) needs always to bear in mind, both items of information will always be tempered by the application of the factors relating to a specific wall, considerably downwards in most cases. Even without taking those factors into account the direction the wall is facing determines the basic 'airfield spell index'. A direction other than into the prevailing wind will always produce a lower figure. For example in Brighton the least strong wind blows from NNW and the airfield spell index for that orientation is 35.6 litres per m² per spell which is in the exposure category of 'Moderate', but not so very far off 'Sheltered'.

So far, the airfield index has been determined on the basis of the location and the orientation of a surveyor practising in that location but equally, of course, it can be determined for one of a dwelling's walls. To temper the 'airfield' spell index to a specific 'wall' spell index the factors which need to be applied relate to Terrain Roughness, Topography, Obstruction and the size, shape and other characteristics of the wall itself. Applying these to an example of a wall to a prominent two storey, flat roofed, house with a sea view on rising land in Brighton produces the following figures.

There is only one Terrain Roughness factor which is above 1.00 and this is where land facing the wall, i.e. upwind, offers no effective shelter at all and is within 8km from the coast or a large estuary. A factor of 1.15 needs to be applied in this instance but for the three other situations described, and illustrated by photographs in the Standard the factors are either 1.00 for no effective shelter, but more than 8km away from the coast, 0.85 for where there are frequent low obstructions and 0.75 for where there are closely spaced obstructions, such as trees, or the location is built up. In the Brighton example quoted, the wall is facing out to sea and there is no effective shelter so that the factor to apply to the airfield spell index is 1.15.

The Topography factor to be applied allows for the effect on wind speed of local topographical conditions. Its basis is that the land in front of the wall does not slope down steeper than 1 in 20 (there is however an Appendix to the standard providing details of how calculations can be made if required to produce a factor for walls forming part of dwellings on the crest of a hill or on the edge of an escarpment where slopes are in excess of 1 in 20). There is only one situation where the recommended factor exceeds 1.0 and this is where the wall faces down a valley or faces a group of buildings liable to produce a funnelling of the wind where 1.2 should be used. All other situations are factored at 1.0, except in the rare case of a wall sheltered by steep sided enclosed valleys, known to be sheltered from the wind, when it may be taken at 0.8. In the Brighton example cited, the location is not in a valley either likely to produce funnelling or provide shelter and therefore the factor to be applied is 1.0. The factor to be applied to the airfield spell index for an obstruction in front of the wall such as

another building, trees or fences a mere 4 to 8m away from the wall is 0.2, but the farther away the obstruction the less effect it has, so that for an obstruction 100m to 120m away, it is 0.9 and where more than 120m away it is 1.00, because it has no effect at all. The example in Brighton is envisaged as having the advantage of a seaward view from all floors but the converse disadvantage of no obstructions at all to give it shelter, necessitating an obstruction factor of 1.00 to be applied.

Although the shape of the wall under consideration e.g. gable, eaves or a wall supporting a flat roof, and also actual positions on the wall e.g. base, eaves level, top of gable, as illustrated by diagrams in the Standard, provide different factors for the adjustment of the airfield spell index if that should be required for special cases, the Standard indicates that the average factor values as set out may be taken for the basic three types of wall, depending on their size. The average factor values recommended are 0.4 and 0.3 for two and three storey gables, the same respectively for two and three storey eaves walls with pitched roofs of 20° pitch or over and typical overhangs of 350mm and 0.4 and 0.2 for two storey and 10 storey walls supporting flat roofs, although it should be noted that in the latter case the figure should be increased to 0.5 for the top 2.5m. In the envisaged Brighton example a two storey wall supports a flat roof and accordingly the factor to apply would be 0.4.

It can be seen, therefore, in the example that the airfield spell index for the location of Brighton on the basis that the wall under consideration faces the prevailing wind is bound to be reduced substantially when the factors for Terrain Roughness, Topography and obstruction but, particularly, when those concerned with the size and shape of the wall are applied. In the case of the example envisaged the airfield spell index was 143 l/m² per spell and the factors were respectively Terrain Roughness 1.15, Topography 1.0, obstruction 1.0 and for the wall itself 0.4. These when applied to the airfield spell index produce a wall spell index of 65.8 litres per square metre per spell, which falls into the category of 'Severe' according to Table 11 in BS5628:Part 3 2001. If there are problems or rain penetration through the wall the surveyor then can check whether the construction as ascertained meets the recommendations provided by Table 12 of the same Standard for walls in the 'Severe' category of exposure. For solid brickwork this would be at least 328mm, one and a half brick, in thickness and rendered in accordance with BS5262:1991 Code of Practice for External Renderings. Even a two brick thick solid wall ought to have been rendered as should a dense concrete wall, though here a thickness of 250mm should be satisfactory. The minimum thickness of 90mm for the outer leaf of a cavity wall should be adequate, but if rain penetrates to the interior it may be because the cavity itself was not increased from the minimum required of 50mm, as recommended for 'Severe' conditions in the Standard. It is thought that something like 20 to 30% of rainfall driven against a wall can penetrate the outer leaf and a wider cavity could help to prevent rain being blown across the cavity, should pointing be neglected or the perpends be inadequately filled.

The site research carried out to produce BS8104:1992 confirmed what had been previously assumed from hearsay evidence about certain aspects of exposure conditions and enabled the factors recommended to be given appropriate relative weight. For example the walls of dwellings on the edge of an estate facing open countryside, a park or playing fields are at a far greater risk of rain penetration than the walls of dwellings within the estate,

even one row in, except where walls have a clear view down a road or are exposed above their surroundings on a knoll or a gradual slope. It also confirmed the value of overhangs, projecting sills, string courses and the like as protective features for walls, determining that water running off, falls vertically to the ground rather than being blown back on to the wall lower down. As an example, the success of rendering or cladding the top half of an exposed two storey gable wall, as a repair, is cited, as long as the lower edge of the rendering is finished with a bell mouth or any cladding is tilted forward at that point.

However, the research also cast doubt on the effectiveness of often used features such as copings at the top of parapet walls and overhangs at the top of gable walls. Eaves overhangs to sloping roofs where the pitch exceeds about 25° are effective because wind flow patterns at the wall face tend to act as though the pitched roof extended the height of the wall and, to an extent, the eaves overhang acted like a long sill. However, with copings to parapets and gable projections the high wind uplift tended to a degree, to increase the volume of rainfall which collected below the projection, in many cases to a greater extent than would have been so if there had been no projection at all. This effect could be a contributory cause of the problems of excessive dampness appearing in the parapets of the dwellings shown in 209.

The Standard for assessing exposure, BS8104:1992, advises that if, when using the maps, the location lies on a contour line or on the boundary line between sub regions, the worst case figure should be used and similarly when there is doubt about the precise orientation of a wall, the highest of the possible rose values should be taken. Even so, it is emphasised that the index figures produced are not precise enough to enable fine distinctions

Table 2: Ascertaining the Construction of the External Walls of Dwellings and their suitability to cope with Categories of Exposure as defined in BS5628: Part 3: 2001

External Material	Date of Building	Sound when tapped Externally	Internally	Overall Thickness mm	Likely Construction	Suitability to cope with Exposure Category
Brickwork in bonds other than stretcher	Pre 1980	Solid	Solid	230–240	Solid 1 brick thick wall with plaster internally	Sheltered
	"	"	"	350–360	Solid $1\frac{1}{2}$ brick thick wall with plaster internally	Sheltered
	"	"	"	460–470	Solid 2 brick thick wall with plaster internally	Moderate
	"	"	"	570–580	Solid $2\frac{1}{2}$ brick thick wall with plaster internally. Note: No plain brick wall of any thickness sounding solid on both faces is suitable to cope with exposure in the Severe or Very Severe category	Moderate
	Post 1980	Solid	Hollow	320 upwards	Solid 1 brick wall with insulation on the inside face. Overall thickness depends on type of insulant, gap and lining	Sheltered
Brickwork in any bond mostly in header	Late 1700s and early 1800s	Hollow	Hollow	170–200	Timber framed wall clad externally with mathematical tiles hung and pointed to look like brickwork with lath and plaster internally	Sheltered
	"	Solid	Solid	300–310	Solid 1 brick thick wall clad with mathematical tiles bedded in mortar with plaster internally	Sheltered
	"	"	"	410–420	Solid $1\frac{1}{2}$ brick thick wall clad with mathematical tiles bedded on mortar with plaster internally	Moderate
Brickwork in stretcher bond	Pre 1960	Solid	Solid	290–300	Cavity wall of two leaves of brickwork separated by a 50 mm cavity left unfilled or filled with injected insulation	Moderate
	Post 1960	"	"	275–285	Cavity wall of two leaves the outer of brick, the inner of blockwork separated by a 50mm cavity left unfilled or filled with injected insulation	Moderate
	Post 1980	"	"	300–330	Cavity wall of two leaves the outer of brickwork, the inner of blockwork separated by a 50mm cavity with insulation bats fixed to the inner leaf	Moderate
	Post 1940	"	Hollow	240–300	Cavity wall with brick outer leaf attached across a 50mm cavity to a frame of either timber, steel or pre-cast reinforced concrete (PRC). If post 1980, most likely frame of timber. Roof space check should determine	Moderate
Rendering	Pre 1940	Solid	Solid	260–270	Solid 1 brick thick wall rendered externally to improve weather resistance with plaster internally	Moderate
	"	"	"	370–380 and 480–490	Solid $1\frac{1}{2}$ and 2 brick thick walls rendered externally to improve weather resistance with plaster internally	Severe
	"	"	"	450–upwards	Solid wall of rubble stonework rendered externally to improve weather resistance with plaster internally	Severe
	Post 1940	Solid	Hollow	260–320	Cavity wall of two leaves with an outer leaf of cheap bricks or concrete blocks attached across a 50mm cavity to a frame of either timber, steel or pre-cast reinforced concrete (PRC) with a lining internally of plasterboard or other boarding. If post 1980 most likely frame of timber. Roof space check should determine	Moderate
	Post 1990	Hollow	Solid	270–280	Solid 1 brick thick wall of cheap bricks or concrete blocks with insulation fixed to the outside face and with plaster internally	Severe

Table 2

External Material	Date of Building	Sound when tapped Externally	Internally	Overall Thickness mm	Likely Construction	Suitability to cope with Exposure Category
Rendering (continued)	Pre 1980	Solid	Solid	290-310	Solid construction of no fines dense concrete, rendered externally and plastered internally. Only likely to be found on comparatively large local authority developments	Severe
	1950s	"	"	240-260	"	"
	Post 1960	"	"	190-210	"	Moderate
Stonework in regular courses	Post 1920	Solid	Solid	265-280	Cavity wall of two leaves the outer leaf of stone, real or artificial, the inner leaf of brick or concrete block, separated by a 50mm cavity left unfilled or filled with an injected insulation	Moderate
	Post 1940	Solid	Hollow	290-310	Cavity wall of two leaves the outer leaf of artificial stone tied across a 50mm cavity to a frame of either timber, steel or pre-cast reinforced concrete (PRC) with a lining internally of plasterboard or other boarding. If post 1980 most likely frame of timber. Roof space check should determine	Moderate
	1840-1920	Solid	Solid	450-500	Solid wall of stone with a backing of brickwork maintaining regular courses, plastered internally	Moderate
Stonework uncoursed	Any date	Solid	Solid	500 upwards	Solid wall of stonework throughout, plastered internally	Moderate
Rendering but always above a stone or brick base	1600-1800s	Solid	Solid	350-900	Solid wall of unbaked earth i.e. clay mixed with straw and water, rendered externally and plastered internally. 'Cob or clom' 600-900, 'Witchert' 350-450, 'Clay lump' 390-410, 'Pise de-terre' 350-450. Location usually determines	Severe
Plain tiles, slates, shingles or weather boarding	Pre 1980	Hollow	Hollow	200-300	Wall of timber frame clad externally with plain clay tiles, natural slates or timber shingles usually above ground floor level or timber weather boarding throughout, or to ground floor only	Severe
	Any date	Hollow	Solid	280-390	Solid 1 brick thick wall of cheap bricks or concrete blocks, clad with plain tiles, either of clay or concrete, or slates, either natural, stone or artificial timber shingles or weather boarding, either of timber or plastic. If tiles or slates used then usually to upper floors only leaving ground floor to be either rendered or weather boarded.	Severe
	Post 1980	Hollow	Hollow	200-300	Wall of timber frame with weather resisting cladding of plain tiles, clay or concrete, slates, natural, stone or artificial shingles of any material or weather boarding, timber or plastic, backed by a moisture resisting layer and plasterboard or other boarding internally. If tiles or slates used, then usually to upper floors only leaving ground floor to be weather boarded.	Very severe

Note: The table while covering many forms of construction is by no means exhaustive and others may be encountered, though less frequently. For example, solid walls of $\frac{1}{2}$ brick thickness of around 90-115mm, or dense concrete about the same is incapable of coping with any degree of exposure unless rendered and then only in 'Sheltered' positions. In areas of 'Severe' or 'Very Severe' exposure cavity walls to dwellings from the 1920s onwards may be found where the overall thickness may be greater by 25-115mm on account of an increase in the thickness of the outer leaf or an increase of the lower amount in the width of the cavity. Impervious external claddings, with sealed joints, of glass, metal or plastic may be encountered to cope with 'Very Severe' conditions.

between degrees of exposure to be made. BS5628:Part 3:2001 also stresses that because the indices are based on inherently variable meteorological data, and are not precise, local knowledge and experience should be used to determine which exposure category should apply when an assessment produces an index near the borderline between one category and another.

The lack of precision inherent within the indices however, in no way reduces their value. It enables judicious rounding up into easily memorised figures and allows an inspector to familiarise himself with the wall spell index and exposure category of the type of wall likely to be most severely affected by facing into the prevailing wind for his own locality. It also enables him to check that the correct procedure has been adopted and the appropriate recommendations followed when it comes to advising on whether near or near new construction is likely to be adequate in the future or not. This can be done for any location in the UK and using the Brighton example quoted, it ought not to be too difficult for a local inspector to memorise a rounded up airfield spell index of 150 l/m² per spell and the product of the worst possible factors at 0.5 (Terrain Roughness 1.2 x Topography 1.15 x Orientation 1.0 and the highest of the average wall factors x 0.4). These would produce a wall spell index of 75 l/m² per spell equating to the exposure category 'Severe'. For a wall in the worst possible circumstances, not far off the 65.8 l/m² per spell produced by the detailed calculation on the specific wall and sufficient for the purpose, as it is in the middle of the range of amounts of wind driven rain which define that category. Rounded off figures for exposure for any location can be applied when considering the type and construction of any of the walls or details described in the remainder of this section, though of course

other considerations may apply if defects are found. For example a soundly designed wall may fail to keep out the rain if essential maintenance is neglected.

Load Bearing Walls

Cavity Construction

As demonstrated by 174 and 175, the external appearance of brickwork can be a guide as to whether the wall is of load bearing cavity construction or load bearing solid construction. Stretchers alone indicate cavity work whereas a mixture of headers and stretchers signify solid construction. Snapped headers could, of course, be used in a wall of cavity construction to imitate Flemish bond or some other bond where headers are visible but would be expensive and laborious as the half bricks would have to be cut flush, only justifiable rarely perhaps to match other existing work. Unfortunately the appearance of other facing materials, as stated previously, provides no ready guide in a similar way as to type of construction. Rendering is just rendering and natural stone could be in any form. Artificial stone, on the other hand, might provide a clue if a difference between real and moulded joints can be detected and the dwelling is of an appropriate age and in a traditional stone building area.

Although by no means a new idea, cavity construction, ie two leaves of brick or block, of whatever type, separated by, but held together across a gap ideally of a 50mm minimum, by ties began to be much favoured in the late 1800s as a way of preventing damp penetration through walls of one brick thickness. This was all that was required structurally for the two storey dwellings then much more in vogue. It had been found that a wall of this thickness was not really proof against penetrating damp, particularly where exposed to driving rain at first floor level. To overcome the problem, walls had either to be built one and a half brick all the way up, costly and wasteful in the circumstances or, more attractively from the cost point of view, rendered.

As might be expected, the latter option above was usually favoured and indeed this continued to be the practice of builders of new dwellings for many years into the 1900s. As BRE points out much of the development of two storey dwellings by private builders, particularly in the south east in the period 1920 to 1940 was carried out in solid brickwork, albeit with the top storey rendered and the bay window tile hung. This practice continued even into the 1950s when private development was allowed to start up again after the Second World War. Against this commonly adopted method of construction, however, a number of developers catering for the upper end of the market and perhaps employing a designer, did indeed turn to cavity wall construction, as did others in the north and west where the weather was more severe. Where adopted cavity walls were generally provided with brick outer leaves although natural stone was also available in blocks of varying lengths and heights and thin enough for the purpose.

In the public sector, however, cavity construction was thought of as best practice by the designers, either employed or engaged by local authorities, and its use will be found to have been adopted in many of their developments from the 1900s onwards. Sometimes the decision would be strongly influenced by location. For example, local authority housing in the north of Scotland built in the late 1920s was provided with cavity walls. Most public sector development

202: (above left) Although cavity construction was well known about in the 1920s and 1930s, most speculative builders particularly in the south east, continued with solid construction, rendering the first floor and providing bay windows which helped to distinguish their developments from those of the local authority, a vital selling point.

203: (above right) While most private builders continued to build in the 1930s in load bearing solid construction, those conscious that rough weather could be experienced in their area, such as here in the North, turned to building in load bearing cavity construction, thus avoiding the need to render at first floor level except on the retained bay window, where the construction would usually be framed up in timber with a panel of rendering. Unless there is evidence of the use of a long lasting wall tie, the inspector would have to recommend opening up to check type and condition if reassurance is to be given, as those here must be at least 60-70 years old. This is particularly so as while there is no evidence of failure in the brick joints or the use of black ash in the mortar, the dwelling is exposed on a hill on high ground and on the outskirts of a city notorious at one time, for industrial pollution.

of low rise housing up to three storeys has been built in cavity work, mainly with a brick outer skin but for cheapness, to meet cost yardsticks, alternatives such as concrete blocks and rendering could be used. In the private sector from the 1950s onwards when cavity

204: Stretcher bond on this short terrace of two storey local authority houses of the period 1920 to 1940 indicating load bearing cavity construction thus avoiding the need for rendering at first floor level. Only two of the original windows remain but apart from re-roofing nothing much else seems to have been needed externally. What advice should the inspector give to a seller or buyer? There is no evidence of the use of aggressive or soft mortar, no undue pollution in the area, the dwelling is not exposed and the generous eaves overhang shields the walls to some extent. However, the wall ties must be at least 60 years old and even though there are no signs of cracks in the mortar joints suggesting wall tie failure, a tie needs to be exposed to check on type and condition in the absence of other evidence.

205: It should not be assumed that all modern housing is built with cavity construction. Problems which continue to come to light when it is used and the increased levels of insulation required over the years have turned some designers back towards solid construction as here, in the late 1990s, where Flemish bond has been used and the insulation provided internally.

work became almost universal, brick remained the favoured material for the outer skin except in areas where stone was the traditional building material. Here, town planners and the conservation movement pressed strongly for the use of either natural or artificial stone, the appearance of the latter being much improved from the 1960s onwards.

When soundly built, cavity wall construction solved the problem of damp penetration through one brick walls and could be cheaper than a one brick wall rendered wall. While it also improved the insulation qualities of the dwelling, always provided ventilation to the cavity was limited to weep holes at the base and over window and door openings, its detailing was far more complicated and required more careful workmanship than that

involved with a solid wall. Such demands are those which are all too easy to skimp on construction sites and recent inspections of work in progress on a large number of sites continue to reveal a lengthy catalogue of potential problems through poor workmanship, in most cases to be left uncorrected and covered up all too readily by completion. Typical of these included mortar being left on wall ties and collecting at the base of the cavity, damp proof courses laid incorrectly, weep holes omitted and wall ties sloping downwards towards the inner leaf. Only the omission of weep holes from among those listed above, and there are many others of varying degrees in importance, will be visible to the inspector once the dwelling has been completed. On the other hand, the consequential patches of damp penetration may well be all too obvious on walls not recently decorated or perhaps hidden behind items of furniture strategically placed. Suspicions might well be aroused if walls have in fact been recently re-decorated and detection of damp might involve touching with the palm of the hand and judicious use of the damp meter, particularly on a wall exposed to the prevailing wind.

The surveyor needs to be attentive to detailing around windows and doors and at the base of walls where faults common to cavity construction in both design and workmanship are prone to allowing damp to penetrate. General surfaces of walls need to be scanned for the same reason. Other faults to walls which also cause damp penetration such as inadequate projection of sills, copings, eaves and verges and, for example, defective pointing are not restricted to cavity wall construction and, if present, should obviously not be over-looked but also not confused with those typical of that form of construction. Other faults such as those related to the performance of materials or defects due to ground movement, both probably causing cracks, also apply to all forms of construction and will be dealt with under their appropriate headings. However, there are two defects, wall tie failure and sulphate attack which are more properly dealt with here as the first is more or less unique to cavity work and the second, usually affects cavity construction in a more pronounced way.

Wall Ties

It is perhaps an exaggeration to say that cavity walls contain the seeds of their own destruction since the outer leaf is mainly an envelope attached to the structure, often carrying little load other than that from lateral wind pressure and it can be replaced or even re-attached if required without much disturbance to the rest of the dwelling. However, it is inevitable that one or other of these operations will be necessary since, at least for dwellings built before the early 1980s, it has been shown that the vital component, the wall tie, is likely to fail well before the expected life of the leaves for which it forms the attachment.

It is an obvious fact that the more exposed an outer leaf of a cavity wall is to the weather in

the form of driving rain the more rainwater will pass through. All being well the bulk will drop down inside the cavity so that no dampness appears internally. However, a proportion will remain on that part of the wall tie embedded in the outer leaf. That proportion will be greater and remain longer where a cavity has been provided with insulation, either injected some time after construction or provided originally. The outer leaf will, in consequence, remain colder being deprived of a proportion of the heat escaping through the cavity thus restricting the rate of evaporation. The life of a metal wall tie depends on its ability to resist rust in the presence of that dampness along with the quality of the mortar in which it is embedded and the shape of the tie and in this lies its vulnerability.

Wall ties have had a chequered history. In the late 1800s comparatively heavy ties in shapes not too dissimilar from those covered by today's British Standard might be made of cast or wrought iron, the former tarred and sanded, the latter in some cases galvanised. Both lasted quite well, often giving 100 years of service. Hollow clay blocks were also available in the early 1900s, some glazed to shed any moisture and cranked down in shape from the inner to the outer leaf. In the absence of movement in the structure which would break them, these ties are practically indestructible and it is conceivable that these were used in some of the early local authority developments of cavity construction in the periods 1900 to 1920 and 1920 to 1940.

By the 1920s, cheaper mild steel ties were being manufactured in twisted flat bar fishtail and butterfly shapes. Sometimes these were left unprotected, sometimes galvanised or alternatively coated with black bitumen paint. These types became subject to a British Standard, 1243 in 1945, which curiously required differing levels of galvanising for the two types. Even more curious, the specified levels of galvanising were reduced in 1964. It was only when wall tie failures attributed to the omission of a galvanised coating or the corrosion of an inferior protective layer came to light in the 1970s that the Standard was again revised in 1981 to bring the galvanising to a much higher level than before and for the same level to apply to both types and to the double triangle shape tie which had been added to the two other acceptable shapes. Under the Standard, ties could be made in two lengths, 150 and 200mm, and besides galvanised mild steel, could be made in copper or copper alloys and from 1964 in stainless steel. A draft performance based Specification has been available since the mid 1980s enabling ties made of other materials and of different shapes to be tested so that the inspector examining plans of a new or near new dwelling should not be surprised at what he might come across.

To enable some assurance to be given of at least a 60 year life, BRE recommend that for new dwellings ties to BS1243 should be of either mild steel, with the thick galvanised coating of 940g per m^2 prescribed in the 1981 amendment, austenitic steel of grades 302, 304 or 316, or copper, aluminium bronze or copper bronze. Alternatively ties tested in accordance with the draft performance Specification can be used which can be in a variety of shapes in stainless steel or plastic. The latter are considered by BRE to be quite acceptable if limited to dwellings or no more than two storeys in height, but only considered satisfactory provided they are of an appropriate length to span the cavity and be embedded for at least 50mm in each leaf. The inspector of a new or near new dwelling needs therefore at least, to find out what was purported to have been used as wall ties where it has been ascertained that cavity construction has been employed. BRE has

pointed out that some developers were still using inadequate ties in the late 1980s. As to workmanship, the specification should have provided for ties at the rate of 2.5 per m^2 with extra ties around openings and along the top edges of gable triangles, although whether the specification was followed cannot, of course, be checked on site during a visual inspection, once the dwelling has been completed.

In its investigations into defective wall ties, BRE found that corrosion could be accelerated in certain circumstances. Ties embedded in black ash mortar corroded much more rapidly because of the high sulphur content in the ash, particularly where dwellings were situated in areas where exposure to driving rain was the norm, in areas where there was a high level of industrial atmospheric pollution and in coastal areas where the atmosphere contained a high level of salt. It also found that ties positioned in gable walls were much more susceptible as were those set in the weaker lime mortars. In consequence BRE recommended that in conditions of extreme exposure only ties of stainless steel were considered sufficiently durable and where there was salt contamination only grade 316 would do with an alternative of plastic.

As a result of all their investigations, BRE concluded that any dwelling built before 1981, the year when the Standard of galvanising for mild steel ties was raised, is at some degree of risk unless it is known that stainless steel ties were used. A part of the BRE investigation recorded that 'many' outer leaves were sucked out in the gales of 1987 and 1990 due to corroded ties. If these leaves had been built using twisted butterfly wire ties in ordinary Portland cement mortar, there would have been no indication in advance that this might happen. This is because the increase in bulk from the corrosion of a wire tie is insufficient to disrupt brick joints, unless they are exceptionally thin. On the other hand, in cases where vertical twisted bar ties have been used and corrosion causes a substantial increase in bulk, there should have been some indication of cracks in the mortar joints adjacent to where the ties had been positioned. It was said at the time that most collapses of this nature involved inferior ties, the use of permeable lime or aggressive types of mortar, such as black ash, or exposure to severely polluted or marine environments.

In cases where there is cavity construction to a gable wall in a coastal or heavily polluted area, visible evidence of aggressive or soft mortar and signs of cracking in horizontal mortar joints at regular intervals the inspector must, in the light of the BRE advice, give an urgent warning of the potential danger of collapse. This is particularly so if the dwelling is near a railway or a main road and possibly subject to heavy vibration. The provision of safety screening must be advised to keep persons well away unless the leaf can be taken down immediately.

In the above situation, but where there are no visible signs of cracks in the horizontal mortar joints on a pre-1981 dwelling, there is either the possibility of collapse at any time should wire ties have been used and have corroded or, in the case where vertical twisted bar ties have been used, corrosion may not have reached the stage where cracks appear. Here again in both cases safety screening is required as a matter of urgency until a tie can be located and taken out for examination to determine which type of tie has been used and its condition.

However, there are plenty of dwellings of cavity wall construction in areas well away from the coast, where exposure categories as defined in BS8104:1994 are classed as

'Sheltered' or 'Moderate', there is little or no atmospheric pollution, no evidence of the use of aggressive or soft mortars and no visible signs of cracks in the horizontal mortar joints. Does the inspector just say that there are no visible signs of defects and leave it at that for a dwelling known or ascertained to have been built before 1981? He will have no clue as to what type of wall tie has been used, so that while BRE indications that galvanised mild steel vertical twisted bar ties from before 1981 have a life of 35 years as against 20 years for the twisted wire type (15 years where the leaf is attached to a timber frame) are of vital interest, they are not a great deal of help when the type of tie used is not known. What the inspector must say, for a dwelling built before 1981 in cavity construction, is that there is a possibility it will shed its envelope, at least in part, at any time. Should such a thing happen a few months after the report has been submitted, a client would be entitled to say that he should have been told of the possibility and at least advised that a check on the type and condition of a tie in an exposed wall be made. This is particularly so in view of the possibility of collapse and injury to occupants or passers by but also because of the aspect of re-instatement costs. BRE puts these at 1998 prices at a few hundred pounds for a small centre terrace house rising to £3000 for a large detached house, not small amounts to someone who has exhausted his resources on the purchase of a dwelling.

It is conceivable that a seller will have got wind of the problem of wall tie corrosion and if brick joints had become cracked or rendering had begun to show signs of a pattern of horizontal cracks at places where wall ties were likely to have been positioned, steps might have been taken to have walls repointed or re-rendered. This should not, however, lead the inspector to change his advice in any way. Where a dwelling is of cavity construction either load bearing or framed with masonry outer leaf and built before 1981 then even if there are no signs of defects associated with wall ties, the unequivocal advice must be, in the light of the BRE findings, that a wall tie be located at one of the more exposed positions of the dwelling and examined for type and condition.

What of the dwelling built before 1981 where the seller maintains that the wall ties have been renewed? A number of different techniques are available whereby ties can be replaced from the exterior without the need to demolish the outer leaf, either leaving the old ties in place or having them removed, depending on what type was originally used.

The specialist firms which carry out this type of work usually provide a guarantee and in the first instance the inspector would wish it to be made available for examination. It would carry more weight for the firm providing it to be substantial, well known in the business for many years, of sound reputation and for the guarantee to be backed by insurance. For a dwelling to be mortgageable, a life of 60 years must be anticipated and this is the life at least expected from the mild steel wall ties covered by the 1981 amendment to BS1243 for use on entirely new work. Guarantees on wall tie replacement work of less than this length of time are therefore not ideal as problems could arise on a resale at a later date. This is the disadvantage of a method of joining the two leaves together, where ties have failed, by injecting a heavy duty polyurethane foam into the cavity. The method, which even so cannot be used in areas of high exposure because of the risk of damp penetration to the inner leaf, is reckoned only to have a life of around 20 years and is therefore of very limited value. Wall ties for replacement work are, of course, different from those used for new work

because they need passing through the outer leaf for fixing within the inner leaf by a secure method. There is no Standard available and the stainless steel and plastic types of tie used are of as wide a variety as are the fixing methods which range from expansion, screwing, using special tools, and grouting with resin or other cementitious material. The ideal is that ties are subjected to an *in situ* pull test before the outer leaf is made good.

Even though the inspector considers a satisfactory guarantee is available, he should still ask to see the specification of the work carried out to ensure that it is appropriate to the site conditions and also see the report which would have been prepared after the inspection of the original ties. He needs also to be sure that the specification covered the whole of the external walls of the dwelling and not just those which were thought to be more exposed than others.

206: One of a terrace of early 1900s two storey houses built of load bearing cavity wall construction in a coastal town where the atmosphere was not only salt contaminated but also polluted by industrial processes until the recent past. The wall ties must be nearly 100 years old and the evidence of staining from a leaking rain water pipe makes it imperative to advise the seeking out and exposure of a tie to establish type and condition.

207: Gable walls in load bearing cavity wall construction are at greatest risk from wall tie failure particularly as here where the dwelling is from the early 1920s and the ties must be around 80 years old. To be assured that collapse is not likely, even though conditions are favourable as regards mortar and exposure and there is no evidence of defects, advice to expose a tie needs to be given to ascertain type and condition. It is surprising that the 3.5m tall stack, far in excess of current Building Regulation limits, is not showing more signs of distortion than it does.

208: Load bearing cavity wall construction for a two storey detached house built in the 1980s. It ought to be possible for a house of this age to establish by enquiry what was purported to have been used by way of wall ties originally. There is evidence of damp staining on the gable wall and the surveyor would wish to know the type and that either stainless steel or copper was used or if mild steel, that it was galvanised to the higher standard required by the 1981 amendment to BS 1243 if assurance is to be provided.

Sulphate Attack

A defect, the indications of which could confuse the inspector with those of wall tie failure is sulphate attack on the jointing and pointing of mortar of brickwork and any applied rendering. Although sulphate attack can affect brick walls of solid as well as cavity construction, it only occurs on those dwellings built in the last 100 years or so where Portland cement has been used in the mortar mix. Since there are fewer dwellings of solid construction from that period, sulphate attack is more likely to be encountered on dwellings of cavity construction. Against this, however, is the fact that much rendering with Portland cement in the mix was applied to dwellings at first floor level in the 1920s and 1930s and, when applied overall, constituted a feature of those purporting to be in the 'modern style'. Prior to the 1920s, most jointing, pointing and rendering was based on lime or older forms of patent cement which were not subject to sulphate attack.

Sulphate attack is caused by the presence in the bricks of soluble salts. When brickwork remains wet over long periods as it can do where dwellings are situated in exposed positions, particularly in gable walls and where there is little protection from overhanging eaves, rainwater dissolves the salts and carries them into the mortar of jointing and pointing. The salt solution attacks the tricalcium aluminate in the Portland cement causing it initially to expand but eventually reducing it to mush. Signs that this is happening are cracks in all the horizontal joints of the pointing in the damp areas of the brickwork, a whitening of the mortar and often signs of a whitish staining around the edges of the bricks. In contrast, the visible signs of wall tie failure are limited to cracking in the horizontal joints in the immediate area of the wall ties themselves. If neither of these defects can be seen at low level or from windows, an examination of walls at high level through the inspector's powerful binoculars may reveal such defects in the brick joints and enable the two types of defect to be distinguished.

Where an external wall has been rendered with a mix containing Portland cement over bricks with a high soluble salt content, the fine shrinkage cracks which often accompany the drying out of such a rich coating, allow moisture to penetrate through to the brickwork and for it to be in contact with the bricks for far longer periods than would have been the

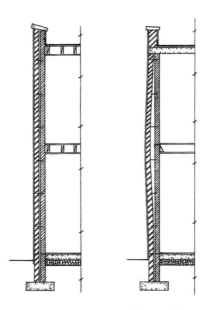

Figure 22: Typical box-like dwellings of the period 1960-1980 with their flat roofs frequently have inadequate damp proof courses in the walls at roof level and, if rendered, allow yet more damp penetration through shrinkage cracks. Wall tie failure and sulphate attack cause different problems depending on whether the roof load is transferred to the foundations through both leaves or merely via the inner leaf. This applies irrespective of whether roofs are flat or pitched. On the left where there is no roof restraint to the outer leaf, the expansion in the joints tends to be progressive towards the top, tilting the wall ties towards the inner leaf and probably causing damp penetration to appear on plastered walls internally. On the right, the roof restraint to the outer leaf causes it to bulge away from the secure fixing provided by the ties, eventual collapse being a possibility. Horizontal cracks to all brick joints, or on the rendering on the line of the joints, indicates sulphate attack. If the horizontal cracks are intermittent, often at about 450mm vertical intervals, then wall tie failure is the defect. There could, of course, be both types of defect happening simultaneously.

Figure 21: At first floor level on the walls of this typical semi-detached house of the period 1920 to 1940 there is the characteristic crazy pattern of shrinkage cracks caused by the use of a rendering mix too rich in Portland cement. The more pronounced horizontal cracks on the leaning chimney stack, too tall for its thickness, are, in this case, caused by the heavy deposits within the stack from the use of solid fuel, inducing sulphate attack on the brick joints and rendering. There is obviously a danger of collapse. Should the bricks in the main wall contain sulphates, damp penetration through the shrinkage cracks, will bring about a more extensive sulphate attack on mortar joints and the rendering and the more pronounced horizontal cracks will appear. If the walls are of cavity construction, wall tie failure will cause the same type of horizontal crack to appear even if the bricks are free of sulphates but the extent of the cracks will be limited to the joints where wall ties are situated, but then probably only if the vertical twisted bar type of tie had been used.

case if the wall had been left uncovered. Damage can then be even more severe than on unrendered walls and a pattern of frequently occurring cracks in rendering along the line of the majority of brick joints will appear. The problem can then become compounded as yet more rainwater is allowed through to the brickwork and eventually areas of the rendering become loosened and bulged. However, if the defect is confined to wall tie failure then cracks in the rendering will again be limited to the horizontal joints where wall ties occur, but still, of course, indicating the need for attention. These horizontal cracks should not be confused with the crazy pattern of cracks running in all directions caused by the shrinkage of a rendering mix too rich in Portland cement, although these sometimes precede sulphate attack by allowing dampness through to the brickwork.

Both wall tie failure and sulphate attack can lead to the expansion of the brickwork and, in a cavity wall, can cause disruption at the position of the ties and possible bulging of the outer leaf when there is sufficient loading at the top to prevent an upward movement in the brickwork.

Remedies for sulphate attack can range from renewing defective bricks, adding a protective flashing to a band course or a canopy over a porch to, in serious cases, renewing all the bricks on an elevation. Other methods, all of which are primarily concerned with preventing rainwater reaching the brickwork, can extend to slate or tile hanging or covering elevations with weatherboarding.

Insulation

Mention has already been made that the presence of insulation in the cavity causes the outer leaf to remain colder and wetter for longer periods and can therefore aggravate defects. In its basic original form of two leaves and a ventilated cavity, the design was better than a one brick solid wall for insulation purposes, but was still well below what could be considered reasonable. This situation was thought satisfactory when domestic central

heating was practically non-existent but when all-embracing warmth became desirable and was achievable in the 1950s and it was found that much of the heat went out through the walls, steps were taken to improve the insulation properties of the wall. For new building this, initially, consisted of sealing off the cavity from the outside air, but retaining, of course, enclosed trunking for the apertures to provide underfloor and other necessary ventilation, and constructing the inner leaf of load bearing insulation blocks. These measures only produced a slight improvement and attention then turned to providing insulation within the cavity itself even though this compromised the basic principle of that form of construction.

For new work, to meet the insulation requirements of the Building Regulations, first introduced in 1966 but increased in 1975, 1982 and 1990 in that respect, so that they are now nearly four times as stringent, the provision of insulation to the cavity could be in the form of bats or slabs or various forms of insulating material. This might also involve the use of longer wall ties as it was essential to maintain the recommended 50mm clear cavity. However, incorrect or a lack of proper fixing could cause subsequent damp penetration to the interior when, for example, the cavity became blocked by slipped or loose sections.

For existing cavity wall construction, which by then might already be 30 to 40 years old, the method of injecting urea-formaldehyde foam, introduced from Denmark in 1959, was usually adopted and was extensively employed following the oil crises of the 1970s. In some dwellings, this produced problems from fumes leaking into the interior as the foam cured and set, foam oozing into the roof space, out from weep holes and air bricks for floor and other ventilation, blocking them up in the

209: There are clearly problems with the parapets of these two storey flat roofed terraced houses from the period 1960 to 1980 built in load bearing cavity construction. Brick on edge copings are inadequate on their own and need something more to keep damp at bay. If the bricks used have a high sulphate content, there is a considerable danger of sulphate attack developing. Extreme care would need to be taken inspecting these dwellings, and others nearby in the same form of construction, to check what has gone wrong and whether any remedial measures taken are, or are likely, to be effective.

210: The more innovative designs which began to appear from around 1980 can produce their own problems and the inspector needs to check that sound principles have been followed. Here on this three storey terrace, the parapet and semi-circular feature to the front wall needs a sound coping and good bricks and pointing as like all parapets, it is exposed to the weather on both sides. Although there may be no defects apparent internally it is clear that damp at least is beginning to affect the parapet, staining the brickwork and eroding the pointing and if the bricks have a high sulphate content there can be a danger of sulphate attack. It may be possible to access the parapet gutter and peer over the top for a closer look.

process, and water penetration across the cavity where fissures occurred in the set foam causing damp staining on walls internally.

There was adverse publicity and in the US and Canada where most houses are of framed construction and the fumes in consequence escaped more readily, urea-formaldehyde foam was banned because of fears of nasal tumours. The ban was lifted in 1983 when the fears were shown to be unjustified but in the UK the Building Regulations in England and Wales had already been amended in 1975 to require notification, since a proposed installation of cavity fill was deemed to be an alteration to the dwelling and a breach of the requirement that the cavity should be clear.

Approval was normally given provided the installation was carried out by a contractor registered under a scheme introduced by the Agrément Board. The British Standards Institution took over the running of the

registration scheme of firms of assessed capability in 1979 after the publication in 1978 of two British Standards, 5617 Specification for Urea-formaldehyde foam systems suitable for thermal insulation of cavity walls with masonry or concrete inner or outer leaves and 5618 a Code of Practice for the work covered by 5617. Specific guidance was not included at this time for installation in dwellings of framed construction and because of problems where this had inadvertently been carried out, it was found necessary to amend the Code of Practice in 1982 so that before any work was carried out the initial survey identified dwellings unsuitable for the purpose, unless brought up to suitability in advance.

The requirement of BS5618 for an initial assessment as to suitability before urea-formaldehyde foam was installed was extended to the proposed provision of all types of insulation material by the publication of BS8208 Guide to Assessment of suitability of external cavity walls for filling with thermal insulants Part 1: 1985 Existing Traditional Cavity Construction. Other forms of cavity wall insulation can include expanded polystyrene beads or granules, man-made blown mineral fibre, for which a British Standard Specification was issued in 1982 and a polyurethane foam similar in appearance to urea-formaldehyde foam. The British Standard Code of Practice 5618: 1982 covers only the urea-formaldehyde type but Agrément Certificates were issued in respect of the other types. All provide insulation of a similar standard but urea-formaldehyde foam has remained the cheapest and is therefore the most likely to be encountered. On the other hand, because it shrinks on curing it is, along with polyurethane foam, the most likely to allow rain penetration across the cavity. Although more expensive, polyurethane foam could well be encountered where it has been used also to combat wall tie failure.

Since the only way that the surveyor on a visual inspection can himself tell whether a cavity wall has been insulated or not is by catching sight of hardened foam or some other insulating material at the top of the cavity in the roof space, he has to rely, in general, on statements made or documents produced by the seller. For a new or near new dwelling, the specification approved under the Building Regulations will hopefully be available and the surveyor will have his own opinion on the way in which the insulation requirements have been met. These typically include the maintenance of the normal 50mm cavity sealed from the outside air but with insulation either within or fixed to the inside face of the inner leaf, raising the overall dimension to between 300 and 330mm. Also typical are cavities filled with mineral wool or fibreglass, both of which are thought unlikely to provide a passage for the transmission of

water to the inside face of what is in effect a solid wall about 280mm in overall thickness. Insulation by the use of foam, either urea-formaldehyde or polyurethane is not likely to be encountered for new work because of the reputation it acquired when used in existing dwellings and this is also presumably the reason why other regulations apply in Scotland where exposure conditions are more severe.

For existing dwellings where the cavity has been insulated since 1980, the inspector will hope to see documentation to show that the work was carried out by a registered contractor under the British Standard scheme for firms or Assessed Capability, along with the initial report, a description of the proposed work and receipts for the work actually carried out. By this paperwork, if it is complete, the inspector will have obtained all the assurance he can that the dwelling was considered suitable at the time for the provision of cavity insulation and be informed of what was done. His inspection may either show or disprove whether the correct decisions were taken and that the workmanship, so far as he can see, is satisfactory.

Where work to provide urea-formaldehyde cavity fill insulation was carried out between 1960 and 1980, there may be documentation if it was carried out after 1975 by a contractor registered under the earlier Agrément Board scheme and the proposed work had been notified to the local authority. Even so, and particularly so where no documentation whatever is available and there are damp penetration problems, the inspector needs to give some initial consideration to the factors determining whether a dwelling is suitable for such work. At the time, the 1982 amendment to BS5618 covering this aspect had not been issued.

The need for the 1982 amendment had come

about because of earlier damp penetration problems and it was primarily concluded that, along with other factors to be taken in to account, dwellings in exposed positions were unsuitable for cavity wall insulation. If there are problems of damp penetration in a dwelling situated on a hillside or in a coastal region, unscreened by neighbouring dwellings and in an area subject to heavy rain and gales, it is fairly elementary to conclude that undue exposure is at least one of the causes. On the other hand, the situation might not be quite so straightforward.

Having regard to the high incidence of reported penetrating dampness in the UK, affecting around 15% of the total stock, the inspector needs to be aware, as already indicated earlier in detail under the heading of Assessing Exposure, that much work has been carried out to measure rainfall and the speed and direction of the wind. An inspector needs to know for his own particular area, for example, from which direction rain is driven at its maximum. While for most of England, Wales and Scotland it is from the West, there are, in fact areas in both England and Scotland where it is from the East.

Driving rain is of course the problem. Rainfall, wind speed and direction held on computer over 50 years or so has been analysed to produce national maps of two types of directional driving rain index, an average annual index and a once in three years spell index, i.e. peaks and lows. The first is useful for dealing with weathering and staining problems while the second is essential for assessing the risk of rain penetration through masonry walls. The information is provided in a BRE Report of 1985, Directional Driving Rain Indices for the UK – computation and mapping – and there is also a British Standard 8104:1992 which provides a Code of Practice for assessing the exposure of walls to wind

driven rain. These provide the basis for the inspector's local knowledge of weather patterns and exposure for his own area and to which he will need to add the local site conditions for the dwelling under consideration. Most dwellings have at least two exposed elevations, some have three and detached dwellings have four, so that it is probable that one is affected more than the others. This elevation needs isolation and detailed consideration if the inspector is to provide a reasonable explanation of the faults.

Other aspects to comply with BS8208 Part:1 1985 which have to be considered before a dwelling is deemed suitable for the provision of cavity insulation are the form of construction and its age, the condition and extent of the cavity to be filled, the condition of the two leaves and any services within, whether there are any gaps between the cavity and occupiable parts of the dwelling, the condition of joints if the inner leaf has been left unplastered or has dry lining and whether air bricks for ventilation to sub-floors are enclosed across the cavity.

Depending on the outcome of the initial survey, a range of works may need to be completed before the cavity fill is installed and the inspector will need to see the specification for these and receipts if possible or alternatively form his own judgment on what he can see.

All the considerations in the Guide need to be taken into account and need to be related to any evidence of subsequent difficulties found on the inspection. These could range from isolated patches of damp on walls to cases of rot in wood floors where the ventilation has been cut off.

Wall tie failure, sulphate attack and problems of damp penetration to the inner leaf either due to faults in design or workmanship, particularly around openings for windows and doors, or from the subsequent addition of insulation to the cavity where none existed before, are defects particularly associated with load bearing cavity wall construction. Of course, such construction is not immune from other defects due to causes originating above ground level which also affect load bearing solidly constructed walls such as roof thrust, overloading, movement due to temperature changes, expansion and shrinkage which will be dealt with under the next heading. They are also not immune from defects originating from below ground level such as inadequacy of foundation design in relation to the subsoil, leaking drains and tree root action, among others, which will all be dealt with under the heading of Foundations. The mismatch of materials normally used in cavity work, the fact that ties are set a mere 50mm in the outer and inner leaves together with the use, over the last 100 years of the stronger cement based mortars as against the weaker lime based, often makes the effect of these more dramatic and serious when they do occur. Cavity construction is not able to accommodate move-ment in the same way as earlier solid brick construction which was usually more substantial even though built in the weaker lime mortar.

Solid Construction

Brick Walls

For ascertaining the construction of the external walls, the inspector will have taken a measurement of the thickness at each floor level and, if possible, on each elevation. Determining at the design stage what thickness of brickwork ought to be provided for an external wall of solid construction has always been governed by its ultimate height and also in more recent years by its length. This indeed is how the Building Regulations set out the

thickness of walls for dwellings up to three storeys in height in Approved Document A in the 1991 edition of the Regulations for England and Wales and the Small Buildings Guide of the Building Standards (Scotland) Regulations 1990. For larger dwellings at the present time, calculations need to be made and it is conceivable that for small dwellings, walls thinner than those set out could be shown to be sufficiently strong. However, usually the practicalities of building and performance requirements preclude the use of walls less than one brick's length in thickness, usually just termed as a one brick wall or when required to be thicker a one and a half or two brick wall etc.

Of course, over the years dwellings have been built far taller than three storeys in height and earlier Regulations allowed for such multi-storey dwellings to be built in brickwork by setting out a far wider set of circumstances than the current Regulations with appropriate thicknesses for each.

The Schedule in the current Building Regulations provides a useful check in the case of a dwelling up to three storeys in height should there be a doubt in the inspector's mind as to whether defects in the form of leans and bulges can be attributed to inadequate thickness for the wall's height or length or alternatively whether the defects seen are due to some other cause. Experience suggests the latter to be far more likely in the case of such low built dwellings.

For the convenience of considering the suitability of the thickness of walls of dwellings taller than three storeys in height built in solid brick construction but before the advent of the first Building Regulations in 1965, the Schedule from the 1991 edition but translated into terms of brick thicknesses has been extended as Table 3 by the addition of

Table 3: Suitable thicknesses for domestic solid load bearing brick walls of various heights and lengths as derived from the Building Regulations 1991 for England and Wales, the Building Standards (Scotland) Regulations 1990 and extended from the London Building (Constructional) Amending Bylaws (No1) 1964

Height of Wall	Length of Wall	Thickness of wall in brick lengths
Not exceeding 3.5m	Not exceeding 12.0m	1 brick for the whole height
Exceeding 3.5m but not exceeding 9m	Not exceeding 9.0m	1 brick for the whole height
	Not exceeding 9.0m	1 brick for the whole height
	Exceeding 9.0m but not exceeding 12.0m	$1\frac{1}{2}$ brick for lowest storey, 1 brick for the rest
Exceeding 9m but not exceeding 12m	Not exceeding 9.0m	$1\frac{1}{2}$ brick for lowest storey, 1 brick for the rest
	Exceeding 9.0m but not exceeding 12.0m	$1\frac{1}{2}$ brick for lowest two storeys, 1 brick for the rest
Exceeding 12m but not exceeding 15m	Not exceeding 10.5m	2 brick for lowest storey, $1\frac{1}{2}$ brick for next two storeys, 1 brick for the rest
	Exceeding 10.5m but not exceeding 13.5m	2 brick for lowest storeys, $1\frac{1}{2}$ brick for the rest
	Exceeding 13.5m	$2\frac{1}{2}$ brick for lowest storey, 2 brick for the next two storeys, $1\frac{1}{2}$ brick for the rest
Exceeding 15m but not exceeding 18m	Not exceeding 10.5m	2 brick for the lowest storey, $1\frac{1}{2}$ brick for the rest
	Exceeding 10.5m but not exceeding 13.5m	2 brick for the lowest two storeys, $1\frac{1}{2}$ brick for the rest
	Exceeding 13.5m	$2\frac{1}{2}$ brick for lowest storey, 2 brick for the next two storeys, $1\frac{1}{2}$ brick for the rest
Exceeding 18m but not exceeding 21m	Not exceeding 13.5m	$2\frac{1}{2}$ brick for lowest storey, 2 brick for the next two storeys, $1\frac{1}{2}$ brick for the rest
Exceeding 21m but not exceeding 24m	Not exceeding 13.5m	$2\frac{1}{2}$ brick for lowest storey, 2 brick for the next three storeys, $1\frac{1}{2}$ brick for the rest
Exceeding 24m but not exceeding 27.5m	Not exceeding 13.5m	3 brick for lowest storey, $2\frac{1}{2}$ brick for the next 2 brick for the next three storeys and $1\frac{1}{2}$ brick for the rest

Note: Where walls over 18.0m in height exceeded 13.5m in length, thickness was to be increased by a $\frac{1}{2}$ brick except in the top two storeys extracts from the London Building (Constructional) Amending Bylaws (No 1) 1964 with the old Imperial heights and lengths rounded off in metricised terms.

In the absence of any visible defects in the form of leans and bulges, observable by looking along the wall from the ends, upwards and along both ways from the mid-point at the base, over the top of a parapet or by looking out of windows, there is usually no need to take other measurements in addition to thickness coupled with a check on Table 3 to see whether that thickness is adequate. Most often it will be, but on occasions from times when the enforcement of Regulations was less stringent, builders for the sake of cheapness would cut corners and walls will be found to be half a brick less in thickness than they should have been. The margin of safety in the bricklayer's practice of the time incorporated in the Regulations was such that the gamble often paid off, but not always if other factors in the construction such as poor and, or, insufficient mortar or soft and misshapen bricks came in to play.

The Regulations governing the thickness of solid brick walls have been refined from experience down the years. For example the Building Act of 1774 categorised buildings into seven rates, in four of which appeared dwellings and prescribed the thickness of party and external walls for each. The categorisation took into account various factors including floor area, the number of storeys, the anticipated construction cost and the height above ground level. In some ways the definitions were contradictory and open to various interpretations, resulting in some proposed dwellings being tagged with what could be considered an incorrect rate. It is said that a fourth-rate house could comply by its anticipated cost and floor area yet still be built higher and have more storeys than it should.

211: The Building Act of 1774 categorised proposed terrace dwellings into one of four different rates according to a number of factors. These included proposed floor area, anticipated construction cost, the number of storeys and the height above ground level, prescribing the thickness for external and party walls of each. This photograph shows a first-rate house of five storeys built in 1789.

212: A terrace of four storey houses built around 1800 and categorised as second-rate under the Building Act 1774. The use of different variously coloured bricks in the rebuilding of the top storey of one is unfortunate for appearance but why was it necessary? Should a surveyor inspecting one of the others in the terrace consider whether the need to do so might not also be approaching in the one he is viewing. Do these rainwater pipes hide where the joint of new to old was made, leaving, it would seem at parapet level, a break in the face of the brickwork?

As a result some builders took advantage and built with walls of a lesser thickness than required. Some dwellings had to be demolished on becoming a danger but many still remain, often with parts rebuilt.

213: A short terrace of four, four storey houses built in the late 1700s and categorised as third-rate under the Building Act 1774. As with the first and second-rate houses shown in 211 and 212, the need to rebuild at top floor level in the past, either wholly or in part, should remind the inspector that it is not only adequate thickness that has to be considered but also essential restraint. Usually floor joists span from party wall to party wall, the shortest distance, leaving the front wall unrestrained for much of its height. The outcome of the practice at the time of building the party walls before the front walls of terrace dwellings has necessitated the provision of restraining discs at mid height on the first and second floors to secure the front wall at its centre.

214: Rows and rows of three storey fourth-rate houses as categorised under the Building Act 1774 were built, often very cheaply, in the latter part of the 1700s and the early part of the 1800s. Many have long since gone by way of re-development with those remaining usually very flimsy and much altered as here.

Contemporary pattern book illustrations show houses of various rates, but even when built in accordance with the Act a third-rate house built after 1774 of basement and three upper storeys could be built with walls of $1\frac{1}{2}$ brick thickness in the basement and a mere 1 brick thickness for the three upper floors. Lacking restraint from floors but carrying a roof load, it s not surprising that the front walls of upper floors will often be seen to have been rebuilt and this will be found to have been even more of a requirement for dwellings built in the hundred years or so before the Act was passed.

215: It is not often that the inspector will see such clear indication of the front wall of a terrace dwelling leaning outwards. The line of red facing bricks delineates the party walls between the dwellings of this terrace built around 1720 of basement and four upper floors. The top storey of the front wall of the house on the right has been rebuilt independently, leaving the brickwork of the house on the left at the same floor level proud of the new vertical face. A surveyor on a visual inspection of the house on the left, and possibly of any other house in the terrace where the top of the front wall has not been rebuilt, would need to advise the necessity of further investigation to establish whether it is safe to leave such a lean uncorrected. It is thought that regular monitoring to detect whether there is any further movement, at least, would be necessary. There is only a narrow basement area between the dwelling and the public pavement so that any collapse could prove to have serious consequences.

If not actually rebuilt, the inadequacy of thickness and lack of suitable restraint also accounts for the numerous metal ties seen on many a dwelling of the 1700 and 1800s.

216: A further example of a very visibly partly rebuilt front wall to a four storey terrace house built about 100 years later than the one shown in 215. The Building Act of 1774 did not really provide for the necessary restraint to prevent tall walls leaning outwards. This one has to be rebuilt from parapet down to first floor level leaving the adjoining building's front wall at least 100mm out of plumb at the top.

217: This late 1700s terrace house of basement and three upper floors would have been categorised as second-rate under the Building Act of 1774 but has needed to be provided with metal cross ties to the front wall probably to restrain movement due to inadequate thickness of brickwork and lack of restraint. A check on whether the ties are continuing to function or failing to hold the wall in position can, and needs to, be made by looking out of the windows, provided it is safe to do so. The dwelling on the left has had to have roughly half the top storey rebuilt.

218: It has been necessary to restrain movement of the front wall at the top two floor levels of this early 1700s house of basement and four upper floors. Measurement of thickness at each floor level and a comparison with Table 3 will help to determine whether this has been necessary due to inadequate thickness or it may be due to a lack of restraint. Whether the metal disc plates are fulfilling their function needs checking by looking out of the windows.

219: While it is possible to appreciate an owner's reluctance to rebuild a front wall, and in the case of this early 1700s terrace house of basement and three upper floors, much character, including the wood cornice and door casing, could be lost in the process. However, restraint of movement in this form does cast doubt on structural stability, seems ugly and heavy handed and must affect value to an extent even if, from appearance, it is doing its job.

220: Flank walls of the end houses of terraces are notorious for movement lacking the support provided by adjoining property. The flank wall of this basement and three upper floor late 1800s terrace dwelling is bulged at top floor level. Other dwellings in the terrace have butterfly roofs of the type shown at Figure 27, prone to causing a lean or bulge in the flank walls of the end houses. Here in the refurbishment, the old roof has been replaced with a new structure to provide additional accommodation, stacks built higher and the flank wall tied in with steel channels and disc plates, a belt and braces approach. Internal inspection should reveal whether progressive movement has been restrained, now that the original cause has been removed.

As time went on, the proposed height and length became the sole governing factor for determining the thickness required for a dwellings external and party walls and reference to Table 3 should be made if there are any doubts about adequacy. The third-rate house mentioned above if it had been built in accordance with Table 3 would have had the first storey above basement level of one and a half brick thickness in addition to the basement but, even so, would also need to have been provided with sufficient restraint by attachment to roof and floors.

The problems of inadequate wall thickness, lack of restraint and, as will be seen, poor construction in the speculative building of terrace housing in the 1700, 1800 and the early part of the 1900s was often dealt with either by rebuilding or the provision of metal tie rods. These would be secured to disc plates on the face of the wall and fixed to a structurally sound feature internally or through to a wall, via the floors on the other side of the dwelling, moving in the opposite direction. Nevertheless, there are many tall terrace dwellings existing where recourse to either of these procedures has not been considered necessary, yet walls are clearly leaning outwards or bulged.

Such dwellings have, in effect, reached a new position of repose and given continued domestic use in a quiet location free of undue vibration, any undermining and an avoidance of major alteration, either externally or internally, will, to all intents and purposes, remain in repose for many years to come. An inspector can be more assured of this if decorations have not been renewed for some years and the signs of previous making good to cracks remains undisturbed other than perhaps from minor signs of shrinkage in the form of hairline cracks. This is so whether bulged or leaning walls have been provided with tie bars and disc plates or not and even whether adjoining walls of another terrace dwelling have been rebuilt vertically.

It is not thought that many of the restraining features such as disc plates, flat bar crosses or 'S' shaped braces on many older dwellings are really doing anything very much at all. In contrast with long channel sections, suitably packed out, they are often inadequate for the purpose being too small and too far apart in position and if movement had been continuing the plates or bars would have been left behind in the brickwork as the wall continued to move. The explanation for the apparent success of many of the restraining features is that they were an unnecessary provision in the first place, the wall in question already having reached a new position of repose and not requiring restraint.

Faced with the situation outlined above, there is no reason why the surveyor in his report on a visual inspection should not relay his view that on the evidence as described of what he has seen, age of dwelling, quality of brickwork, degree of lean or bulge if this can be ascertained with safety by looking out of windows, oldish decorations and absence of cracks, the dwelling has reached a new position of repose, the movement is of long standing, structurally acceptable and further movement is unlikely of a progressive nature. In addition he must stress the absolute avoidance of major disturbance to the present condition unless proper advice is taken. All too often, if such a warning is not included, a buyer will assume that all is well and that he can bring in his builder and interior decorator and start to pull the dwelling apart without professional advice. When things go wrong, he will complain and say that he should have been warned while looking around for someone on whom he can pin the extra costs.

Dwellings are sometimes redecorated before being put on the market for resale, not necessarily to deceive unwitting buyers. If the inspector encounters new decorations in a dwelling where there is visible evidence of leans and bulges, irrespective of whether any means of restraint have been provided, he cannot on a visual inspection, consider himself reasonably assured as he can with old decorations and sound making good. Reasonable evidence in that form that movement is not continuing is lacking and in the circumstances he should advise the need for further investigation, probably to include monitoring for a period of time. Advice of this nature will be considered unfortunate by both seller and buyer. The former might conceivably be able to produce documents in the form of earlier reports and details of work done. Whether these will be sufficient for the surveyor to conclude that it amounts to adequate

evidence to show that movement is not progressive will depend on the circumstances of each case.

Of course, in the case of older decorations, where the making good of cracks provides the surveyor with the evidence that the movement is progressive and of a continuing nature, his report can be more succinct. Considerable expense is likely to be incurred and any idea of the amount involved would need to await further investigation.

While what has been noted above, as to advice to be given in the case of leans and bulges refers to overall defects in brick walls which can be identified as originating above ground level, it also applies to more localised defects but, in addition, more particularly to defects presenting similar characteristics but originating below ground level. The latter are more often, however, likely to be of a progressive nature involving diagonal slanting cracks, probably extending to ground level and will be dealt with in the paragraphs under the heading of Foundations.

Defects in the form of leans and bulges originating from above ground level can also be caused by poor workmanship in the form of inadequate bonding even if thicknesses are adequate. Up to the 1700s English bond was commonly used but thereafter Flemish bond became much more common. It is thought to have been first used at Kew Palace, Surrey in 1631. It was thought at one time that English bond was the stronger of the two but it would appear that there is no evidence for this.

It is possible, of course, to build a wall entirely of headers, the very opposite of a wall built entirely in stretchers, but it is very rare to find a dwelling built throughout in header bond although some do exist.

223: An example of brickwork in English bond.

224: It is unusual to find a dwelling built entirely, as here, in header bond. As at a time before being altered this two storey dwelling was originally what passed for in the 1700s as a post office, it maybe that the original builders thought it would be more secure.

221 and 222 : These two photographs, taken a good few years apart, show the flank wall of a two storey end of terrace late 1800s dwelling provided in 221 at first floor level with cross tie plates to restrain outward movement. In 222 the cross tie plates have been removed, yet the bulge at first floor level and the wide mortar filled gaps at the sides of the window and door frames remain. Their removal can only be the consequence of one of two possible decisions. In the first place it could have been decided to provide a substitute form of restraint not apparent from the exterior. Alternatively it could have been decided that the ties were no longer required as the wall had reached a new position of repose. Either way an inspector would need to consider carefully what form his advice should take. This would depend to an extent on the condition of the mortar filling the gaps around the windows and the door and whether there is any evidence of a tie being provided to the feet of the hip rafters in the roof space.

However, headers alone are useful for taking brickwork around fairly sharp curves, on bow windows for example, and around the bends in walls as was popular in the 1930s. The introduction of stretchers would produce too many sharp angles which would look ugly if not covered over by a rendering.

Not so strong as either English or Flemish bond are some of the other bonds which may be encountered. English Garden Wall bond, one course of headers to three or five of stretchers, for instance is less satisfactory having a long continuous vertical joint between each three, or five, courses of successive stretchers causing a deficiency in transverse strength while Rat Trap bond has too many voids,

225: Headers alone are useful for taking brickwork around fairly sharp bends, as here on this two storey brick bow window to a 1920s semi-detached house. On a curve of this radius stretchers in the bond would introduce too many edges, although this is usually overcome by rendering on bow windows at ground floor level and timber framing with either rendering or tile hanging at first floor level.

226: English Garden Wall bond, comprising five courses of stretchers to one of headers, used in the construction of this two storey terraced dwelling containing two flats built in the style typical of the early 1900s. As this bond has fewer headers than either English or Flemish bond it is easier to obtain a fair face on both sides of a one brick wall, hence its primary use in the past as a division wall between gardens. Although it is considered deficient in transverse strength because of the continuous longitudinal vertical joints running through each of the five successive stretcher courses, it is normally thought suitable for dwellings not exceeding two storeys and is cheaper than using either English or Flemish bonds. Nevertheless a surveyor inspecting either the block or one of these flats would need to look carefully to see whether the front 'layer' of stretchers is parting company from those behind and bulging outwards particularly, as here, the pointing has been drastically neglected at first floor level.

looks awful and, as its name implies, can provide a refuge for vermin. It is reckoned, however, to provide a saving of about 25% over a solid wall.

Jointing of the brickwork should, of course, be properly carried out with the mortar applied to all surfaces of the brick for the full thickness of the wall and if the bricks have frogs, these should have been positioned uppermost. To an extent all this has to be taken on trust once the wall has been completed. However, the sight of brickwork in the roof space on the internal face of gables and on party or separating walls might provide a clue if it is of poor construction as to the situation found elsewhere should this be thought of as the possible cause of defects. Nevertheless this also has to be somewhat in the realm of speculation as the bricklayer could have adapted a much more relaxed

attitude to construction at this level compared with elsewhere.

So too, to some extent, must be a suspicion that the wall in question although supposed to be of solid construction might perhaps be in two skins. Much brickwork in terrace construction built speculatively presents a fine face of quality brick to the street. This could, however, have been applied to far less satisfactory bricks behind and, to save yet

227: One brick thick walls built in Rat Trap bond, where all bricks are laid on edge, as here, the stretchers with a 75mm gap between are about 25% cheaper to construct than walls in English or Flemish bond. The temptation to save money presumably led to their use here in this attached two storey dwelling situated in a rural area and built perhaps in the late 1700 or early 1800s in response to the introduction of the heavy tax on bricks first imposed in 1784 and increased in 1794 and 1803. The problem with this bond as the name implies is haven for vermin making it barely acceptable even when separating roof spaces and not seen as shown on 72. Problems with attachment to the adjacent dwelling do not arise here because of next door's construction in flint.

more money, the headers seen on the face could in fact be half bricks. Such practices were the subject of much complaint even in the 1800s and it took a long time before legislation, but more importantly enforcement, curbed speculative builders from adopting them. The strength of a wall would depend very much on the backing brickwork and its workmanship but very often the bricks used for backing would consist of misshapen and under burnt examples despite an Act of 1764 requiring bricks to be 'good, sound and well burnt'. Furthermore even the mortar might be of poor strength and thinly applied leaving many voids. Bulges when the two skins separate

are not uncommon but hefty and hasty ill-considered tapping to see if there is a hollow sound is not to be recommended. The inspector might disappear in the subsequent clatter as the two skins part company completely.

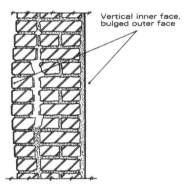

Figure 23: There are usually consequences to building solid one and a half brick thick walls with a facing skin of good quality bricks and a backing of those of poorer quality, a practice common in the 1700 and 1800s. Often headers would be snapped to save more money and only a few facing headers carried through to provide a bond. Poor quality bricks and mortar in the backing compounded the problem resulting in sections bulging forward and causing even those headers intended to provide the bond, to snap.

The practice of building solid walls in two skins for the sake of appearance in speculative building is of long standing and continued to a greater or lesser extent up to around 1900. To keep the rougher quality backing brickwork more or less in line with the courses of the finer quality facing bricks and to achieve at least a modicum of bond at intervals between the two the use of bonding timbers was common. Because these are usually within a half brick's length from the face of the wall, they are inevitably affected by damp over the years which, coupled with the absence of ventilation, causes them to rot and lose their strength to hold the brickwork above in position. The differential movement as the inner skin drops down leaving the outer skin at its original level, causes what headers there are to snap and the outer skin to bulge outwards.

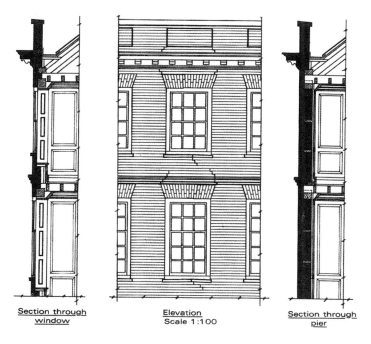

Section through
window

Elevation
Scale 1:100

Section through
pier

Figure 24: The multi-storey brick dwellings of the late
1600s and 1700s usually incorporated a considerable
amount of timber within the walls, originally to hold
the structure together but to cause problems later
when damp penetrated, perhaps due to lack of
maintenance, to the unventilated areas causing the
timber to rot. As the section through the window
shows there was little brickwork at this point but a
continuous large ring beam at each level would be
connected to floor and roof structures and there would
be panelling below the windows and shutter boxes at
the sides. The section through a pier shows the ring
beams and the bonding timbers which were used to
maintain the levels for the rough backing brickwork
and to provide a fixing for wall panelling, as here, or
later for lath and plaster. When the bond timbers rot,
the brickwork in the piers settles disturbing the arches
over the windows and producing cracks in sills and
brickwork. The house illustrated was perhaps built
between the two Acts of 1707 and 1709. The first
banished wooden cornices so that this dwelling was
provided with one either in stone, or built up in tiles
and mortar, and a parapet. The window frames,
however, are still set level with the face of the brickwork and although the Act of 1709 required them to be
set back a distance of half a brick's length enforcement
took a while, and outside the centre of London,
a length of time to filter down.

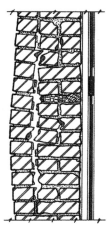

Figure 25: The outer edge of the unventilated bonding
timbers found in the walls of many brick built
dwellings of the 1700 and 1800s is usually within a
mere half brick's length of the outside air and therefore
subject to penetrating damp causing rot and loss of
strength. In consequence the bricks above drop down
fracturing any through bricks and causing the outer
face to bulge.

228: *The contrast between the thickness of the external envelope at windows and elsewhere can clearly be seen in this photograph taken in a very early 1700s dwelling where window frames are set flush with the face of the brickwork and the inside faces of walls are timber panelled. The inside of shutter boxes and the panels above, below and at the sides of windows must be inspected for any signs of rot.*

229: *The exterior of one of the windows of the very early 1700s four storey dwelling shown at 228. The brickwork is typical of the dark shade popular at the time but relieved by bright red dressings and window frames are set flush with the face of the wall. Subsidence at the corner has caused the brick courses to be out of alignment and the sill to fracture.*

Another feature of speculative building of the 1700 and 1800s for terrace housing was the practice of building all the party walls before the main front and rear walls. The time lag did not make a great deal of difference since most terrace dwellings of these periods were provided with basements dug out of the ground and foundations, such as they might be, were at a low level which avoided many of the problems of differential settlement. Bonding between party and other walls was often rudimentary and at times a straight joint would be left with perhaps reliance on strips of iron to hold the front and rear walls upright. This practice can be a further cause of bowing of brickwork as the strips fail to hold and pull out. Even if not unduly serious the movement can produce vertical internal cracks to the plasterwork in the corners of rooms and is probably the reason why it was found necessary to provide restraint to the front wall on the party wall line of the terrace in 212.

Current requirements for dwellings lay considerable stress on the need for restraint to be provided for external walls from both roof and floors, an aspect to which insufficient attention appears to have been paid before the 1960s. The building in to external walls of floor joists provided a degree of restraint to those walls as long as other defects of a more serious nature did not cause the joists to pull away, since there was usually no fixing other than the occasional nailing to a wall plate. Walls, however, parallel to the way floor joists spanned were usually left without any restraint whatever with a consequential risk of leaning or bowing outwards. Even more serious consequences arose from situations where staircases were placed against a substantial area of an external wall where there was no balancing thrust from an adjoining dwelling. The end flank walls of terrace dwellings are particularly prone to bulging, particularly on a downhill slope, either when staircases are positioned against them or when floor joists span from front to rear. The defect is also not uncommon with semi-detached dwellings although not usually on such a dramatic scale.

230: (below left) There is visible evidence of a differential movement in the front walls of these four storey early 1700s terrace dwellings. A surveyor would need to advise further investigation to ascertain whether movement is progressive if he was inspecting either dwelling unless he was entirely satisfied by the presence internally of cracks in plaster, clearly made good many years ago but showing no signs of further opening. The comments made earlier apply here as they do to many dwellings more than 100 to 150 years old.

231: (below right) Timber built into the external walls of dwellings of 1700 and 1800s as lintels and for bonding purposes causes much disruption to the brickwork when it becomes defective through damp penetration. The problems here are compounded by the soft and crumbling brickwork unsympathetically renewed in places. A surveyor's facile advice to rebuild would be unwelcome since a buyer would no doubt be seeking charm and the caché of living in a dwelling nearly 300 years old. Permission to do so would in any event be difficult to obtain unless potential danger could be demonstrated. A buyer needs well nigh a bottomless purse before embarking on the ownership of a dwelling of the age and type such as this, a point on which they sometimes need reminding if not themselves knowledgeable. Careful restoration seeking out suitable matching material but retaining as much of the original as possible would be the requirements of the conservationists, work which is always expensive.

232: The artificial stone hood mouldings to the windows of these four storey 1840s terrace dwellings were not protected by flashings and have clearly been a source of damp penetration causing rot in the timber lintels and disturbance to the brickwork above. Flashings seem to have been added later to the windows of the house on the right which may have protected it from more serious damage. However, lacking ventilation at these points, rot can spread to adjacent floors and in the absence of good evidence that the problem has been investigated and any necessary work carried out, an inspector would need to advise further investigation.

Figure 26: (left) The effect of positioning the staircase against the flank wall of the end house of a terrace and supporting floor joists on a central partition leaves the flank wall with virtually no restraint as here shown on Section 'A' and the key plan. The butterfly roof provides no restraint and probably aggravates the situation. The consequences are often a substantial lean developing in the flank wall and long vertical cracks at the junction with the front and rear walls, 'C' and 'D' on the drawing, widening at the top. This is not an uncommon defect in houses built during the 1700 and 1800s, not only at the ends of terraces but also later when semi-detached dwellings became popular. Scale 1:200

Section of tall thin wall with no restraint 'A'

Key plan for 'A' Top floor

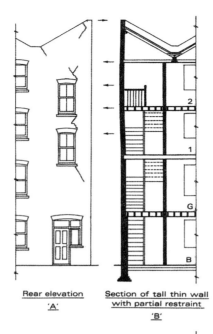

Rear elevation
'A'

Section of tall thin wall
with partial restraint
'B'

Figure 27: Where the staircase is positioned against the flank wall of the end house of a multi-storey terrace but floor joists are alternately running from front to back or from side to side as here on Section 'B' and the key plan, movement in the flank wall can be complicated. If the bond of flank wall to front and rear walls is sound, cracks in the rear wall may be induced at the weakest points, namely the window openings, as panels of brickwork pull away as shown on the rear elevation 'A'. Displacement of the flank wall in this instance may take the form of a horizontal bow as shown on the key plan. Scale 1:200

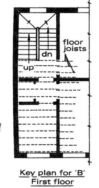

floor
joists

dn
up

Key plan for 'B'
First floor

Gabled flank walls can also be subject to leaning outwards often necessitating rebuilding, due to a lack of restraint. Floor joists below may be spanning from front to rear as will the ceiling joists acting as ties to the feet of the rafters, leaving the gable wall virtually free standing. Of course, if the ceiling joists are

not acting as proper ties through a lack of nailing there could be roof thrust at the top of the front and rear walls as well, causing them to lean out in addition, an unsatisfactory situation to say the least, likely to involve much work of correction.

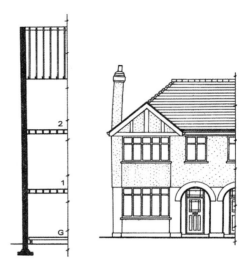

Figure 28: (above left) Expecting a one brick thick wall to do too much, as here, where no restraint is available from the roof or the floor joists which are spanning from front to rear with a load bearing partition for intermediate support, will almost certainly induce a substantial lean. Such a fault is not uncommon in dwellings of the mid to late 1800s and early 1900s.

Figure 29: (above right) :The spread of the private sector semi-detached two storey house in the period 1920 to 1940 with its typically one brick thick walls, bay window and hipped roof slope also saw the introduction of the small solid fuel boiler to produce hot water positioned in the kitchen at the rear, Flue linings were not considered necessary and the result was often a tall leaning boiler stack due to sulphate attack. If this was combined with a lack of restraint to the flank wall, often the case, an unhappy effect could be produced as the latter leant out and the stack leaned inwards. Many stacks have had to be rebuilt with a lining although the advent of balanced flue boilers has meant that many have since been taken down to roof level. Should this be the case the inspector needs to check that ventilation to the flue has been maintained. Scale 1: 200

In comparatively rare cases severe deflection in a floor due to light construction and overloading can exercise an eccentric sideways thrust on a wall causing a bow to develop, instead of the floor load acting vertically downwards as it should be in normal circumstances.

Most of the taller terrace dwellings of the 1700 and 1800s were provided with heavy ornamental cornices near the top of the front wall with only a comparatively light weight parapet to hold down the back edge. If there is no restraint from the roof on the front wall, there is a tendency for the cornice and the brickwork of the top storey and parapet to lean outwards. This could be dangerous if steps are not taken either to rebuild or tie back the feature. Such a defect will be found in dwellings where there is little or no bond between the front and the party or separating walls and the inspector should look for vertical cracks at the junction of these walls, particularly where there is evidence of earlier making good which has subsequently reopened indicating the likelihood of further movement.

 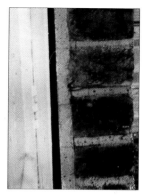

233: (above left) Looking upwards and along brick walls will detect bulged or leaning areas, as here, in this rear wall and chimney stack of an early 1900s house. With such comparatively slight movement and in the absence of cracks or fractures in the brickwork the inspector needs only to record their presence in the report in most instances.

234: (above right) Examining the sides of windows where the frames are set back and in recesses will often indicate that brickwork leans or that there is a bulge. In this example the brickwork is about 100 years old and the movement even now is no more than about the 25mm which in a storey height BRE have said normally requires no repair on structural grounds alone. It is estimated that the making good was probably carried out some 30-40 years ago in a fairly rich cement based mortar yet there are no shrinkage cracks in the mortar itself so that the gap now visible between making good and frame would seem to indicate very slight movement over a considerable number of years. The surveyor could indicate with reason that there is no cause for concern but that to make the gap weather tight repointing with a mortar not subject to shrinkage would be advisable and yearly observation made to see if there is any further movement.

Figure 30: Heavy ornamental cornices of stone or made up from oversailing bricks, tiles and cement were a feature of dwellings of the 1700 and 1800s, particularly after 1707 when wooden cornices were banned by statute because of danger from the spread of fire. If the top of the front wall is not weighed down by the roof, as here, there is a tendency for the upper part of the wall to lean outwards often necessitating rebuilding and tying back.

Towards the end of the 1800s and into the early part of the 1900s the smaller terrace dwellings tended to have splayed bay windows. These were limited to the ground floor only unlike the bow windows of the early 1800s which like the later splayed bays on larger dwellings built either side of the 1850s, were carried through at least one more floor and sometimes to the top of the dwelling. Wherever a bay or bow window finishes, there has to be a beam to carry the walling above and this is often found to be a source of weakness. Either the beam, invariably of timber is inadequate for its purpose or neglect of the condition of the flashing, sometimes only a cement fillet, at the junction of the roof and the front wall allows dampness to penetrate and causes the beam to weaken through either wet or dry rot. The consequences would be a settling in the wall above accompanied by distortion to window openings at a higher level and to sills. Depending on which way the joists are running there could be movement in the floor above and possibly dry rot.

Figure 31: Inadequacy in the timber beam supporting the wall above a bay window or neglect of the flashing at the joint of the roof to the bay window, whether pitched or flat, allowing dampness to penetrate and rot to develop is a fault and defect frequently found in the cheaper two storey terrace dwellings of the late 1800 and early 1900s. The walling settles, windows are distorted, sills slope downwards and a crack appears in the ceiling of the room below. If there is no evidence of effective repairs having been carried out, the inspector needs to advise that the beam be exposed to check for rot.

Of course, such faults and defects are by no means limited to beams providing support over the openings for bay windows. Every opening needs a beam to carry the walling

235: The single storey bay window was a popular feature of the smaller late 1800s and early 1900s terrace dwelling. Lack of a proper flashing or poor maintenance can allow damp to penetrate at the junction of the front wall and bay roof. If not attended to promptly, rot can affect the beam holding up the front wall over the bay, causing it to sag and allowing the brickwork to settle downwards. The rot may also extend to affect the floor. Signs of the defect are brick courses and sills out of alignment, as here, and a crack across the ceiling internally. The current state of the flashing, which here has now been properly provided, the internal ceiling decoration and the sound of the floor above, together with any change in level, will be of help in deciding whether the matter has been satisfactorily dealt with in the past.

above. While inadequacy is often compensated for to an extent by strength in the window frame itself, in the case of timber lintels these are often positioned within a half brick's thickness of the outer face of the wall and in consequence, over time, are damaged by penetrating damp. This can induce wet or dry rot and ultimately the beam ceases to be effective in supporting the walling above, reflected to begin with by disturbance and movement in any brick arch. Concrete lintels, much more likely to have been used in dwellings from the 1930s onwards are not prone to such defects but do sometimes expand causing cracks to appear in plasterwork at either end. This is a once and for all defect not to re-appear if adequate making good is carried out.

236: A surveyor instructed to inspect either of these three storey semi-detached houses of the mid 1800s will have food for thought. The front wall of the top storey of the one on the left has clearly been rebuilt leaving the one on the right with its front wall at the same level still leaning out at least 50mm in front of the rebuilt wall for a height of about 800mm from the top. By the line of rebuilding visible on the left hand property the maximum amount of lean must have been at the centre between the two windows not on the line of the party wall as might have been expected. The question arises as to why the rebuilding was necessary. Parapet gutters if neglected, or even if badly laid out, are frequently a source of damp penetration affecting the feet of roof timbers, the ends of joists, and sometimes the lintels as well. An inspector would need to consider in the case of the house on the right whether, in the light of internal evidence, it is necessary to advise opening up to assess the true condition of hidden timbers and whether or not the front wall should here also be rebuilt if movement continues.

Figure 32: Point loads on walls were not always adequately spread over the structure below by stone or nowadays by concrete padstones. A case in point is at the ends of the beam carrying the valley gutter and the feet of the rafters forming the butterfly roof over many of the three to five storey terrace dwellings of the 1800s. One quarter of the total loading from the roof is applied as a point load in the centre of both front and rear walls. At the front this can often aggravate the situation shown on Figure 30, by adding to the pressure pushing the wall outwards. At the rear, usually because of the presence of a staircase, the end of the beam may be exerting a point load quite near to the head of a window causing fractures. If the bond to party walls either side is reasonably sound a bulge could be induced in the pier between windows as above on the left.

Bricks and Jointing

Well selected clay dug out of the ground, weathered and suitably fired in a kiln can last an astonishingly long time and bricks made of such material are no exception, witness those of Roman origin incorporated in buildings constructed much later but still in sound condition. On the other hand, soft underburnt bricks made from unsuitable clay at the same time, and even much later, have long since crumbled away. Most familiar to the surveyor are the poor quality red bricks which seem to have been made in great quantities from the 1880s to the 1920s. They are vulnerable to weathering and their disintegrating faces are a depressing sight in urban and country districts alike.

The satisfactory performance and the length of life of clay bricks in a wall, whether of cavity or solid construction, is governed by the bricks selected to withstand the degree of saturation likely to be experienced, not only in the general area of walling between damp proof course and eaves but also at other positions in the dwelling. In addition, the composition of the material used and the standard of workmanship adopted for jointing the bricks, the profile of the completed joint along with its subsequent maintenance have a strong bearing on the durability of the bricks.

Bricks used in parapets and chimney stacks, as copings, string courses or in cornices but, particularly those below ground and up to the level of the damp proof course, within the 150mm splashing distance of impervious paving, are most at danger of becoming totally saturated with water. If this happens, it can lead to breakdown by frost action and, perhaps, sulphate attack. An examination, through binoculars if necessary, of the dewlling's vulnerable positions where bricks have been used plus any other positions, such as in free standing or retaining walls, where they may have also been used, can provide crucial information. This applies for a dwelling of any age. The current state of the bricks can be assessed and, perhaps, advance warning obtained of possible future problems even if the condition of the walling between damp proof course level and eaves is generally satisfactory.

Where there are no visible signs of deterioration and the surveyor is inspecting a new or near new dwelling where the brickwork has not been subjected to weathering by the elements for very many years, he should endeavour to see the specification. This is so that he can not only comment on the method of construction but also check on the way the selection of the bricks themselves was handled.

The process following on from the assessment of the exposure category for the worst affected wall by way of BS8104:1992 Code of Practice for Assessing Exposure of Walls to Wind Driven Rain, through the selection of appropriate construction to cope with that exposure, as recommended in the British Standard Code of Practice for the Use of Masonry BS5628:Part 3: 2001 Materials and Components Design and Workmanship,

proceeds in Table 12 of that Code to a consideration of the factors influencing durability. These are principally the location in the dwelling of the brickwork under consideration, the protection provided by the features of the dwelling to limit saturation and the characteristics of the masonry units and the mortar selected. Table 13 in part 3 of that Code sets out, over eight A4 pages, recommendations on the quality of masonry units and appropriate mortar for use in specific locations of the dwelling under conditions where there is a high or low risk of saturation. Four separate columns cover construction in fired clay and calcium silicate units and concrete bricks and blocks.

For fired clay units, the six designations of durability provided in Table 3 of BS3921:1985 Specification for Clay Bricks are used together with the five designations of mortar mix given in Table 15 of BS5628:Part 3: 2001. For example, for unrendered external walls (other than chimneys, cappings, copings, parapets or sills) where there is a high risk of saturation, clay units designated FL or FN, ie frost resistant with a low or normal salt content are recommended in mortar designations (i) or (ii) cement : lime : sand, 1: $^1_4$: 3 or 1 : $^1_2$: 4 – 4 $^1_2$ respectively.

It is possible to make these recommendations because the characteristics of clay bricks, in the main, are defined in the Standard and the bricks accorded a durability designation while the characteristics of the designated mortar mixes are set out in Part 3 of the Masonry Code.

As to the frost resistance, all bricks offered as complying with BS3921:1985 are classified into one of the following categories:
Frost Resistant (F)
> Bricks durable in all building situations including those where they

are in a saturated condition and subjected to repeated freezing and thawing.

Moderately Frost Resistant (M)

Bricks durable except when in a saturated condition and subjected to repeated freezing and thawing.

Not Frost resistant (O)

Bricks liable to be damaged by freezing and thawing if not protected as recommended in BS5628:Part 3 during construction and afterwards e.g. by impermeable cladding.

Although it is hoped to produce a single Pan European standard test for clay masonry units there is, however, at present no British Standard test to determine the frost resistance of clay bricks. Manufacturers are left to decide, if they wish, whether their products are in fact frost resistant by one of two methods. A very severe test devised by the British Ceramic Research Association involves a panel of bricks, kept saturated and subjected to 100 cycles of freezing and thawing. There is no doubt that a variety of brick which can survive a panel test of that severity is frost resistant, a view acknowledged in the Standard. Not many bricks, however, are able to and manufacturers can alternatively claim that a brick is frost resistant on the basis of experience in use and satisfactory performance over a period of at least three years. Clearly, for a location where frost resistant bricks are required a designer needs to know more than that a brick is classified (F) under BS3921:1985. He needs to know whether it has survived the severe panel test of the British Ceramic Research Association and if not, or not subjected to the test, to be shown built examples which have coped with similar conditions in similar locations as those proposed, for a number of years. While the three year period mentioned in the Standard might be considered adequate in some circumstances,

(after all bricks are used in all sorts of buildings not just dwellings), a purchaser of a dwelling would wish to be assured, as far as is possible, of a much longer period of likely satisfactory performance. At least one major manufacturer offers facing bricks which are covered by a 60 year guarantee against frost failure. Some are sold as complying with the Standard and some as not. Others are offered in accordance with the Standard but without the benefit of the guarantee. There is a wide range available. In the mid 1900s it was reckoned that there were about 2000 varieties on the market even though by then more than half the total output was concentrated in the hands of three companies. There is accordingly plenty of scope for palming off less than ideally satisfactory bricks on an unsuspecting public.

When advising on a new or near new dwelling and there is an opportunity to examine the Specification, an inspector would expect to see that bricks to comply with BS3921:1985 as amended in 1995 had been specified but coupled with a guarantee of 60 years against frost failure, such guarantee backed by insurance. Dwellings put up in the public sector are expected to have a 60 year life and this usually applies when money is also allocated when refurbishment schemes come up for approval.

As to soluble salts content, this is more straightforward and an amendment effective from the end of 1995 to BS3921:1985 Specification for Clay Bricks sets down the limits for all bricks, not just those classified as Low (L) when the Standard was first published. Whereas previously the classification Normal (N) set no limit at all on soluble salts content, sulphate content is now restricted to 1.6% in contrast to the 0.5% for bricks in the Low (L) classification and there is a limit on the combined level of other salts of 0.25%. Tests to determine levels are set out and bricks which do not meet the appropriate levels cannot accordingly be offered as complying with the Standard, even though they may do so in other respects. The two categories are defined in the Standard as follows:

Low (L). The percentage by mass of soluble ions measured as described in Appendix B, shall not exceed the following.

Magnesium	0.030%
Potassium	0.030%
Sodium	0.030%
Sulphate	0.500%

Normal (N). The percentage by mass of soluble ions, measured as described in annex B, shall not exceed the following.

The sum of the contents of sodium, potassium and magnesium	0.25%
Sulphate	1.60%

The categories for frost resistance and soluble salts content are combined to produce durability designations thus:

Table 4 Durability designations for bricks		
Designations	Frost Resistance	Soluble salts content
FL	Frost resistant (F)	Low (L)
FN	Frost resistant (F)	Normal (N)
ML	Moderately frost resistant (M)	Low (L)
MN	Moderately frost resistant (M)	Normal (N)
OL	Not frost resistant (O)	Low (L)
ON	Not frost resistant (O)	Normal (N)

BS3921:1985 as amended in 1995 Specification for Clay Bricks is at pains to point out that bricks complying with the Standard are given designations of durability according to their classification by frost resistance and soluble salts content rather than being placed in any defined order as can, for example, exposure conditions by severity or mortar mixes by strength. This is to avoid any suggestion that bricks of one designation are of better quality than bricks of any other. In other words the anomaly is recognised that some bricks which appear to have the characteristics that should provide for a very long life do not always live up to expectations and that what seem to be poorer quality bricks can sometimes survive for a very long time. Hence the advice in both the British Standards for bricks and for the use of masonry, as well as in the NHBC Standards, that there is no substitute for visible evidence of satisfactory performance over many years.

Of course, much depends on conditions in use, particularly contact with other materials as well as manufacture and, of course, on the extent of other features of the dwelling protecting the brickwork. However, it does mean for a surveyor inspecting a new or near new dwelling without a sight of the specification a degree of difficulty. In the absence of visible defects, it is well nigh impossible to tell from appearance alone the quality of a brick. How, therefore, can a surveyor express a view as to the durability of brickwork in a new or near new dwelling? Even relatively poor quality bricks will last up to the expiry date of any warranty and often quite a few years longer so a less than scrupulous developer might well be tempted and choose to invest in fancy trimmings to improve a dwelling's saleability rather than better quality bricks. Not every surveyor will be fortunate as to be able to see defects, as in 238, since generally they take longer to become apparent. It is accordingly understandable should a surveyor decline to express

237: Recessed jointing to brickwork no more than 15 years old on a three storey block of flats. The designer has taken a chance in that such jointing is not recommended except where the exposure category is sheltered but here it is moderate. The use of a hollow wall will probably help to reduce the likelihood of rain penetration to the interior but the bricks in the outer leaf are likely to become saturated and subject to repeated freezing and thawing. An examination of the Specification will enable a surveyor inspecting, on behalf of either a buyer or seller, to find out the durability designation of the bricks used. If the specification cannot be produced, the surveyor in this instance is fortunate in being able to see what is likely to happen to the bricks as time goes on. Already at the base of the wall there is evidence of the brickwork disintegrating because of saturation and consequential frost action due to external paving being taken above the damp proof course level. Even if such dramatic evidence was not available, examination of the bricks by the window would alert the surveyor to the likelihood of future disintegration. A purchaser of a near new flat in this block is going to be faced with a hefty bill for his share of the remedial work which if not carried out soon could lead eventually to entire replacement of the outer leaf with a better quality of brick. A negligence case, Hood v Shaw (1960), where a surveyor failed to notice a number of defective, flaking bricks on a house about 30 years old, left him facing the cost of providing a new outer leaf of more durable bricks.

238: The disintegrating brickwork at the base of the wall referred to in the caption to 237.

employed. With its microscopic labyrinth of voids, the mortar in the joints is the most vulnerable part. This is particularly so where the masonry units lack a degree of absorption and rain is inclined to run down the face of the unit finding minute cracks in the jointing and allowing water to be carried through, by capillary action, to the interior. BS5628:Part 3:2001 sets out the mortar mixes which are provided with the designations used in the large table, to which reference has already been made, giving guidance on the choice of units most appropriate for particular locations in the dwelling. The mortar mixes and their designations are set out in Table 5 (p220) which has been adapted from the Standard.

The Standard further recommends that before deciding on the mortar to be used, the designer should pay particular attention to local practice and to any mixes that have been developed to cope with special conditions found in the locality.

As to jointing it is axiomatic that all joints should be fully filled with mortar but as this is a matter of workmanship it may not be achieved unless there is the closest of supervision on site. Bed joints are usually filled, as the bricklayer will recognise that not to do so could affect strength. It is the

an opinion on durability unless a specification is made available.

As important as the masonry units themselves in keeping the weather at bay is the quality of the mortar in which they are bedded and jointed coupled with the workmanship

Table 5 Mortar mixes

Mortar designation	Type of mortar, all in proportions by volume		
	Cement:lime:sand	Air entrained mixes	
		Masonry cement: sand	Cement:sand with plasticizer
(i)	1:0 – $^1\!/_4$:3		
(ii)	1:$^1\!/_2$:4 to 4$^1\!/_2$	1:2$^1\!/_2$ to 3$^1\!/_2$	1:3 to 4
(iii)	1:1:5 to 6	1:4 to 5	1:5 to 6
(iv)	1:2:8 to 9	1:5$^1\!/_2$ to 6$^1\!/_2$	1:7 to 8
(v)	1:3:10 to 12	1:6$^1\!/_2$ to 7	1:8

Note: The stronger the mortar i.e. (i) or (ii) the better its durability but this has to be balanced against its lesser ability to accommodate movement. The traditional types in the left hand column provide better adhesion to the units and consequently better resistance to rain penetration but need more protection during construction.

239: What happens to brickwork after many years of frequent saturation and subsequent freezing and thawing. Mortar stronger than the bricks often gets left behind.

240: Inadequately filled perpend joints to a 1960s hollow external wall of sand faced Fletton bricks, the surface of which can be scraped off. While in this case, bad workmanship will allow more water to penetrate to the cavity, perhaps seeking out flaws in the drainage channels around windows and doors, than would otherwise be the case if the joints had been properly filled, the effect would be much more serious if the wall had been solid.

perpends that are often neglected to the extent that only the visible ends are filled, 'topping and tailing'. In the case of a new or near new dwelling where the specification is available, the surveyor will be able to check that all matters of mortar mix and workmanship have been covered but once the dwelling has been completed it will be impossible to see such detail unless the bricklayer has been so careless as to leave visible joints only partly filled.

What can be seen and checked is the profile which the joints present to the outside. This can have a pronounced effect on keeping the rain at bay. With new construction it is better to finish the joint as the work proceeds since this avoids the need for any raking out of the bedding mortar. If pointing is carried out after bedding and jointing, however, it should be in a mix similar to that of the jointing material which, in turn, should have matched the strength of the bricks. A tooled finish is better than a flush joint which leaves the surface more open since mortar is pressed into the joint which assists adhesion to the bricks, while the tooling tends to compact the surface. Thus bucket handle or struck weathered joints are preferable rather than merely striking off the mortar with a trowel to produce a flush joint. Recessed joints providing a ledge for water, possible rain penetration and a risk of frost damage are to be avoided. While many consider that brickwork can be made to look more attractive with recessed joints, their use is best confined to interior fair faced work.

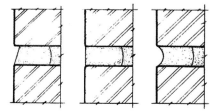

Figure 33: Profiles of three types of pointing. On the left struck weathered, in the centre flush and on the right bucket handle. Those to the left and right are much to be preferred as the tooling compresses the mortar to provide better adhesion to the bricks and closes up the pores on the surface.

Where the surveyor is inspecting a dwelling beyond that stage in its life of being considered new or near new and there is no specification available, attention should be given to what can be seen of the pointing, preferably at the highest level on the most exposed wall, perhaps at the side of a window. BRE suggests that pointing should last 30-40 years but depending on the quality of the original mortar, the profile adopted and exposure conditions much will last a lot

longer. Nevertheless, the age of the dwelling will determine in broad terms when repointing is required. However, this is seldom carried out when it should be. Loose, crumbling or soft mortar at the position mentioned above is an indication that it really is required. The trouble is, of course, that lower down on the most exposed wall and on other less exposed walls pointing is usually in better condition and, accordingly, it all gets left unattended for many more years. This can be a false economy should rain penetrate solid brickwork and cause damage internally or in the case of both solid and hollow walls allow the bricks to become saturated with the danger of frost action and sulphate attack.

241: Badly eroded pointing to one of the corner piers of an angled bay window to a late 1800s house. The brickwork does not appear to be unduly affected at the moment but the danger is that if not soon repointed, rain could penetrate the top of the bricks with the danger of frost attack, possibly upsetting stability, and, through the open joints, to affect the woodwork of the window's box frame. Repointing is urgently required.

Repointing of brickwork on dwellings built before the early 1900s often ignores current advice to match the strength of the bricks and the strength of the lime mortar with which they were originally jointed. Many examples will be found of pointing carried out in a mortar based on too much Portland cement, on the theory that it must be 'stronger'. In practice, it usually shrinks considerably, not only producing fine cracks on the surface which can allow water to penetrate but also shrinking away from the original jointing mortar and the bricks themselves. After a few years it is often easy to lift sections of the pointing out of the joints and, if the joints have not been prepared properly i.e. not raked out sufficiently and wetted, the pointing will often fall out of its own accord.

In careless or unskilled hands, the raking out of joints in preparation for repointing can cause much damage to the edge of the bricks. It should be done carefully by hand to a depth of 19-20mm but now all too often a mechanical disc grinder is used for speed with dire results. In consequence a much wider band of mortar has to be used for the repointing than would otherwise be the case with a detrimental effect on the appearance and poor adhesion to the bricks. Even when raking out is done by hand, carelessness can cause damage if the bricks are soft or unduly brittle. Brickwork that has been repointed many times in the past can often be seen to be affected in this way to a lesser or greater extent.

If bricks do become damaged in this way, tuck pointing might sometimes be adopted. This type of pointing was used for new work in the 1700s when window openings and quoins were dressed with a softer, smoother red brick to contrast in colour with the rougher, less well shaped, bricks used for the general walling. So as to match the finer jointing of the dressings the rougher bricks would be flush jointed in a mortar to match the colour of the brick used and a groove left to be subsequently filled with a narrow projecting band of contrastingly coloured mortar, usually white, but occasionally black. In repair work, this type of repointing can successfully disguise damage to the bricks caused by careless raking out. It is, however, not normally very durable and costs at least five to six times as much as ordinary repointing.

242: Tuck pointing on yellow brickwork of the mid 1800s used to disguise the damaged edges of the bricks and to produce a fine precise appearance. The white 'tuck' contrasts with the base mortar matching, and flush with, the face of the yellow bricks.

243: Three varieties of brick on an early 1700s wall. On a pilaster, smooth red rubbed bricks with fine jointing on the right, ordinary greyish brown stock bricks for general walling in the centre and dressings of red bricks around a window on the left.
The technique of tuck pointing for the dressings and general walling would have produced a more precise finish to match the rubbed bricks on the pilaster but the slight setting back of the repointing is a reasonable alternative and more durable than a thick flush joint or the ribbon effect which a struck weathered joint would have produced.

As to the bricks themselves, where the surveyor is inspecting an older dwelling or where there is no specification available to provide information on their origin, the evidence of the eyes must be relied upon. Those locations mentioned in paragraph 3 of this section on Bricks and Jointing on p216 where the bricks

244: *An example of gross ribbon pointing on the left, badly matched bricks both in size and colour on the right. Rebuilding might be the answer to restore a brick appearance or the application of rendering to disguise. It is, however, a listed building so that the views of the conservationists would have to be taken into account.*

are liable to become saturated need to be viewed from as close a distance as possible or through binoculars for any signs, however slight, of failure in the form of spalling or erosion. Areas where the brickwork is a darker colour than elsewhere are often an indication of damper conditions and repay even more detailed examination and the judicious use of the penknife to scrape the surface if that is at all possible.

245: *Areas of darker coloured brickwork are usually indications of where damp is concentrated more than elsewhere and are likely to be the areas where future problems from frost damage or sulphate attack will occur.*

Where there are no visible defects, the surveyor might be able to say a little more about the variety of brick used although, as mentioned previously, the likely durability cannot be gauged by appearance alone, hence the very essential need to see the Specification if at all possible in the case of newish brickwork. What further can be said depends to a great extent on the age of the dwelling and the surveyor's local knowledge.

The age span of dwellings for inspection is obviously very considerable, from say the 10 years provided in the warranties for new dwellings, upwards. Where there are no defects visible and the dwelling is on an estate with an age assessed at up to 60-70 years it would be sensible to view the exterior of a dwelling of the same age in a more exposed position if at all possible. This might show defects likely to become apparent on the subject dwelling in time. If the dwelling itself has an assessed age of 60-70 years or more and there are no visible defects on either it or similar dwellings nearby then it could with reason be said that it is unlikely to develop faults in the brickwork unless the ground is disturbed or maintenance is neglected. This comment could be reinforced if the surveyor's local knowledge enables him to be reasonably certain that the bricks are of a variety commonly used in the area with a reputation for long life. There might be evidence from brickwork of the same type on older dwellings in the area to justify the comment. A surveyor, a bit long in the tooth, might himself have noticed new development in his area during his years in practice and taken note of the make and type of brick being used. The desirability of labelling mentioned in BS3921:1985 can be useful in this respect. Such local knowledge is invaluable.

Clay suitable for making the facing bricks under consideration here, normally selected for their appearance and in contrast to common bricks used internally or as backing, is now obtained from many areas in the United Kingdom but this was not always the case. When the only way to make bricks was by hand, only the softer clays with good plasticity from certain areas were suitable. This was the situation up to about 1850 and had lasted for 600 years from the re-introduction of brick making from the Netherlands after a lapse of 800 years following the departure of the Romans.

246: Bricks from Roman times were more in the form of thick tiles than the sizes of bricks we know today. These are incorporated in a much later stone wall but have lasted very well.

The early processes to produce hand moulded bricks were very time consuming, usually carried out near where the bricks were destined for use and involving at least one winter when the clay was weathered. Even so, the results were variable since the firing first by wood, up to about 1700, then by coal was by rule of thumb and relied on the brick maker's expertise. Initially shades of red towards brown depending on the proportions of iron and lime in the clay were produced and these remained in favour until about 1750 when paler colours of grey, yellow or even off white began to be considered much more fashionable. Underburnt bricks came out softer and paler coloured than properly burnt bricks and even more so than those which had been over burnt which came out a very dark colour and were often misshapen

and brittle. Some depending on their position in the kiln, might be vitrified either wholly or in part by the heat, turning a very dark purple in the process. The underburnt should, of course, have been rejected but the colour of the overburnt might be exploited in the creation of walls with the perennially popular diaper pattern.

247: Bricks from the hottest part of the kiln usually came out harder and darker than the remainder, but were often smaller and brittle, easily broken if handled carelessly. They were often incorporated as headers in a diaper pattern, as here in a wall from the 1500s. The overburnt bricks are clearly wearing better than the red bricks but even most of these have lasted already over 400 years. More effort should have been made to obtain a better match for the replacement bricks used at the sides of the windows. It is evident that major repair is required. Desirable snow guards have been fitted.

It is the irregularities in shape, texture and colour due to clamp burning and often the presence of small cinders which make old hand moulded bricks attractive to many and for which some prospective owners are prepared to pay at least three times as much as for machine moulded bricks. These characteristics are not entirely achievable by processes relying on the machinery developed since the late 1850s and early 1860s. This included grinding machines which could process the harder and less plastic clays of the Midlands and North, drying and moulding machines and continuous burning kilns all making uniformity, much prized at the time, easier to achieve.

Perhaps the greatest undermining of local brick making came later towards the end of the 1800s with the extraction of the shale like Knotts clay with its higher percentage of lime than iron from the vast Oxford clay bed formation running through Bedfordshire and into Lincolnshire. This has properties, which without the need for weathering or drying, enabled it to be pressed into shape by the new machines very quickly and subsequent kiln firing which only required a third of the fuel needed for processing other clays. Even with comparatively high transport costs, local brick makers could be undercut particularly when, from the late 1920s, with additives, texturing the surface and controlling the amount of air in the kiln to determine colour, passable facing bricks could be produced by the big manufacturers along with the basic sharp arrised pasty looking pinky yellow smooth surfaced Fletton brick, not helped in appearance by the striped effect produced by kiln burning. For a while only those areas inaccessible by rail remained free of the products from Bedford, Peterborough and North Buckinghamshire but cheaper and more reliable motor transport and now shrink wrapping means that bricks are available if, and wherever, required, not always with advantage to the landscape if insensitively used.

As far as age is concerned, it is therefore quite safe for the surveyor to say that the original brickwork of any dwelling assessed as having been built before 1850 is hand moulded. For dwellings built subsequent to that date, both hand and machine made brickwork is possible and other evidence will be needed to suggest which is the most likely. Size and quality of dwelling can be an indication since the larger and better quality dwellings tend to have hand moulded bricks while at the other

end of the scale cheapness will have directed attention to the machine made product.

Surveyors will, hopefully, have made themselves familiar with the main colour and character of the bricks traditionally used in their locality for the building of dwellings from the 1700s through to the early part of the 1900s. There were building booms in each of these centuries, in the 1730s, 1780s, 1820s, 1880s and 1930s and correspondingly most localities will have a variety of dwelling types from each. For example, reddish shades for brickwork are common but particularly bright from the marls in a triangular area bounded by Tewkesbury, Grantham and Birkenhead and even brighter from the shales of the coal measures in the Northern counties, inducing adjectives for the red such as tomato, lobster and even blood for those produced from the mid 1800s around Accrington in Central Lancashire. Bricks are often named after the towns from where they originate, for example Leicester, Coventry, Ruabon and Fareham reds. Many towns are distinguished for their fine sober reddish brickwork, Blandford, Farnham, Lymington and Bewdley are examples. Descriptions of shade are numerous and a good exercise for the vocabulary, salmon, orange, tan, terracotta, pink, tangerine, crimson, deep plum, port wine, mulberry through to dark blue, purple and near black for those a bit overburnt on the surface, as distinct from the true dark blue, all the way through, of the hard engineering brick, the Staffordshire Blue, wholly vitrified.

Attractive brownish coloured bricks, containing a balance of iron and lime, came and still come, from the clay in the Weald of Sussex near Lewes, the towns of Rye and Tenterden exhibiting good examples. The area surrounding the Weald came into its own for brick making when fashion turned against the reddish and brown shades and whitish bricks from gault

248: White bricks of gault clay containing lime but no iron and red bricks where the clay contains much iron.

249: A little judicious scraping reveals that these silver grey bricks, very fashionable in the first half of the 1800s when this house was built, are in fact the same red bricks used as dressings around the windows, the colour change produced by a coating of sand to which salt has been added before firing. Natural weathering has had very little effect over 150 years but the position is comparatively sheltered.

clay were produced north of Maidstone, near Midhurst, between Worthing and Littlehampton along with those from gault areas at the foot of the Chilterns and the Berkshire downs into Wiltshire.

Areas around Stamford, Peterborough and Cambridge are notable for their 'white' bricks containing a proportion of lime but no iron but most renowned of all are those produced in East Anglia. Cossey Whites, which were in fact light yellow because of a trace of sulphur in the clay, were very popular around Norwich in the 1830s and Woolpit bricks from Central Suffolk were exported to the USA in 1827 for building the Senate wing of the Capitol in Washington DC.

Pale buff, bordering on yellow, bricks were also acceptable in fashionable circles but above all grey with as little yellow as possible. True greys came from around Luton, from the Thame and Wallingford areas, Reading and Newbury. Sometimes salt was added to a coating of sand on red bricks to produce an attractive grey; there are examples in Lewes in Sussex and Wickam in Hampshire has houses of the late 1700s built of silver grey bricks, highly prized at the time.

The washed out or dusty looking yellow bricks

with which much of Central London was built in the 1800s originated from either side of the Thames estuary and bricks of the type are still produced from the same areas today.

The true 'blue' brick, which is usually nearer slate grey in colour, as distinct from the over-burnt or part surface vitrified red brick, is of comparatively recent origin and comes from the tough clays of certain coal measures; South Staffordshire producing for example the well known, hard, strong Staffordshire Engineering brick. Another area from where blue bricks originate is Almondsbury, near Bristol. Such bricks are all machine made and classified, depending on compressive strength and degree of water absorption in BS3925: 1985 as Engineering A or B.

Using a higher proportion of fine sand, increasing the time for weathering and specially washing loamy clay produces the denser textured and hard brick suitable for cutting with a bricklayer's saw and rubbing down to produce the fine precise gauged brickwork with its very thin joints used in

250: *Fine sawn and rubbed brickwork of the early 1700s surrounding a blanked off window with carved stone impost and brick pilaster.*

252: *The difficulty with the use of the better shaped red bricks for arches and as dressings around windows is that they are softer and consequently erode quicker than the bricks used in the general walling. It is possible to obtain matching bricks with perseverance so that here renewing the bricks in the arches and around the windows coupled with sensitive repointing of the general walling should restore this elevation nearly to its original appearance of around 300 years ago. It is a pity, however, that an earlier renewal of windows incorporated glazing bars which are strictly speaking too thin to be historically correct.*

251: *Moulded panels of red clay used for decorative effect on a large mansion of the late 1800s, parts of which with its towers and gables resembles more a 'folly', the effect currently spoilt by staining, dirt accumulation and the filling in of windows with a different variety of red brick complete with light coloured pointing in no way matching the original.*

sober coloured bricks used for general walling. Such bricks will not generally be confused with the much harder bricks used in gauged work and will often be found to have become eroded by the weather much sooner than the brickwork of general walling and require replacement.

The muted but variously coloured matt, slightly creased, wrinkled, open textured surface of the hand moulded brick represents the ideal and while the wooden moulds of machine moulded bricks can be sanded before the clay is squeezed in by machine, the resulting effect is never quite the same as achieved by throwing the clay into the mould by hand. When built into a wall, the slightly open texture allows a certain amount of rainwater to be absorbed into the brick and provided the pointing is of a similar strength and texture to the brickwork the 'overcoat' effect is achieved. When the rain ceases, evaporation enables the wall to dry out.

arches and quoins and also for carving capitals and the bases of pilasters.Clay can also be pressed into specially prepared moulds to produce ornamental panels with patterns in relief either of geometrical shapes or leaves and flowers, popular from the middle to the end of the 1800s.

While many varieties of red facing brick have excellent qualities of durability, sometimes the less well burnt but better shaped bright red bricks would be used in arches and around window openings to contrast with the more

The quality of ideal surface texture can be achieved by methods of shaping the brick other than by hand or machine moulding. The wire cutting process was invented as long ago as 1841 but the first machine is only reputed to have been made and used at Bridgewater in 1875. A continuous column of clay is pushed, extruded, through a set of wires or knives which are set at right angles to each other to produce bricks of the correct size and shape but from a stiffer clay than that suitable for hand moulding. The product is a dense brick without a frog and initially one of poor appearance and poorly shaped arrises. To produce facing, as distinct from common bricks, further processing by pressing, rolling and sanding is carried out.

Even harder drier types of clay, described as 'stiff plastic' or 'semi-dry', can be shaped by a strong force applied in a mechanically operated press which delivers the clay into steel lined boxes of the correct size and shape. While the bricks are of a much more regular shape, usually with one or perhaps two frogs, the clay which has to be used for the process produces bricks of poor appearance unless specially treated by sand facing or provided with a mechanically patterned textured finish. The Fletton brick is a pressed brick made from the semi-dry Oxford clay of poor appearance but numerous dwellings will be found to have been built since the 1930s of sand faced Fletton bricks, see 240, the true nature of the brick revealed by a bit of surface scraping by the surveyor in an unobtrusive spot, unless the weather has already done the revealing for him. Sometimes the textured pattern will be sufficient to compensate the eye for the poor colour, the bricks being passed off as 'Rustic' but the surveyor should have no difficulty in identifying their use in a wall as the texturing is exactly the same on every brick.

253: The end house of the terrace of which 249 forms part adjoins a narrow road entrance where the corner is curved. The headers used in the wall to turn the bend were those partially vitrified in the kiln with a hard dark blue surface, not quite hard enough, however, to resist scraping by carts and motor vehicles over 160 years.

The direct opposite to the ideal open textured surface for a brick is the dense close textured smooth shiny surfaced brick often very fiercely or very darkly coloured, almost glazed in the firing of ordinary facing bricks but, sometimes, with a glaze deliberately applied before a second firing.

Sometimes such a surface was introduced in the mistaken belief that it prevents rain penetration. Raincoats are better without joints *viz.* a cape, and it is the joints in a wall of such bricks that are its cardinal weakness since rainwater streams down the face of the brick to seek out any minor cracks and open pores in the mortar of the joints, of which inevitably there are many. It is rather like water running down the glass of a window and seeking out weaknesses in the lower section of the frame and sill. In other circumstances the firing of the hard stiff local clays produced bricks with hard dense surfaces and the problems they created in the Midlands and the North from rain penetration probably led to the more ready adoption of the hollow cavity wall in those areas where such bricks are more common than elsewhere.

Another reason for glazing bricks was the belief that they would better resist the salty atmosphere of coastal regions. In fact bricks are not generally affected by salts in the atmosphere unlike stonework which readily succumbs to acids in any atmosphere, be it from the sea or industrial processes in whatever the locality. A further use for white glazed bricks was as facings for the enclosed areas of the tall blocks of flats popular in the late 1800s and early 1900s to reflect light and in other positions where natural light was limited.

254: *Excellent bricks of an attractive red shade and sand faced texture from the mid 1700s, of regular shape rubbed to sharp arrises and meticulously pointed with fine near matching joints, now slightly eroded in places.*

'Brick' or 'Mathematical' Tiles

It is appropriate to mention at this point what are known as 'brick' or 'mathematical' clay tiles, the existence of which is the reason why it is necessary to tap the outer surface of a brick wall to see if it might sound hollow. The tiles which look, when properly fixed and pointed, exactly like bricks in a wall were developed for reasons of fashion in the 1700s and were still being produced in the early 1980s. Even though an owner might be comparatively well off, the fact of occupying a timber framed house was considered very infra dig in the late 1700s. If the owner could not afford to rebuild, a make over by covering the dwelling with mathematical tiles provided what appeared to be a brick dwelling in the latest fashion. It is gleefully recorded elsewhere that even the Inspectors compiling the Statutory List of Buildings of Special Architectural and Historical Interest were deceived by the construction in a number of instances, believing it to be brickwork. To be fair, the exact moulding and narrow joints can deceive the most experienced surveyor.

For makeovers of timber framed dwellings, it was a fairly simple matter of covering with softwood boarding, battening and then nailing

255: *Compare with 254 which is located close nearby, but here the 'brickwork' is in fact mathematical tiling of the late 1700s in header bond but sufficiently convincing to deceive an Inspector preparing the Statutory List of Buildings of Special Architectural and Historical Interest. Tapping the tiles would have enabled him to discern the difference from solid brickwork when such tiles are used as a make over to bring an older timber framed dwelling up to date in appearance. Not so, of course, if the tiles had been bedded on to solid brickwork as can be the case in other instances. The slight unevenness is probably due to movement in the timber frame which can also have a tendency to loosen the pointing.*

and bedding the tiles to the battens and each other in lime plaster and pointing carefully the fine joints between the tiles with lime putty. However, it was also possible to bed, nailing where possible, the tiles on to an external rendered or plastered surface applied to brickwork or stonework and it is recorded that at least two large country houses were covered over with mathematical tiles of a yellowy white shade in the 1780s to conceal the, by then, highly unfashionable red brick of the original construction.

It is sometimes said that mathematical tiles came to be used as a means of avoiding payment of the tax on bricks first introduced in 1784 at the time of the Napoleonic Wars. However, the tiles were well in vogue by then for the purposes mentioned above and their use was first recorded in the 1740s. Although tiles were not included when the tax was first introduced, and still not when increased in

1794, they were when it was again raised in 1803. The tax on tiles was wholly repealed in 1833 but on bricks not until 1850. The tiles use was widespread across the southern and eastern counties of England until it became fashionable again, around the mid 1850s, to show off the old oak timber frame. The tiles were produced in all the colours normally available for clay bricks in the locality required, both as stretchers and headers. Some in the latter part of the 1700s were provided with a lead based powder before glazing to produce shiny black tiles thought to be a protection in a salt laden atmosphere. Generally these were laid in header bond since it was a time when bow windows were fashionable and headers fitted better together round the curve.

Where the bedding of the tiles is in lime plaster rendering they often continue to adhere even though the nails have rusted. Even so, unless there is evidence of recent refixing they must be considered a danger to occupants and

256: A dislodged section of mathematical tiles enabling their shape to be seen and how they fitted together. These from the tail end of the 1700s were bedded in a lime and sand rendering on comparatively soft red brickwork and the fine joints pointed up in lime putty. It is conceivable that the tiles came from the same makers as the red bricks, in which colour they would have been highly unfashionable at the time. These tiles were, however, given a coating of lead based powder and a second firing to produce a shiny black glazed surface. This was believed to be a protection against the salt laden atmosphere of the coastal town in which this and the similar dwellings in the terrace are situated. While this dwelling has been neglected, the remainder of the terrace has been well maintained and the tiles have lasted well for 200 years. There is evidence of some renewals elsewhere but the perfection of a modern black ceramic glaze is at odds with the remainder, not helped by the incorporation of some headers whereas all the original tiles at this location were in header bond.

257: The upper part of the front elevation of the neglected 1790s four storey dwelling of which 256 is part. The settlement in the window head at first floor level has allowed the mathematical tiles to slip forward loosening their bond to the lime and sand rendering to which they should be attached. Although held in place at the moment, it is clear from 256 that the underlying brick is so soft that a loosened tile could pull off the face of the brick below, adding to the danger of falling on the owner, a member of his family, a visitor or a passer by.

passers by and suitable warnings need to be included in the report. Of course, if they do fall off they are almost certain to break and replacements may need to be specially made, no doubt at a very special price as well, unless it is possible to achieve a reasonable match with brick slips or, in the case of those which were originally glazed, with ceramic tiles.

258: Typical moulded and cast clay Coade stone details around the main doorway of a 1770s house. Such details could be bought from a catalogue much more cheaply for a whole terrace of houses than if carved in stone, with the added advantage that they were practically indestructible.

Terracotta

A further use of clay in building is terracotta, from the Latin meaning cooked-earth. The fine clay from which it is made contains a higher proportion of silica than that used for brick making, 75% as against 50-60%, and less alumina, 10% as against 20-30%, proportions which permit a hard compact vitrified skin to be formed when fired at a high temperature and less shrinkage than is the case in the making of bricks. The material can be cast and moulded to produce crisp ornamental details for cornices, string courses and decorative panels and, because it is so hard, it is also ideal for casting sectional window and door surrounds and the components for balustrades, anything which requires repeating and would have to be carved, if in stone.

Italian craftsmen were brought in to produce ornamental details in the 1500 and 1600s to adorn some of the brick dwellings of the time, Cardinal Wolsey's early 1500s palace at Hampton Court is one such example. However, it was not until 1722 that a factory in Lambeth was established to produce a range of designs to be bought off the shelf from a catalogue. Other factories were set up, as many as 400 at one time, but when the Lambeth works fell on hard times it was taken over by a Mr and Mrs Coade who, by eliminating the iron oxide from the clay which gave terracotta its shades of red, adding a further material nobody is quite sure what but

possibly the white mineral felspar and increasing the firing temperature, produced what was marketed as 'Coade stone'. This proved to be superior to other makers products and from 1767 to about the middle of the 1800s was much favoured for architectural details being virtually indestructible in whatever weather conditions, unlike much of the natural stone used previously for carving similar details. Because it is a cast product, mouldings for the production of the details can be prepared in any particular architectural style, from classical to Gothic revival.

If terracotta is found to be in good condition and on a dwelling built from the late 1700s towards the middle of the 1800s, it is probably from the Coade works. Unfortunately the raw material used by others was not always well selected or well prepared and underburning was fairly common. Consequently some terracotta had a tendency to disintegrate fairly rapidly, particularly if the near impervious skin was damaged by knocks, cutting to make the pieces fit and becoming chipped in the process of fixing, thus allowing damp to penetrate to the softer interior.

259: Light buff shaded terracotta used as blocks for dressings and sills around windows, with brickwork as general walling on an expensive house, looking well and long lasting after around 100 years of exposure.

The proportion of other material, particularly iron oxide determines, as with brick making, the final colour from red through pinkish shades to the later more common buff. In the mid 1800s terracotta was also cast in large sectional hollow blocks, usually filled with a light concrete on fixing. These were intended to resemble and were used as a cheaper alternative to the finer work in stone employed on the more expensive buildings as ashlar. In this form it was popularised by A W Pugin and Sir Charles Barry, both heavily involved in the rebuilding of the Palace of Westminster following its destruction by fire in 1834, together with the influential writer and traveller John Ruskin in the mid 1800s. Their advocacy was taken up by many architects such as Alfred Waterhouse who used terracotta for the Natural History Museum in South Kensington and the RICS Headquarters in Parliament Square.

Terracotta was also sometimes provided at this time with an enamel glaze and given a second firing at low temperature to produce large tiles and the smaller details in different and often vivid shiny colours, when it is known as faience. By the late 1800s such vividness had gone out of fashion but the more subdued shades of terracotta continued to be much used on commercial buildings well into the 1900s. For domestic purposes at this time, terracotta if used at all was confined to the more expensive houses and blocks of mansion flats, generally left unglazed and employed in combination with brick. While terracotta from this period, manufactured by the larger reputable firms, in the main, seems to have worn well, it can bring about the same weather resisting problems as brick with a dense impervious surface, producing the 'raincoat' effect as distinct from the 'overcoat' effect, and permitting water to run down the surface and enter through the joints. Accordingly there can be a build up under the impervious surface causing it to spall. This can be severe if the water is carrying salts from the atmosphere and can cause blistering and powdering to the surface. Of course, such salts can be derived from other material forming the backing to details, from brickwork for example and from deposits of the products of combustion in chimney flues. Such defects will become apparent first around the edges of blocks where joints occur. If fixing of the details is by metal cramps and these rust due to water penetration at the joints, major cracks can be induced by the expansion of the metal.

While the impervious surface of sound and well produced terracotta resists most aggressive atmospheres, the less sound and much lighter in colour underburnt surfaces will become soft and pitted because the surface will not be properly formed. Lamination can also occur if the material is not pressed adequately into the mould and warping and cracking can become apparent if the material is not sufficiently dried before firing.

While it is no longer possible to buy terracotta details and panels from a catalogue, there are still firms who will produce replacements to

260: For repetitive work, as here, in this balustrade and as coping stones to the one and a half brick thick walls, terracotta was often used as a cheaper alternative to stone in the latter part of the 1800s and the early part of the 1900s for the more expensive dwellings.

match old and damaged sections, though inevitably, of course, the price is high and probably only economic if details are required in quantity.

261 and 262: These two storey Airey Duo-Slab no-fines in situ concrete semi-detached and terraced houses were built in the 1920s, the top pair refurbished in the 1950s, the lower terrace in the 1960s.

Concrete Walls

In situ Mass

In the second half of the 1700s a technique called *pisé*, of building walls by ramming a mixture of clay soil, straw and water between temporary shuttering was developed in France and found favour in Scotland for use in dwellings on the east coast, south of Aberdeen and on the south side of the Moray Firth, east of Inverness. So the technique was by no means new when Sir Robert McAlpine used it again to build dwellings in Scotland of dense *in situ* concrete in 1876 and there are other recorded instances of dwellings built of this material before the 1914-1918 war.
In the 1920s after the war, the first single leaf no-fines concrete systems using removable shuttering were developed by Corolite, Unit Construction, Wilson Lovatt, Airey and Laing, whose Easiform system evolved from 1928 into two leaf cavity construction. Boswells of

the late 1920s were similar and were provided with pre-cast concrete corner posts to facilitate the setting out of the shuttering. Another type from the 1920s with external walls of cavity construction were Forrester Marsh.

Other systems utilising no-fines concrete, but cast with permanent shuttering, were also developed in the period 1920-1940. For example 2000 Fidler houses are recorded as having been built in Manchester and London, which had walls of *in situ* concrete poured between two leaves of 50mm thick clinker concrete slabs. Plastered inside and rendered externally, the overall thickness was roughly the same as an internally plastered one brick thick solid wall. Over another 1000 Universal houses are known to have been built in this period with solid external walls about 180mm thick of *in situ* no-fines concrete, cast between permanent shuttering of asbestos cement sheeting and clearly posing a problem these days if repairs are required.

The period 1940-1960, with its post war shortages of bricks and skilled labour and the urgent need for new homes to replace those lost to bombing and slum clearance, provided a boost to the construction of dwellings using no-fines concrete. Putting up, removing and re-positioning temporary shuttering, mixing and pouring concrete made good use of what was available and many local authorities, but particularly the Scottish Special Housing Association in conjunction with the construction company Wimpey, soon had large programmes in hand, usually based on negotiated rather than competitively tendered contracts. Wimpey alone built

263: This pair of two storey Wimpey no-fines concrete semi-detached houses were built in the 1950s but demonstrate the effects of the right to buy legislation on the street scene. The dwelling on the left has been purchased while he one on the right has been refurbished by the local authority in 1992, though both clearly have had their roof coverings renewed at the same time. The new windows to the right hand dwelling of the same size as the old are set in their openings at the same level as those in the dwelling on the left, generally necessary on window renewal, but appear to be set further back because of the increased thickness of the newly applied insulating rendering. The increase in thickness over the original can be seen to the left of the rainwater pipe.

around 300,000 dwellings of this type up to about 1980, claiming that their no-fines technique, although there was nothing proprietary about it, was used to build more homes for local authorities than any other system. In the 1950s the company was building at the rate of 18,000 a year with a limited number, in addition, for the private sector. Including a small contribution from other types such as the Incast and the BRS Type 4, with their flat reinforced concrete roofs, in excess of 1% of the United Kingdom's total dwelling stock is provided with external walls of no-fines concrete.

Besides being an admirable use of resources at the time, the no-fines technique provided homes of a comparable standard to those being built of cavity brick construction. With external cast *in situ* concrete walls based on a mix containing no sand and aggregate from 10mm to 20mm in size, a honeycomb structure was produced, so that when plastered internally and with external rendering, often sprayed on, a good level of thermal insulation for the time was provided. This can be easily improved on refurbishment up to current requirements by re-rendering with a thicker insulating material and possibly dry lining to the interior. An added advantage was that, in the main, if the rendering on general walling should degrade from exposure and become cracked, any water that penetrated to the concrete would trickle down through the comparatively open texture and run out at the base without finding its way to the inside face. Even when damp penetrates in any cracked areas above windows, it tends to be diverted out through the bellmouth invariably formed in the rendering immediately above the windows. With no-fines concrete dwellings, cracks accordingly in the rendering tend to have little significance other than on appearance, but if the bellmouth becomes cracked it may need renewal or the cracks repaired with mastic.

264: These two storey Wimpey no-fines concrete houses from the 1960s, in a stepped terrace form, have panels of weatherboarding between the windows. The joint between the general wall surfaces and the weatherboarding and the area around, needs examining to ensure that there is no damp penetration as such joints are often a weak spot. Although the rendering appears to be in fair condition it is stained and appearance would be improved by the application of a couple of coats of masonry paint.

Even though steel reinforcement was incorporated in the structure to support roofs and floors and over windows, later to be substituted by pre-cast units of denser concrete set into the shuttering, the rendering provided increased protection to the reinforcement. As a result the corrosion problems of other post 1940 dwellings, mainly those of prefabricated reinforced concrete construction, were not experienced to any great extent and even where they were could easily be repaired. Thus, because of their satisfactory construction and the evidence of structural soundness it was not found necessary to designate no-fines concrete dwellings under the Housing Act 1985 and therefore those bought, or about to be bought, under the right to buy legislation can be treated for inspection and report purposes in the normal way.

It ought to be mentioned here that since the no-fines technique is very adaptable and structurally the same as a traditional house, in that loads are taken on external and internal

load bearing walls, the latter providing the necessary lateral support, it was not only two storey semi-detached and terrace houses, based on two lifts of shuttering which were built. Also built were bungalows and low rise blocks of flats and maisonettes from one to six storeys in height, although for the latter usually a reinforced concrete frame would be provided, the no-fines concrete being restricted to the infill panels. Roofs and floors, were of traditional construction in timber, the former in hipped, gable, monopitch and duopitch types and dwellings were sometimes provided

265 and 266: What can happen to BRS Type 4 cast in situ no-fines concrete local authority semi-detached houses built in the early 1950s in a south coast town. The top examples as originally built still with their flat roofs but now with much degraded external rendering, the lower examples transformed with new pitched and tile covered roofs and with walls encased in brick laid to stretcher bond. The surveyor would need to ascertain the type of wall tie used and other information from the specification if asked to report on condition.

with porches and bay windows.

The normal rendered appearance of no-fines concrete dwellings might be varied with sections of tile hanging or weather boarding and, of course, a no-fines concrete wall can form the inner leaf to a masonry constructed outer leaf. This was another way of introducing variety to the appearance, particularly for gable walls and where the considerations of wall tie failure would need to be taken into account in any report. The problems of detailing to prevent damp penetration where different materials meet need also to be borne in mind when tile hanging or weather boarding have been introduced and eventually as in all cases where rendering is used, it will need renewal, the timing dependent on initial quality, exposure conditions and maintenance.

Pre-cast Concrete Block

Building a dwelling with concrete blocks made to give the impression of stonework is not a new phenomena, only brought about by the conservationists keen that new dwellings should fit in with their surroundings in areas predominantly stone built. Blocks about 100 years in age, cast using a mix of cement and stone aggregate and provided with a textured finish to one face, are fairly easy to identify. There was no call at that time for blocks in a wide range of sizes so that a fair representation of a particular bond could be achieved and, it seems, that little conscious effort was made to obtain a particular texture or a particular shade of colour.

The need to match, or at least blend, in with the surroundings provided a boost to the concrete block industry, or 'reconstructed stone' as the makers prefer to call their product, from the late 1970s onwards. For example, one manufacturer supplies blocks in eight

267: Known locally as 'the stone house', this single storey cottage was built in the early 1900s of concrete blocks. The blocks, all looking exactly the same with a lightly textured finish on one surface only, were laid in courses, stretcher bond style. They have worn well but have subsequently been painted, probably brightening up the appearance but, nevertheless, unfortunate.

268: Concrete blocks in walling at least 100 years old. The blocks here, simulating coursed squared rubble, are at least in three different sizes, providing a slight variety, although not much effort has been made to break, in some places, the long near straight vertical joints. There is a fair amount of erosion to the top of the texturing on each of the blocks, but no doubt they will last for many more years.

different sizes with a consistent bed width of 103mm to provide the outer leaf for cavity walling, but which can be in a variety of coursed and random designs following the traditional styles. The blocks can be obtained in three different colour shades, rather uniform greys and buffs, and all can be in three different textured finishes, tumbled, split and pitched. While it is said that water absorption is low, the blocks are not impervious to driving rain, the incidence of which is more likely to be experienced in areas where stone building predominates, so that deep recessed joints are cautioned against and all joints required to be completely filled, avoiding any 'tipping and tailing'. The latter aspects of workmanship are not, unfortunately, items which can be checked on site unless so grossly neglected as to be visible in places, but, certainly, for new or near new work every effort should be made to check the Specification for such instructions and to see that the blocks obtained have been manufactured to BS1217:1986 Specification for Cast Stone. There may also be an Agrément Certificate for the particular make and type employed.

269: Concrete blocks to simulate squared rubble built to courses in a small rural town, on a two storey development, built about 10 years ago in an area where exposure conditions border on the moderate to severe. The joints are perhaps a little too recessed and high up on the gable walls facing south west, water could penetrate the top edge of the blocks, freeze and cause delamination or crumbling, particularly on blocks which have already suffered physical damage, of which there are quite a few.

For new, or near new, cavity work all the usual features need checking from the Specification if possible, width of cavity, how the insulation is dealt with, the type of wall tie used, damp proof course details and the presence of appropriate weep holes. An

important aspect where reconstructed stone blocks are used, is to ensure that the units match the characteristics of those used for the inner leaf, otherwise differential movement can cause delamination on the face and possibly cracks. Being a concrete product, movement joints need to be given consideration where length of walling exceeds 6m and where panels exceed in length, twice their height. Sometimes differential moisture movement in the face of the blocks will cause crazing and occasionally spalling, though naturally the coarser textures are less prone to this type of defect. However, severe exposure will undoubtedly take its toll over the years.

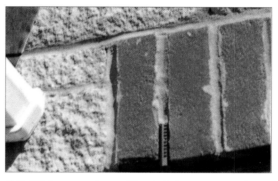

270: With any cavity walling it is important to check that drainage weep holes are provided above any horizontal damp proof course. Here for this cavity wall with reconstructed stone concrete block outer leaf and flat red brick lintels, often referred to mistakenly as 'soldier arches', a red plastic fitment is provided for the purpose, correctly at 3m intervals. A glimpse of the lead flashing to the adjacent porch roof can be seen along with its plastic gutter.

Pre-cast Concrete Panel Walls

Of the many low rise systems for houses and small blocks of flats which were developed to boost the provision of new housing after the conclusion of the 1939-1945 War, were some which incorporated storey height pre-cast concrete panels as the enclosing walls along with reinforced concrete columns and ring

271: A 1950s two storey pair of semi-detached dwellings enclosed by storey height pre-cast concrete panels which, with in situ reinforced concrete columns and ring beams to support the floor and roof structure and an exposed aggregate finish to the panels, enabled rapid construction.

beams to support the floors and roof structure. One such was the Reema hollow panel system of the 1950s which had wide full height panels, clearly seen on 271 which with their exposed aggregate finish on the external surface enabled, with the use of a small crane, dwellings to be put up quite quickly.

Although the Reema hollow panel system did not utilise metal angles to bolt the sections together, other systems did and the combination of large pre-cast panels with bolted connections was developed and became the norm for the high rise developments of the 1960s, when the demand for more new housing became even greater. Most of the systems for low rise housing, including the Reema, eventually suffered deterioration from carbonation, the use of chloride accelerators and inadequate cover to the reinforcement resulting in the houses, but not the flats, being designated as known to be 'defective by reason of their design or construction' under the Housing Act 1985. The names of the systems designated are set out in Table 6 on p 261 and, as they all comprised a frame in one form or another, they will be dealt with in the subsection dealing with framed, as distinct from load bearing walls.

For high rise developments, as well as the low and medium rise blocks of flats which continued to built but now on the large panel and bolted connection principle, panels would be cast in the main off site, complete with an external finish and with window and door frames incorporated while tower cranes would hoist them into position, there to be fitted with joint seals and baffles and to be bolted together. One completed box of walls and floors would bear the weight of the ones above and transfer the load downwards to the ones below. The Reema system, the panels not hollow in the 1960s version but a mere 160mm thick, was but one of around 30 different systems among them being Bison, Wates, Taylor Woodrow, Anglian, Jesperson, Carnus, Crudens, Skarne and Tracoda. In their original state it was not easy to differentiate the systems visually but this is not important now, as long as they are recognised as of 'large panel' construction and this may only be deduced by enquiry in

272: The systems with full storey height pre-cast concrete panels for low rise housing of the 1950s became, with bolted connections, the types mainly used for the high rise developments of the 1960s and early 1970s, as here for this 19 storey block of flats.

view of many now being overclad. It is estimated that over 1% of the total United Kingdom housing stock is enclosed by walls of prefabricated reinforced concrete, but it is not possible in the statistics to separate the number of dwellings of large panel construction from those with smaller panels fixed to a frame.

The panel systems suffered a setback in 1968 with the collapse of part of one of the blocks, Ronan Point, following a gas explosion in one of the flats, with some loss of life. The blocks had not been designed for a gas installation in the first place and an extensive programme of modification and strengthening had to be put in hand for existing blocks and changed requirements for new blocks put in place. BRE report that no major failures have occurred

273: It was said at the time that the blocks of flats constructed of storey height pre-cast concrete panels with bolted connections in the 1960s were never designed to withstand a gas explosion, yet gas installations were provided to most blocks. The result when the gas for a cooker was left turned on in a flat at Ronan Point in 1968, is shown here. A modification and strengthening programme was put in hand at the time for existing blocks and changes made in the requirements for new blocks.

since but recommended that all blocks intended to serve longer than 25 years from the date of construction should be the subject of a full appraisal for structural safety and durability and that the initial appraisal should be followed by a visual inspection after one, two and five years and subsequently at no more than five yearly intervals.

That a surveyor inspecting and reporting on a flat in a block of large panel construction clearly has a duty to have regard to the reports recommended by BRE and to consider what effect their content might have on his conclusions was determined in the Court of Appeal case, *Izzard* v *Field Palmer* 1999. The surveyor was instructed to carry out a mortgage valuation in 1988 of a two floor maisonette in a four storey block constructed on the Jesperson system, which the surveyor either identified or ascertained, built in the 1960s. The block was one of a number on an estate of 807 dwellings built by the Ministry of Defence but sold in 1985 to a developer who did some renovations and then sold the flats on 999 year leases from 1986. The flats were said to be attractive to first time buyers even though it was known that they were expensive to maintain. It was held that the surveyor, who had valued the flat at the asking price of £42,000 and said it was readily saleable, was negligent in that he left blank a box on the report form which asked for details of 'matters which might affect value', gave no consideration to the literature which was available on the type of construction and, accordingly, not advising that further investigation should be made into the possibility of high service charges being incurred. Damages of £28,000 were awarded, being the difference between the purchase price and a valuation produced by the plaintiff, on the basis not of defects missed but because no warnings were given and no account taken of the effect on value.

All three Appeal Court judges agreed with the finding of negligence, but one disagreed on the amount of damages. In a strongly argued dissenting judgment he considered the plaintiff's valuation should be discounted thinking, perhaps, it was artificial and based on hindsight. Instead he preferred to base his judgment on another at £40,000 given in evidence, which was on the lines of other comparable valuations on the estate at around the same time of between £38,995 and £41,000 put in as a schedule. The £40,000 valuation took into account the possible difficulty of resale because of the type of construction, advised that past accounts should be looked at but omitted taking sufficiently into account the risk of a higher future maintenance liability. For this

274: The storey height pre-cast concrete panels on some system built dwellings developed to speed housing provision will be seen on all types of development from the 1960s and 1970s, low, high and the medium rise as on the four storey block of flats shown here, identified as of the Bison system. As originally completed, individual panels can often readily be discerned but many system built dwellings and blocks of flats have been overclad in the 1990s, to improve weather tightness and insulation, altering their appearance substantially.

omission the Judge took off £2,000 and considered that the proper valuation of the Izzard's maisonette at the time of purchase should have been £38,000, which would have reduced the court's award of damages from £28,000 to £4,000.

Interestingly, in the above case, the dissenting Appeal Court judge considered that in the Court below the trial judge had been rather too impressed by the evidence of a distinguished expert surveyor, brought in from elsewhere, who had said in evidence that all the surveyors who had valued similar flats on the estate at around the same time had woefully underestimated the risk involved and accordingly valued too highly. While this may well have been true, nevertheless what people will pay sets the tone of values and clearly he considered there was a market for the accommodation on offer irrespective of type. The rub would come should the maintenance liability rocket, but there was no certainty that that would happen since, as the judge pointed out, BRE having investigated Jesperson sites in both the north west and south east of England, had given them a fairly good bill of health.

While, when reporting on flats of this form of construction all the aspects brought out in the court case need consideration, as they do for that matter in respect of inspection and reports of flats of any type of construction, other findings by BRE on these specific types also need to be given consideration and most certainly brought to the attention of a prospective buyer, lender or seller. The blocks are now all 30-40 years old, the high rise versions severely exposed to the weather and with construction at a time not noted for its care and attention to detail. BRE found that much of the concrete used was of poor quality, affected now by carbonation and the use of chloride accelerators and providing insufficient cover to the steelwork. These faults have resulted in corrosion to the reinforcement, cracked and spalling concrete, falling surface finishes, deterioration to the seals and baffles to the joints, leading to a lack of weather tightness, along with poor insulation qualities. Even with overcladding, which it recommends only after very careful consideration of suitability and all the implications involved and which would overcome the problems of water penetration and lack of insulation, the view is that in most cases the concrete used should be looked upon as material with a limited life. It is to address this durability aspect as well as stability where the structural engineers with their five year reports are so important.

Having regard to the BRE findings, the surveyor before reporting must see the initial appraisal and all subsequent five yearly reports. He must see the lease to determine the demise and the extent of the joint liabilities and, not leaving it to others, he must see adequate information on expenditure for maintenance purposes and assess for himself from both the reports and his own observations what is likely to be the expenditure required in the future. By doing this he is not, as some have done in the past, condemning all flats in such structures as unmortgageable but treating each case on its merits and ensuring that the limitations of the flat being reported are truly reflected in the purchase price and his valuation, if such forms part of his instructions. If the reports are lacking or he considers those available to be inadequate for the purpose, he should advise the commissioning of one and assist in the formation of the instructions so that he gets the information he requires.

It may well be that lenders will, in fact, turn down an application on receiving a mortgage valuation report where engineers views on durability are conveyed, which suggest that a life of 60 years cannot be expected and that it is more likely to be 20 to 40 years. That is for the lender to decide but there is no reason why the surveyor cannot provide a valuation based on a limited life of an estimated number of years and a reversion to a share of the site value.

The surveyor certainly needs to hold the hand of the purchaser in these circumstances, particularly where right to buy is involved and the client may be dazzled by what seems to be a bargain, overlooking the fact that the discount is personal to him, has nothing to do with the value of the flat and should be left entirely out of the reckoning. The lease to be granted will no doubt be on the local authority's standard form and there is invariably a provision for termination should the lessor, for good reason, decide on demolition and redevelopment or even on a sale for substantial refurbishment by others. Whether the price quoted, exclusive of discount, as ascertained no doubt by the district valuer, takes all the aspects of age, type of construction, likely durability and liabilities fully into account needs to be given some thought. So, of course, does the asking price quoted by any subsequent seller, on this occasion no doubt supplied by the seller's agent in line with recorded sales of similar flats, not necessarily, as in the *Izzard* case, taking full account of all the circumstances. This is one area where the surveyor's advice can be very valuable, though it is not always welcome by buyers in a hurry or desperate for accommodation. However, if asked for and taken at the time of many right to buy sales, the degree of shocked expression at some of the repair and refurbishment bills now being presented might be somewhat less.

As for the surveyor acting for the seller in the preparation of a Home Condition Report, and as the surveyor is liable to a buyer and lender as well as the seller, he has to tread a very wary path. However unpopular, he needs to ensure that on the basis of the facts and the opinion of the engineer he presents a balanced view of risks and liabilities in terms of the clearest, and least ambiguous, sentences as possible, since he will have no valuation in which to present his opinion in financial terms.

Stone Walls

Earlier in the subsection dealing with ascertaining the construction of the external walls where these presented the appearance of stone to the outside world, the differences between the real and the artificial product were considered together with the evidence of thickness in so far as it related to the type of construction. The immediately preceding subsection has dealt with the artificial product over the last hundred years or so since concrete blocks with a stone aggregate began to be made, but more particularly over approximately the last 40 years when they began to provide a greater impression of the real product than had hitherto been the case.

Real stone now needs consideration and, once again, knowing or assessing the age of the dwelling can be a help towards eliminating certain forms construction from the reckoning. A dwelling built in the last 40 to 50 years if presenting the appearance of natural stone blocks laid in regular courses, whatever the finish given to each block, will hardly ever be of other than cavity construction. Real stone rubble walling is just not feasible in an outer leaf of 100-120mm and, for domestic purposes, it would only be in very exceptional circumstances that an outer leaf of greater thickness would be, because of the expense, entertained. Matching a new extension to an existing much loved and valued dwelling, in an area of severe or very severe exposure,

might be one such case.

Apart from defects relating to the type of stone used, which are common whatever the form of construction, natural stone sawn and laid in courses in the outer leaf of a cavity wall will be subject to the same structural conditions as bricks or concrete blocks used in a similar position. These have been discussed in relation to cavity construction where the outer leaf is made up of bricks and need not be repeated here.

A dwelling of real stone constructed more than 40 to 50 years ago will, on the other hand, be in one or other of the forms of solid construction of long tradition, determined to some extent by the type of stone available but much more so by the amount of money available for its working.

Using stone from as near the surface as possible and as near as it comes from the local quarry, produced the type of walling generally used over the centuries for the smaller dwelling in both rural, village and small town situations, particularly in areas where there was little timber or brickearth available. Rubble walling is the term which is applied to many different types of masonry using the stones more or less as quarried and although it may look as though it has been put together somewhat

Figure 34: Types of rubble walling in elevation, the stones used as taken from the quarry with a minimum of working. 'Random rubble' on the left, 'random rubble brought to courses' in the centre and 'coursed random rubble' on the right, the types reflecting the amount of work by the mason on arranging the stones.

haphazardly nevertheless, for stability, should be built to follow certain basic principles.

Using stones in a wall more or less as they come form the quarry and only removing perhaps a few of the more inconvenient corners, produces the appearance of walls shown on Figure 34. 'Uncoursed random rubble' on the left needs, however to be laid to maintain a bond, both across the thickness and along the length of the wall, otherwise the stones will just fall apart. Across the wall, the bond is maintained by the use of 'bonders', longer stones extending into at least two thirds of the thickness of the wall with not less than one for each square metre of wall on face. Viewing an existing wall, it is not possible to tell which stones are bonders but if a wall is unduly bulged an inadequate use of bonders is a possible cause. If so, rebuilding would probably be required. Should bonders be allowed to project right across the full thickness of the wall, it could be the cause of isolated damp patches appearing internally.

More selection of the stones by the mason can produce a wall with sections in more apparent order, 'brought to courses', as shown in the centre of Figure 34. Walls built to courses are always stronger and here the stones are levelled up to courses at intervals varying from 600 to 900mm in height according to the type of stone and, often, to the custom of the locality, the heights usually corresponding to those of squared stones used at quoins and jambs. The same constructional principles for stability apply as for uncoursed random rubble and as they do also in the case of walls where the stones are arranged according to height to form continuous courses, 'coursed random rubble', as shown on the right of Figure 34. Here again, the height of the courses will probably be arranged to correspond with the stones available for use at quoins and jambs, resulting in courses alternately deep and shallow.

More work by the mason on the stones themselves to make them roughly square, can produce walls along the lines of those shown on Figure 35 progressing from the left according to how much arranging the mason carries out on laying. The least amount produces 'uncoursed squared rubble'. The stones are laid as 'risers' and stretchers' with , in general, the risers not being more than

275: Uncoursed random rubble limestone walling about 120 years old.

276: Coursed random rubble limestone walling about 500 to 600 years old.

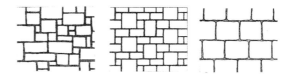

Figure 35: :Types of squared rubble in elevation, the stones roughly squared before the mason sets them out in the wall. 'Uncoursed squared rubble' on the left, 'snecked squared rubble' in the centre and 'coursed squared rubble' on the right.

250mm in height and adjacent stretchers not exceeding two thirds of that height in length. Without those limitations on the size and proportion of the stones in general, but with the introduction of a definite ratio of comparatively small stones not less than 50mm square, called 'snecks' to prevent the occurrence of long continuous joints likely to have a detrimental affect on stability, produces the type of wall shown in the centre of Figure 35 known, not surprisingly perhaps, as 'snecked squared rubble'.

277: Snecked squared rubble limestone walling about 120 years old, somewhat eroded.

'Coursed squared rubble' as shown on the right of Figure 35, involves laying the stones rather like stretcher bond in brickwork in long continuous courses using stones roughly squared to the same height and breaking joint from one course to the next. The courses can be of various depths ranging from 100 to 300mm, but usually averaging out at around 250mm, and the stones can be dressed, if required, to provided different types of finish, split or rock faced for example.

The harder stones are more suitable for rubble walling and in some cases can be roughly worked into irregular polygonal shapes. They can then be bedded to show the face joints running in all directions as another form of rubble walling. In its basic form, where the stones are only roughly shaped and fitted

278: Coursed squared rubble limestone walling about 500-600 years old, with some stones spalling due to face bedding.

with comparatively wide joints, it is known as 'rough-picked polygonal' walling, as shown on Figure 36. The stones can be more precisely shaped and the face edges more carefully formed so that the stones fit close together and then it is called 'close-picked polygonal' walling, as at Figure 37. Kentish Rag is a type of limestone which is particularly hard and difficult to work, frequently coming out of the quarry in irregular blocks, and is often used for this type of walling.

Another form of rubble walling is produced by the use of flints or cobbles. The flints and

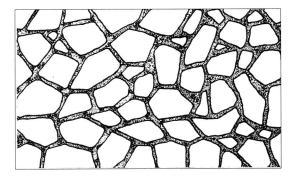

Figure 36: For stones which are hard and come out of the quarry in irregular blocks, polygonal shapes can be adopted for rubble walling. A minimal amount of work on the stones and wide joints produces 'rough-picked polygonal walling', as shown here in elevation.

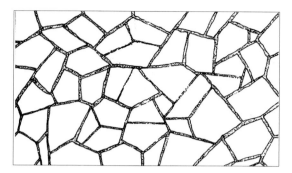

Figure 37: Stones which although irregular are more positively shaped and arranged to fit together, with comparatively narrow joints, produce 'close-picked polygonal walling' as here, shown in elevation.

280: Cobbles from the nearby beach carefully sized and arranged in regular courses and set in the walls of a two storey dwelling with brick quoins and jambs.

279: Dry stone walling of a mixture of local West country sandstones.

281: Cobbles set in a wall and subsequently tarred to contrast sharply with light toned stone lacings, jambs and quoins.

cobbles are rounded nodules of silica, either dug out from gravel beds associated with layers of chalk and limestone or taken from the beach. From the beach, cobbles are sometimes set into panels in walls of other material and appear in dwellings near the coast but more often, the flints used are dug from the ground. While extremely hard and virtually indestructible, flints are brittle and can be snapped roughly in half to expose the inner shiny face, as against the 'rind' of chalk which usually encloses the nodule.

Flint walling was extensively and continuously used in East Anglia and in southern and south east England from Roman times but because of their comparatively small size unwieldy

shape and impervious nature, bonding of flint walling can prove difficult. Traditionally flint walls would be built in two parallel leaves between shuttering with a hearting of flint rubble and mortar and the longest flints used as bonders projecting into the hearting. As sufficiently long flints were scarce, more often the flints, either as dug or snapped, would be set in panels and backed and laced with brick or stonework which would also form the quoins and the jambs around window and door openings. Whole flints uncoursed and used in this way are shown on Figure 38, while snapped and squared flints, termed 'knapped', laid in courses and in conjunction with stone dressings, are shown at Figure 38. Even when used in panels, flints can become

Figure 38: Section and elevation of brick wall with panel of uncoursed flints, as dug, and with lacings and dressings of brick.

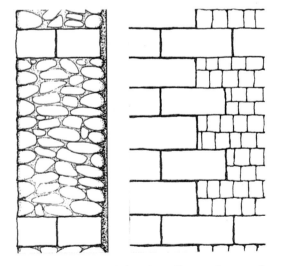

Figure 39: Section and elevation of knapped, squared and coursed flint wall with band course and dressings of cut and sawn stone.

detached from their backing and although there may be no signs of bulging, it is always essential to advise a sounding by gentle tapping over the whole of the surface to detect any hollow areas which could be a potential hazard by way of collapse and possible injury.

282: Uncoursed knapped flint walling at least 100 years old.

283: Knapped ands squared flints laid in courses to line up three courses to each 'V' jointed quoin stone on a two storey dwelling over 200 years old.

The finer grained more compact varieties of both limestone and sandstone, the two most commonly used building stones in the United Kingdom, are both capable of being sawn square and finely dressed to form blocks which can be laid in regular courses with the narrowest of joints to produce ashlar. In the case of most limestones, the softer and more easily worked are also capable of being carved and chiselled into comparatively complicated shapes for cornices, architraves, columns, string courses and the like. This is not always possible with sandstones, or even desirable having regard to exposure conditions in the areas where most of them come from or the incidence of pollution, in the case of the more intractable of sandstones.

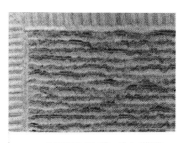

Figure 40: *Apart from the usual finishes of sawn and rubbed applied to the stone for ashlar, to highlight certain features of a facade, the stone is often dressed to a more elaborate finish. Those shown here are 'broached' at the top with 'punched' below, both blocks having 'drafted' margins.*

284: *Ashlar consists of cutting and sawing relatively soft and easily worked stone into precise rectangular and accurately dimensioned blocks and laying them in courses with minimum width joints to produce, as here, a flat uniform surface. Carried to extremes, it often meant that stability could be impaired. To overcome this problem hollow bedding would be used whereby mortar would be spread into a hollow in the base of each stone. As this compressed over time, the stone on the outer edge of the joint would fracture.*

Figure 41: *The dressed stone here is given a 'droved' finish at the top while a 'herring bone' furrowed pattern tooling is shown at the bottom.*

285: *To Emphasise and to give 'appropriate' strength to the lowest storey, the podium in classical architecture, 'V' joints would be used between the blocks of stome for ashlar to produce a pronounced pattern*

For dwellings of good quality, ashlar would be the expected finish for owners with the wherewithal to afford it when building for themselves and also for developers building for the upper end of the market, the cities of Bath and Edinburgh being prime examples where such developments took place.

The surveyor will not be able to tell from the exterior whether the ashlar he is seeing is part of a wall wholly of stone, probably with a hearting of rubble and mortar, or whether it is

286: The importation of Caen stone from Normandy by sea to south east England was comparatively easy compared with transporting the home grown product by land. It is perhaps surprising that there are not more signs of erosion on the Caen stone of this two storey dwelling, built over 400 years ago, but the location is clear of pollution which has been the undoing of much of this type of stone elsewhere.

287: In a narrow street of tall mainly five storey dwellings, this Portland stone fronted house, facing away from the prevailing wind, shows remarkably little erosion after over 230 years. The subtle difference between the dressing to the voussoirs around the door and window openings on this dwelling and those around the door opening on 258, which are in artificial stone can, with interest, be noted. These are 'reticulated', the ridges being linked to form a network of irregularly shaped sinkings (reticules), as against the 'vermiculated' dressing on 258 with sinkings to form a winding snake like (verminous) continuous ridge.

a facing skin blocked on to a backing of brick. From the interior, he might get a sight of the unplastered inner face in the roof space or a cellar, which could confirm a viewpoint either way. Wholly stone construction is perhaps more likely in dwellings built before the coming of the railways in areas where the availability of brick was limited. In the absence of any pronounced defects, it is perhaps a somewhat academic point to seek to differentiate.

Overall in purely structural terms, it is probably better for an ashlared dwelling to be brick built with a facing of stone. Brick walls do not have the disadvantage of walls built wholly of stone where the mortar of the hearting crumbles over the years and drops down as dust with the rubble, to the bottom of the cavity often affecting stability, because the outer leaves are deprived of their bond.

Apart from incorrect bedding of the more pronounced sedimentary limestones and sandstones which is a frequent source of spalling on the face of blocks of stone, another problem with ashlaring was that with the more easily

worked stones, there was always a temptation to reduce the width of the joints to such an extent that there was hardly any mortar to hold them together thus possibly depriving the outer facing skin of a proper bond. To overcome this problem, 'hollow' bedding was sometimes resorted to. However this left such a narrow strip of stone near the face to form the thin joint to the stone below that when the mortar in the hollow was pressed down as the structure settled, the narrow strip would suffer a compression fracture.

Where details such as copings to balustrades and cornices were provided with the thinnest of joints so as to appear in one continuous length, metal cramps were often used to hold the sections together. Over the years, they rust if made, as they usually were, of wrought iron and in doing so expand considerably invariably fracturing the stonework in the process. This is more likely to be found in connection with dwellings built of limestone. Sandstones, being

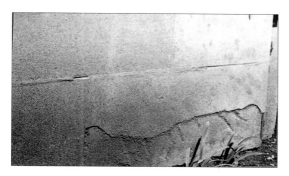

288: When damp from the ground brings up salts to crystallise behind the face of a stone incorrectly bedded, it is not surprising that it starts to spall, as here with Clipsham stone after 60 to 70 years.

289: At the time of cutting and sawing to prescribed sizes and regular shapes, it is not always very apparent where the stones natural bed lies. This does, however, become apparent when years later damp penetrates over a period with its damaging acidic content, as it does to quite a considerable degree with rather porous Bath stone as used here, and then freezes, blowing off the face of the stone.

generally coarser and less easy to work would be provided with thicker joints with more mortar but, even here, for balustrades where there was a danger of them being pushed over, metal cramps would be used.

In general, limestones do not stand up to the weather with its penetrating wind driven rain and possible subsequent frost effect, quite to the same extent as most sandstones and when exposure to the elements is aggravated by the presence of sulphurous compounds

through pollution, then the sandstones win hands down. In the past, such pollution came about through the burning of coal for both manufacturing purposes and domestic use. Manufacturing was more extensive in the north where coal was more readily available and it is perhaps fortunate that most sandstones are found in that part of the country, while most limestones occur in the south.

The absence of smoke in the atmosphere should not, however, fool anyone into thinking that there is now less pollution to affect stonework. There is more than ever, since the burning of gas and oil for industrial processes and the production of electricity, coupled with the exhaust fumes from the internal combustion engines of motor vehicles, releases vast quantities of sulphur dioxide into the atmosphere. This falls as diluted sulphuric acid in rain and, while the element of soot to blacken surfaces may now be much less, the acid dissolves the calcerous matrix, a carbonate of lime, holding the grains together. This is more prevalent in limestones than sandstones, the most durable of which are formed with a siliceous matrix, comprising mainly silica which is virtually indestructible. Even so, dwellings in the south and west are likely to be less affected by pollution since, in general, the prevailing winds blow any pollution from these areas to the north and east. It is of course, the differing composition and characteristics of limestones and sandstones that lead to problems when the two types are mixed in the same building. A run off of calcium sulphate from limestone to sandstone can cause rapid decay in the latter.

Nowadays, for new work, a more informed choice can be made from the stone which is available from those quarries which are working. Most of these have been in operation for a long time, or have been reopened under new ownership, so that the characteristics of the

290: *Facing as it does into the teeth of the south west prevailing winds even more care than usual should have been taken to bed this stone correctly. Continual splashing from below added to wind driven rain has fairly quickly found out that a mistake was made.*

292: *Metal cramps are often used to hold sections of stone together. If of wrought iron and they rust, the expansion of the metal is usually sufficient to fracture the stone. Here, the copper cramp is so much stronger than the stone itself that when structural movement has forced apart the two sections, the cramp has been left behind with the fractured sections of coping*

291: *Fractured and spalling stonework to a balustrade. It may be that the type of stone used here was too coarse grained to be successful where sharp arrises and turned mouldings were required.*

by the dozen where there was a demand with the result that selection would be governed by the market, for appearances sake, and by price, for profit. In consequence much stone was produced which has proved unsuitable in the particular circumstances of its use.

Stones

Faced with a dwelling where it has been established that natural stone is what is being seen, how is the surveyor to ascertain which type of stone has been used? Furthermore, is it likely to provide a new owner with trouble free, and hopefully handsome, walling, given proper maintenance over the years to come?

quarried stone can be demonstrated from long use over the years, though this is obviously not the case with newly opened quarries. Even though a new quarry may be near an old disused or exhausted one which was known to produce durable stone, wide variation can exist in the characteristics of stone quarried from comparatively short distances apart. Petrochemical analysis may be necessary to demonstrate any claim to similarity. In contrast, during the building booms of the past, particularly those brought about by the Industrial Revolution, quarries would be opened

The surveyor's local knowledge is vital towards providing an answer to the above questions. If the dwelling's stone is in reasonably good condition and the dwelling of some age, say more than 40 to 50 years old, it might be thought that there is not much point in saying more, leaving it to an assumption that such circumstances will continue to be the case.

293: Weather worn sill of Clipsham stone after 70 odd years of exposure, revealing the oolitic nature of the limestone, the fossil elements being more resistant to the weather than the main body of the stone.

However, the surveyor ought to be able to do more than this, even if he is comparatively new to the area and lacking that local knowledge which surveyors ought to cultivate. Many, of course, do and this enables them to

recognise on sight a type of stone particular to the local area.

One of the attributes desirable in surveyors mentioned at the beginning of this book was an interest in the older maps of the local area. This was so that the information they can provide about what was on the site before the dwelling being inspected was constructed can be noted, particularly as this could have a bearing on possible foundation problems. It may well be that these maps also show a quarry or more likely the site of a quarry. The inspector himself may know from personal experience such a site as a local hazard, because of steep cliffs and flooded pits or, over the years, one that has become a local beauty spot once the vegetation has grown back. Ordnance Survey 1:50,000 maps note quarries and the 1:25,000 maps differentiate between a gravel pit, a sand pit, other pit or quarry and a refuse or slag heap.

295: Erosion and spalling to the elaborate four flue chimney stack on a Gothic revival dwelling.

294: Although the surveyor's local knowledge might tell him that this five storey attached house is in an area subject to subsidence and he might assume that the defects seen here are due to that cause, closer investigation will reveal that the disturbance to the ashlar is much more likely to be due to defects in the lintels.

Even better, £6.25 of any surveyor's money is very well spent on a map published in 2001 showing the location of the larger quarries in the United Kingdom that achieved some commercial or historic status as building stone producers, even if only on a local scale. The map shows a myriad of quarry locations but it is said that they comprise only a small proportion of the thousands that were worked. However, the map does show those working in 2001 and a wide selection of those which provided stone in the past, giving their names in different styles or indicated by contrasting colours in small circles. This map is produced jointly by the British Geological Survey with English Heritage, Historic Scotland, CADW (Welsh Historic Monuments) and the Stone Federation of Great Britain. It is entitled Building Stone resources of the United Kingdom, and is to a scale of 1: 1,000,000 on the National Grid. The stated prime purpose is that for new work, stone similar to that used already in the locality can be selected or, for repair work, stone reasonably closely matching the appearance of the old but, if necessary, possessing superior characteristics can be chosen.

Regrettably, the map, apart from some panels of general description, which are repeated on an insert within the map's plastic folder, and some photographs round the edge of buildings constructed of different types of stone, gives no information on the colour, characteristics or appropriateness for use in various circumstances of any of the stones from the quarries indicated. Eminently sensible and useful for identification purposes, would be another insert in the folder, providing this latter information for all the stones indicated on the map.

The Appendix commencing at p433, however, extends the information on the map in just such a way for colour and characteristics in respect of around 150 of some of the better known building stones set out by old Counties from north to south with the Unitary Authority quoted under the location if different. In addition, the use of a particular stone in one or two buildings is sometimes given. Inevitably these buildings tend to be large and well known, not all of them dwellings but, if visited will provide the surveyor with an idea

of how the stone looks after weathering over time. This, of course, can be somewhat different from the colour given in the Appendix which, obviously, has to be as quarried since the degree and effect of weathering and pollution will vary considerably.

One or two aspects of the Appendix need mentioning. The vast majority of building stones in the United Kingdom are quarried, but a few, such as a number of the 'Bath' stones are, or were, mined along with Beer, Purbeck, Collyweston and a few others. This is not thought necessary to differentiate in the Appendix. 'Freestone' is not just a stone there to be picked up from the ground as might be thought, but identifies a stone which can be easily worked in any direction, being compact and fine grained and with no well defined bedding plane, examples are Red Corsehill, Lochabriggs and Woolton. There is an absence of information about some stones but, by the inclusion of a quarry address, even approximate, it may be that local information is to hand. Area Geological, Archaeological and Historical Societies can be extremely useful contacts for the surveyor and the interest of local historians is often expressed in papers and pamphlets obtainable from local libraries. In this respect it may be that some stones are known locally by different names, sometimes reflecting their characteristics.

To give an example of how the Appendix and the map might be used, 186 shows a terrace of two storey houses a little over 100 years old in a Northern town, two of which have been cleaned in contrast to a third, on the right, still with its sooty weathered appearance. The Appendix lists a quarry within a few miles where the stone is described as white to brown, a sandstone of medium to coarse texture and durable. Colour, description and evident durability, even on the stonework not cleaned, match well so that there could be a reasonable assumption that the named stone is the one in the frame. However, the map shows two other quarries nearby so that the reasonable assumption would have to remain just that until other available local knowledge is given consideration to see whether the stone from those other quarries can be eliminated, maybe on grounds of colour, texture or known lack of durability.

Clay and Chalk Walls

As BRE points out, more than half the world's population live in dwellings built from the subsoil which is mixed with water and other unburnt local materials. We may feel somewhat superior in that the vast majority in the United Kingdom live in brick, stone, concrete or timber framed structures but a significant, though relatively small, proportion do also, in fact, reside in the former type of dwelling. One of BRE's first tasks in 1922 was to study building in cob and *pisé de terre* and many such dwellings, and others built similarly, have outlasted dwellings built in far more sophisticated ways.

The approximate locations and wall thickness together with the names and the types of construction of dwellings built of unbaked earth are set out on the subsection on ascertaining construction. As stated there, most are likely to have been built in the 1700 and 1800s but all types have certain needs in common and can suffer degradation and possible collapse if these are not met.

The most important need is adequate protection from damp and this was normally achieved in three ways. First, walls needed to be built up on a base of stone or brick to reduce the amount of rising damp which, if in excess, could reduce unbaked earth to mud. If base walls have no damp proof course and there

296: *Unbaked earth walling, minus its rendered coating, comprising clay, chalk lumps and chopped straw at first floor level above uncoursed random rubble ground floor stone walling and with brick dressings to the windows at both floor levels. Some repairs are clearly required before the rendering, to protect it from damp, is renewed. The projecting brick course between the two types of walling will provide a sound base for the necessary bellmouth at the bottom of the rendering.*

297: *Unbaked earth walling, clearly showing the lumps of chalk incorporated in the mix where the internal plaster has been removed in the window reveal of a two storey cottage, in process of refurbishment.*

are signs of damp on plaster at the base of walls internally, the provision of a damp proof course must be strongly recommended and the reason why explained. Preferably, it should be advised even if there are no signs of dampness.

Second, a sound form of rendering to the outer face of the wall construction is needed. This must be permeable and based on lime, not Portland cement, to allow for a certain amount of absorption and subsequent evaporation.

298: If all the rendering, or at least half, is not renewed soon on this already badly repaired unbaked earth gable wall, there is a danger that the moisture level will rise considerably, possibly with the ultimate risk of collapse.

This is so that the interior of the wall can breathe and does not dry out to become dust or become too wet and turn into mud. Any decoration to the rendering must take this need into account. One aggrieved new owner was reported as having settled for a six figure sum of compensation when his surveyor failed to notice that the cob walls of the purchased cottage had been painted with impervious plastic paint. Part of the cottage had collapsed during repair.

The third important need is that there should be sufficient overhang at the eaves to provide as much protection from damp penetration as possible to the top of the wall. A minimum of 450mm is necessary and preferably 600mm. This degree of protection was usually provided in the past by the thatched roofs with which such construction was normally associated. These can be fine but as they do not generally come with gutters, a check needs to be made that the discharge from the thatch is not

liable to splash up against or near to the rendering on the unbaked earth wall, or above any damp proof course provided. However, many thatch roofs were replaced by other forms of roofing in the 1900s with a lesser overhang. If a new thatch roof cannot be provided, it might be possible to add sprockets to the rafters and extend the covering outwards, taking the rainwater away by means of longer swan-neck rainwater pipes. Needless to say, it is most essential to maintain roofs in good repair.

Fractured, damaged or neglected unbaked earth walls can be repaired using the traditional techniques or with stabilised blocks made of the same materials. Burrowing by rats or mice within the mass of the wall, a fairly common event and which has been known to cause collapse, can be deterred by filling any visible burrows with a compatible mix containing broken glass.

299: Stabilised blocks of clay, small lumps of chalk and chopped straw, which can be seen sticking out, awaiting use in the repair of an unbaked earth wall.

Framed Walls

For a surveyor inspecting and reporting on dwellings with framed, as distinct from load bearing walls, there is the difficulty that the framing units whether of reinforced concrete, steel or timber are, with one notable exception, seldom seen. That exception is, of course, the oak framing of houses built roughly from the 1300 to the 1600s but, even with many of those, comfort and, at times, fashion dictated a cover up. Showing off the old oak frame only became fashionable from about 1850 onwards.

What the surveyor does see is the form of envelope applied to the frame and, preferences being what they are, many framed buildings appear from the outside as traditionally brick built. To make their product look as though it was built with load bearing walls was the aim of most of the companies putting up dwellings with framed components in the 1900s particularly those supplying the private sector as well as local authorities.

The procedure for ascertaining the construction of the external walls, wherever this is possible for dwellings, has already been dealt with at the beginning of this subsection but covering this may only provide the information that the structure is framed, not how or with what. That in itself is, however significant as, coupled with a knowledge or an assessment of the age of the dwelling, it provides a pointer to further investigation. Times of shortage, particularly the periods immediately following the 1914-1919 and the 1939-1945 Wars, provided the stimulus for much of the development of the systems used and it is to the information about those systems and there characteristics that the surveyor will often have to turn.

The Requirements for Home Inspectors states that inspectors 'must be able to recognise and identify the types of system building … and should recognise whether they are designated'. Put as baldly as that, it represents an exceptionally tall order but, fortunately to some extent, the gap in the sentence is filled by the words 'found in their area' which enables the surveyor to make some enquiries, for example to the local authority, from publications and from those he knows with knowledge of what was going on in the locality, particularly from the late 1940s through to the beginning of the 1980s. The Building Societies Association has expected as much from the surveyors acting for its members, having stated so in 1987 after doubts had arisen about the long term prospects of certain types of system built housing.

Even so, as mentioned in the subsection dealing with load bearing walls comprising large panels of pre-cast concrete bolted together where no framing was involved, there were over 30 systems developed for this type of construction alone and which were used for low, medium and high rise dwellings. BRE acknowledged that it was difficult to differentiate these in their original form and says the same about many of the framed systems used for low rise dwellings. Of those constructed with prefabricated reinforced concrete (PRC) frames, 34 different types of these were designated by the government, as listed in Table 6 on p261, as 'defective by reason of their design and construction' enabling innocent purchasers of local authority houses built of that type to obtain a substantial grant for their alteration and repair. 'Right to buy' purchasers of flats, however, built with the same systems were given assistance instead by a right to require the local authority to repurchase, a right also extended to house purchasers who did not wish to have any repairs carried out.

BRE in the 1980s identified many other examples of system built dwellings, publishing information in 1986 and 1987 on 30 which were steel framed, extending their studies to 54 timber framed systems built between 1920 and 1975 and publishing this information in 1995. BRE acknowledged at the time that there many more different types and published a book of 940 pages in 2004, entitled *Non-Traditional Houses*, on identifying the types in the United Kingdom built between 1918 and 1975. The book details 450 house types classified by form of construction along with notes for surveyors. Included are the comparatively rare and few in number as well as those which were built in quantity and are better known. Covering so many systems, generally by way of a two page spread, it not so detailed on the constructional aspects and the flaws which have become apparent over the years with some of the systems as was available from BRE earlier. Accordingly, reference might still need to be made to the earlier material. While at £270 the book might be considered expensive, it provides valuable information and ought to be available for consultation when needed by every surveyor. As to comparatively easy and instant recognition of types, this is complicated by the fact that so many were deliberately intended to look like other forms of construction and, for others, is compounded through the changes in appearance brought about not only by the enveloping and overcladding carried out by local authorities but also by private owners seeking to make their homes look different from those of their neighbours.

While the ability to recognise some of the more obvious types of system built dwellings, where these are known to have been built in a particular area, is a useful attribute and would obviously save time, it is far too much to expect a surveyor to carry in his head a mental image of so many different systems

of construction even if it were possible to recognise their differences which, according to BRE, in many cases it is not. Far more useful is the surveyor's routine of inspection which should throw up the fact that a dwelling is not of traditional construction. This will trigger the search for further information which will be essential before the report can be compiled and may even involve a second visit to the dwelling.

For example, following the above recommended procedure would have saved a surveyor having to pay out the £9,000 awarded as damages in a Scottish Court for failing to identify that the dwelling he was inspecting for a mortgage valuation in 1982, and whose report was disclosed to the purchaser, was of non-traditional PRC 'Dorran' type construction, *Peach* v *Iain G Chalmers & Co* 1992. The house had been built privately in 1976 in a good location in the north of Scotland and had a swimming pool. The surveyor described the walls as 'concrete block built, harled' and valued it as a traditionally built house at a mere £1,000 below the asking price of £35,000. A 'Dorran' system built pair of semi-detached houses are illustrated at 300. This shows quite clearly the narrow full storey height concrete panels, the projecting ring beam at first floor level and the fact that the windows are set flush with the external wall face, all features typical of the system. The surveyor described the walls of the house he

300: Original 'Dorran' type PRC houses

was inspecting as 'harled' which could well have obscured the joints of the concrete panels, but the other features were evident, particularly the windows set flush with the external face in walls very much thinner than walls of traditional construction. The BRE report on the structural condition of 'Dorran' system built dwellings was not published until 1984 but evidence was produced to show that problems with PRC dwellings had been well publicised in 1981, that in the locality where this one was situated, they were known for what they were and when sold in 1982 fetched between £19,000 and £25,000, according to location and condition, less than houses of traditional construction so as to allow for possible future problems.

The point to be drawn from the above case is that even if the surveyor had not kept up to date with his reading and on, perhaps, a fairly cursory inspection for a mortgage valuation had not noted the features which distinguish a 'Dorran' type of PRC dwelling, a set routine of inspection ought to have shown up the house as of non-traditional construction and different from other harled houses in the area, if only because of thin walls, thus putting him on enquiry.

As it happens the BRE report of 1984 gave 'Dorran' type houses a comparatively clean bill of health. It did this along with others such as 'Myton', 'Newland' and 'Tarran' based on similar structural principles and differing only to the extent, where visible, on whether steel or timber was used for the roof structure. The report indicated, however, that like all PRC houses they suffered from corrosion of the steel, cracking, spalling and rust staining of the concrete and, accordingly, they were all subsequently designated. Nevertheless, at the time, BRE did not consider there was need for urgent action, just repair and proper maintenance. They were described as of a

301: Refurbished 'Dorran' PRC houses.

type where stability relied on the ability of the wall panel units to support the load and the restraint provided at first floor and eaves level by tie bars connecting the part of the ring beam in the front and rear walls and the party and gable walls. Early signs of any distress would become apparent through cracking in the ring beam units and bowing of the wall panels, defects always to be looked out for, according to BRE, by surveyors inspecting any of these types as might well be the case for a purchaser where a mortgage is not required and it is apparent that the original structural elements are still in place.

What of system built dwellings that no longer look at first glance anything like they did when originally built? The illustration at 301 shows another pair of semi-detached houses built on the 'Dorran' system. The storey height, distinctive, concrete panels have disappeared and so has the projecting beam at first floor level while the windows and the whole appearance up to roof level have a characteristic 1980s look. Another look at the illustration, however, will show in the back-ground a further pair of semi-detached houses which could provide an easy solution for the inspector as to age and type. Most of the non-traditional dwellings are located on fairly large estates so looking around nearby could provide the answer to a problem where the dwelling being inspected looks different from the others as it does here. On the other hand, what if all the houses on the estate are much the same? There are visible clues on the exterior of this pair of houses, both have hipped roof slopes and chimney stacks so that the date of construction in the 1940 to 1960 period is much more likely rather than the 1980s. This should make the surveyor immediately aware that he is dealing with a dwelling which has undergone some refurbishment and was originally built some 20 to 30 years earlier. In some instances the type of dwelling may not be immediately apparent, but there should be sufficient evidence from measurement and soundings and internal details, including features in the roof space, to determine that the house is not of traditional construction, that evidence to be verified from the BRE report when further enquiry reveals the type of system.

Reinforced Concrete Framed Walls

In the domestic field, the surveyor is likely to encounter reinforced concrete framed walls in two forms. The first, and what could be considered now as more traditional, is where the frame would be erected *in situ* as the load bearing structure for a small multi-storey block of flats. In fact unless told, the surveyor would probably not know that the frame was of reinforced concrete. The alternative form of framing for this type of structure is of steel but as this has to be encased in concrete for fire protection, protruding columns and beams look the same. Such a block in either material would be engineer designed and, unless there was any evidence of distress, it is unlikely that anybody would have been instructed to make an appraisal for strength and durability since the day it was built. With no sign of distress, the surveyor would have no reason to advise either a buyer, lender or seller of a flat in the block, to have an

engineers' report. The surveyor's inspection and report can be concentrated on examining what can be seen of the enclosure walls of the flat, the interior and the common parts and the exterior of the block as a whole with a note on the structure being limited to a few comments.

The heyday for such construction was from around 1920 to 1950 before prefabricated reinforced concrete became the panacea for

302: An 11 storey block of flats built about 70 years ago on the coast in the all white 'modern' style fashionable at the time but now looking a bit grey and stained. Visual inspection will not reveal the material used for the frame: it could be either reinforced concrete or steel with perhaps a greater likelihood of the latter. In the absence of any signs of distress to the frame within the flat, in the common parts or on the exterior there would be no need to advise the obtaining of a structural engineer's report on the stability of the frame or its durability. There are, however, ample signs that exposure to the elements has taken its toll on the exterior and the internal inspection will no doubt reveal how the aged metal windows are coping. Maintenance costs and renewals are always high with this type of construction and the surveyor will need to sink his teeth into the service charge accounts and the likely cost of future programmes to provide sound advice to any client whether buyer, lender or seller of a flat.

303: Recognisable as a typical reinforced concrete framed brick clad block of medium rise flats of the 1950s the same remarks about the condition of the frame as in 300 apply to this local authority block. Service charges ought to be relatively modest as there is seldom a lift for a block of this height and each flat has probably, now, its own controllable central heating and hot water installation. Upgrades to windows and entry systems are always a possibility and the surveyor apart from making his own estimate of needs, should press the freeholder for details of any future programmes.

304: In is original condition, probably the easiest to recognise of all the PRC system dwellings of the late 1940s and 1950s and subsequently designated as 'defective by reason of design and construction' is the 'Cornish Type 1' with its exposed frame and mansard roof although some did have tile hung vertical first floor walls and hipped roofs. The characteristic whitish appearance of the dense concrete is due to the use of washed quartz sand, a by-product of the China clay industry.

solving all the housing problems of the time brought about by shortages. Until then *in situ* reinforced concrete construction in the domestic field had been restricted generally to blocks of flats up to five or six storeys, leaving two storey housing to be constructed by traditional means. The change came about in the late 1940s when prefabricated units began to be used for framing low rise housing and, eventually, the panels used for infilling were developed into the large panels with bolted connections for the high rise developments of the late 1950s and 1960s. All told about 1% of the total housing stock comprises dwellings system built using prefabricated reinforced concrete components, 34 of the systems used for low rise houses designed before 1960 being designated as 'defective by reason of their design and construction' through the provisions of the Housing Act 1985, though 18 of these account for 98% of the total stock of PRC houses of this period. These systems are Airey, Ayrshire County Council,

Boot Pier and Panel, Cornish Types 1 and 2, Dorran, Myton, Newland, Orlit, Parkinson, Reema Hollow Panel, Stent, Tarran, Underdown, Unity Types 1 and 2, Wates, Whitson-Fairhurst and Woolaway.

To an extent the designation makes the surveyor's task somewhat easier since it means that unless the dwelling has had its structural components replaced by a system certified by PRC Homes Ltd and provided with a NHBC Certificate, it is not mortgagable. This fact obviously puts off anyone who would require a mortgage and a purchaser who could buy for cash would clearly need advice on value to take account of the need to do the work, the cost of which is substantial as can be gauged by the maximum grant which was available of £24,000 representing 90% of the estimated cost of the work in the early 1990s to the worse affected PRC houses such as Dyke, Hawksley, Orlit, Schindler, Waller and Wessex. Should an unrepaired dwelling come on the market and it was not intended to do any of the structural replacement work and to

305: The exposed frame of 'Cornish' type dwellings at least enables the defects which are common to all types of PRC dwelling, carbonation and the use of chlorides in the mix producing acidity in the concrete, rusting and expansion of the steel work and spalling of the concrete to be seen more readily. Here the defects are evident in one of the structural columns with the spalling concrete rendering the steel reinforcement even more subject to corrosion.

306: In many PRC dwelling the cladding panels obscure sight of the structural frame. Here it has been necessary to open up so that the serious cracking can be revealed at the foot of one of the reinforced concrete columns.

take a chance on durability then the risks would need to be explained in the report. Reference would need to be made to the findings of BRE set out in their report on the structural condition of that type of PRC

307: It was not only two storey terrace and semi-detached which were built using PRC systems. This designated Tarran bungalow mainly as originally built has like its two storey relations, storey height 400mm wide wall panels but here supported on a timber kerb which tends to deteriorate over time as do the concrete elements in the same way as with other PRC dwellings. The roof is of lightweight profiled asbestos sheeting as there is no concrete ring beam to support anything heavier.

308: The cost of removing the structural elements of a designated Tarran bungalow and replacing them with a new structure may not be cost effective and here the enveloping option has been taken. The wall panels have been provided with an external insulating render coat and the original asbestos roof replaced with lightweight profiled steel sheeting to give the impression at first glance of profiled tiles. The useful life of the bungalow could well be extended by 20 to 30 years given adequate maintenance.

dwelling following their 1980s inspections and due allowance made in any valuation which might be required.

The above could well be the situation where a surveyor is instructed to inspect and report to a prospective buyer or seller of an unreconstructed PRC dwelling and is not told, or the seller does not know, the nature of the original construction. While it may be essential to fall back on the advice for establishing whether the dwelling is of traditional or non-traditional construction and the necessary follow up enquiry, it would obviously save time if the surveyor can readily identify some of the more common types. The appearance of 'Dorran' type PRC dwellings has already been discussed with reference to the two illustrations at 300 and 301 and mention was made of the similar principles

used in the construction of the 'Myton', 'Newland' and 'Tarran' types. The latter's system was also used in the construction of bungalows and when combined with some of the features of the 'Newland' type of house produced the 'Tarran Newland' two storey dwelling as merely 'enveloped' at 309 but as reconstructed at 310. At 311 is a reconstructed 'Airey' type, so that both the pairs of semi-detached houses at 310 and 311 would be eligible for acceptance as security for a mortgage advance.

Other PRC dwellings still as originally built are also illustrated here, 312 'Wates', 313 'Airey', 314 'Cornish Type 1' and 315 'Cornish Type 2 while 316 shows 'Orlit' houses with flat roofs and 317 'Orlit' Houses with pitched roofs and new windows.

The discovery of the main problems with PRC dwellings followed a fire in an 'Airey' type

Table 6: Types of Prefabricated Reinforced Concrete (PRC) Dwellings designed before 1960 and designated in the 1980s and 1990s by the government as ' Defective by reason of their Design and Construction'

Airey	Shindler
Ayreshire County Council	Smith
Blackburn Orlit	Stent
Boot Beaucrete	Stonecrete
Boot Pier and Panel	Tarran
Boswell	Tee Beam
Cornish Type 1	Ulster Cottage
Cornish Type 2	Underdown
Dorran	Unitroy
Dyke	Unity Type 1
Gregory	Unity Type 2
MacGirling	Waller
Myton	Wates
Newland	Wessex
Orlit	Whitson-Fairhurst
Parkinson	Winget
Reema Hollow Panel	Woolaway

Note: Some other names of types appear on lists published earlier than the above BRE list of 2004. These are Butterley, Hawksley, Lindsay, Stour and Unit and which may be alternative names or variants of the above types. If encountered, dwellings with these names should be considered in the same way as those above.

309: 'Tarran Newland' refurbished PRC houses.

310: Reconstructed 'Tarran Newland' houses.

311: *Reconstructed 'Airey' PRC houses.*

314: *Original 'Cornish Type 1' PRC houses.*

312: *Original 'Wates' PRC houses.*

315: *Original 'Cornish Type 2' PRC houses.*

313: *Original 'Airey' PRC houses.*

316: *Original 'Orlit' PRC houses with flat roofs.*

house. During reinstatement the structural frame, normally hidden in this type, was exposed revealing spalled concrete and rusting steel reinforcement in the columns due, principally, to carbonation and the addition of chlorides to the concrete mix to accelerate hardening in the rush to produce more prefabricated units at an even faster rate. Carbonation is a process whereby concrete absorbs carbon dioxide from the atmosphere and changes from being alkaline to being slightly acidic. The proportion of chlorides added to the mix tended to be widely variable but in combination with the carbonation produced an acidity far in excess of what would have been considered safe. Unlike the *in situ* concrete columns and beams used in multi storey frames, the prefabricated units

317: 'Orlit' PRC houses with new roofs and windows.

318: A 'Unity' type PRC system built pair of semi-detached houses clearly showing the pattern of panels resembling blockwork laid with continuous vertical joints. The frame is totally enclosed but this does not prevent it being affected to some extent by the atmosphere and by the inherent defect of overuse of chloride accelerators in the mix.

319: A three storey block of flats built with the 'Unity' system of construction where the pattern of the concrete panels resembling blockwork laid with continuous joints, as in 318 can also be discerned. Not designated, the remedy for innocent purchasers of flats under the right to buy legislation would be to require purchase back by the local authority because of the unlikelihood of a mortgage being forthcoming for any other purchaser. If held and sold for cash, perhaps as a buy to let investment, a purchaser would need to be advised strongly about possible future problems and the effect on value. Nothing more than a new roof covering and new windows seems to have been done to this block since it was originally built so that, following BRE advice, a structural engineer would need to be brought in to provide a report on structural stability and durability, if a recent report was not available from the block's freeholder. In regard to 'Unity' system built blocks, BRE considered that added strength was provided by the fact that they had concrete floors, concrete stairs and landings, internal walls of blockwork and brick clad gables.

used for low rise housing were relatively small and, as in many instances provided with little concrete cover to the steelwork in the first place, it was not too long before the expansion of the rusting reinforcement caused the concrete to crack and spall, compromising strength and allowing yet more oxygen and moisture to reach the steel and progress the rusting process even faster. There is no practical way of preventing this progressive deterioration in the PRC units made in the late 1940s, 1950s and 1960s, though its rate will vary according to how much shielding of the units is provided by the design of the dwelling and the degree of maintenance.

It was for the above reasons that the replacement of the original structural elements was decided as being the only sure way of ensuring that the dwellings would have the comparable life span of at least 60 years, as required and as expected from dwellings of traditional construction. For those being retained and not re-structured, BRE recommend regular inspection for any signs of cracking which it says should be seen as a warning particularly at the corners where it considers the columns more likely to deteriorate earlier, along with the ring beam. For blocks of flats above two floors, appraisals of the structural stability and durability should be obtained from a structural engineer, which

320: For refurbishment of this pair of PRC 'Unity' system constructed semi-detached houses the local authority has opted for enveloping rather than the replacement of the structural frame as would have been a necessary choice for a private owner if he wished to sell eventually to a buyer who needed a mortgage.

321; A pair of two storey semi-detached designated PRC houses shown as originally built on the 'Woolaway' system where the structural components are totally hidden by a machine applied roughcast rendering. The design is based on aerated concrete columns at about 750mm centres with wall panels and ring beam of the same material, the only restraint, however, being provided to the front and rear walls by the roof. Wall panels are fixed to the 150mm square columns by bolts, corrosion of which can provide a distinctive pattern of staining, and the columns from the window and door jambs, the wide column division between the two sections of the large windows to the main rooms at ground and first floor levels being a distinctive feature. BRE reported in the 1980s that the columns showed the most advanced degree of carbonation due to the use of aerated concrete and some cracking and that there was some further cracking in the panels but considered that it was more apparent internally since the rendering gave some protection to the surfaces facing towards the exterior.

may involve the need for opening up, analysis of the concrete and examination of the steel.

The structural units of the easily recognised original 'Cornish Type 1' and 'Type 2' dwellings, along with those of the 'Dorran', 'Myton', 'Newland', 'Reema Hollow Panel', 'Stent', 'Tarran' and 'Wates' types are all visible and capable of the regular inspection advised by BRE for signs of cracking and spalling of the concrete. BRE said in late 1984 that if no cracking or signs of movement were seen, it was reasonable to assume that the PRC components of those types were in sound condition at the time of inspection.

Those PRC dwellings where the components are hidden, 'Airey,' 'Ayreshire County Council', 'Boot Pier and panel', 'Orlit', Parkinson', 'Underdown', Unity', 'Whitson-Fairhurst', 'Winget' and 'Woolaway' need opening up for examination of the structural units if any opinion is to be given although the five of these which are rendered 'Boot', Parkinson', 'Underdown', 'Winget' and 'Woolaway', may show early signs of problems in the structural components by the presence

of cracks in the rendering.

With all PRC system built dwellings if cracking or movement is found, then the cause and significance need determining because deterioration is progressive and could be at faster rate than with traditional construction.

A few of the individual characteristics of some of the PRC dwellings may help surveyors immediately recognise that the dwelling being inspected is of non-traditional construction. For example 'Stent' dwellings, some of which were in fact rendered in addition to those left with the structural components visible, had a

322: A pair of two storey semi-detached dwellings built on the Wates PRC system in the 1950s. They have been refurbished and their appearance altered by the addition of an insulating render coat obscuring the joints of the wall panels and the outline of the first floor ring beam formerly visible. A new roof covering has been provided for the original roof structure but, and this may be deceptive, it does not look as though much has been done to the original windows. The chimney stacks in the centre at the rear of each dwelling remain as a characteristic feature but they are here of two flues, one fitted with a capping for a gas fired boiler.

novel way of fixing the roof tiles with clips to a steel mesh which was laid on roof trusses of either steel or timber. Lightweight steel roof trusses distinguished Ayreshire County Council dwellings from the very similarly constructed Whitson-Fairhurst type, which had traditional timber cut and fitted roofs though both, built in Scotland, had the typical timber ground floors and sub-floor spaces even when, elsewhere, construction on a concrete base forming the ground floor was more common.

Another PRC dwelling type could, at least, be distinguished from others although it might, at a quick glance, be confused with a traditionally built brick dwelling. The 'Smith' system type was faced with brick slips 30 to 40mm thick bonded to concrete slabs in a factory and forming a solid wall. The brick slips formed a pattern in stretcher bond with two course soldier arches over window and door openings but, as the panels were joined

together on site, matching was not always carefully done and a mottling effect was sometimes apparent. Because walls were solid, 200m thick at ground floor level but a mere 50mm at first floor, damp penetration was often a problem.

It will be noted that early Wates PRC dwellings as originally built had a single flue chimney stack, more or less in the centre of the rear slope. This was unusual for the time when it was common to have at least a two flue stack and sometimes one of four flues on the party wall line. The single flue was to serve a central warm air unit supplying warmth for the whole house and which had a cast iron flue pipe. The units have probably long since gone but the stacks generally remain on those dwellings which have not had their structural units replaced, either still looking as built originally or merely enveloped to improve insulation standards.

Steel Framed Walls

Like most of the systems used to boost the output of dwellings in the post 1939-1945 War period, those of steel framed construction owed their origins to developments in the 1920-1940 period. A publication by the Department of Scientific and Industrial research in 1951 on 'The Corrosion of Steel Houses' covered eight systems which were built on 36 sites between 1920 and 1927. The most essential report for surveyors, however, on steel framed dwellings is one dated 1987 by the Building Research Establishment on the 'Inspection and Assessment of Steel Framed and Steel Clad Houses'. It runs to 55 pages including a 30 page Appendix describing 35 systems put up in both the public and private sectors and gives some of their locations. The report is specifically directed to building surveyors to

323 and 324: Some steel framed houses were built in the 1920s. These dwellings cased in brickwork with tile hanging to all the first floor of the terrace at the top are of the Dennis Wild type, much as originally built. They employed standard rolled steel sections with stanchions heavier than in many other systems, mostly with no protective coating to the steelwork but some encased in concrete, where their fairly wide spacing is apparent from internal inspection. Although looking like conventionally brick built cavity wall dwellings, they can be identified by their individual patented timber 'Wild Cradle' roof trusses, based on 225mm by 75mm principals with steel tie rods. Cavity widths vary but as the steel is close or even touching the inside face of the outer leaf, driving rain can penetrate and cause corrosion in stanchions necessitating repair and sometimes renewal of the outer leaf. Other versions of the same type were slate hung or were pebble dash rendered, both on battens to the first floor.

325: This and 181 on p 167 are both Dorlonco steel framed semi-detached houses, built originally in the 1920s and 1930s by steel makers Dorman Long. They are distinctive by being double fronted even apart from having the typical steel roof trusses. 181 is more or less as originally built, though with a new roof covering. Walls were rendered on metal lathing. The original rendering will have long since lost, by carbonation, its alkaline protective quality for the steel, which will have corroded, causing cracks in the rendering. This can allow water to penetrate to the structural components and begin the process of corrosion. Earlier repairs usually consisted of either renewing the lath and render, likely to last for another 30 to 40 years, or replacing it with brickwork or concrete blocks. It was often found that the effectiveness of vertical damp proof courses in these circumstances was minimal, because of the difficulty of fixing around the steel. BRE found some cases where the damp proof courses actually directed water on the steel. Later and more recent refurbishments as here, have included repair to the affected steel and recladding the external walls with insulation and new rendering, along with the provision of new windows.

enable them to carry out what was known then as a 'Structural Survey', now known as a 'Building Survey', and to produce advice for prospective purchasers of steel framed and steel clad houses.

The 1987 BRE Report indicates that about 140,000 steel framed houses had been built, about 0.5% of the total stock of dwellings, by the time of writing, the vast majority post 1940. However, BRE also notes that the locations of most are not known and that there were types other than those mentioned in the Report. As to appearance, although a few are distinctive it is probably these which are more likely to have had their appearance altered by recladding with other materials to extend their life, improve their performance and, where privately owned, to make them appear different from their neighbours. Many, however, are brick clad giving them the appearance of conventional dwellings, the disguise of some being nearly perfect, in that it takes more than a superficial examination to detect that they are in fact steel framed. Fortunately an inspection of the roof space reveals that most systems had steel roof trusses, the major exception being the Dennis Wild and Thorncliffe systems. These can be identified by other means, the former by their patented 'Wild Cradle' roof trusses based on 225mm by 75mm timbers and steel tie rods and the latter by their visible cast iron structural panels for the walls, bolted together. This leaves the remainder also identifiable as of non-traditional construction, on account of their steel roof trusses, and accordingly recognisable as steel framed types.

326: The verge and gable of a BISF system steel framed semi-detached house, a pair of which are also illustrated at 195 on p 175. Water tends to collect on the horizontal cladding rails, inducing corrosion not only in the rail itself, causing a section here to fall off, but particularly in the panels leading to disintegration at the foot and distortion of the remainder. It will be noted that BRE classifies the ground floor construction of BISF dwellings with its rendering on metal lath as Type 1 and the panel clad first floor as Type 3 though both, of course, are attached to the same steel frame. These houses were built in the 1950s and after 30-40 years it is to be expected that in many cases both lathing and rendering at ground floor level will have needed renewal. This has probably already been carried out on the pair shown at 195.

In advice on the procedure for carrying out site inspections, BRE suggests in its 1987 Report that where the type of steel famed house is not already known or cannot be ascertained by enquiry, it should be determined from Appendix 1 to the report. Being directed to building surveyors carrying out a 'structural survey' of the type common at the time when instructions were readily accepted, almost as standard, involving a certain amount of opening up of structures to seek further evidence on defects which might only be superficially apparent, the report includes the results of inspections where this procedure was followed. In consequence, from the point of view of surveyors carrying out the visual inspections envisaged in this series of books including, of course, those for 'Building Survey' reports where opening up does not, as is common nowadays, form part of the normal agreed terms of engagement, it is the outward manifestations of those defects closely inspected, usually corrosion in the steel normally hidden from view, which are of initial concern. Each of the systems described in Appendix 1 of the report has been classified, wherever possible, into one of the four major types of steel framed dwelling. Referencing back to the main body of the Report provides information on the major distinguishing features of each type helping to narrow down identification of positions where serious corrosion could be present.

The four broad types of steel framed systems which BRE classifies according to major distinguishing features are set out below. Some systems were offered in variants of two types and accordingly are asterisked and appear in two lists or in one case (BISF) the ground floor in one list and the first floor in another.

Type 1: Framed dwellings with outer leaves of brick or other masonry or clad with rendering on a wire mesh.

Examples are:
Birmingham Corporation
  (House No 1)
BISF (ground floor)
Crane
Cramwell
Cruden*
Denis Poulton
Dennis
Dennis Wild
Dorlonco
Nissen Petren
Presweld*
Riley (ground floor)
Trusteel Mark 2
Trusteel 3M

Notes: Crane system dwellings are all
         bungalows in the Nottingham area.
         Crudens are all in Scotland. Preswelds,
         Birminghams and Nissen Petrens are
         few in number, the latter, a
         development of wartime Nissen hut,
         with easily distinguishable barrel roofs
         are all either in Edinburgh or Yeovil,
         some of the latter listed. Trusteel are
         most widely used, Mark 2 with
         latticed steel roof rafters and 3M with
         pressed steel channel components.

Type 2: Framed dwellings with cladding of
        concrete panels. Examples are:
        Bell Livett
        Coventry Corporation
        Cruden*
        Cussins
        Gateshead Corporation
        Hitchens
        Livett-Cartwright
        Presweld*
        Steane

Notes: Relatively small numbers of Bell Livett,
         Coventry Corporation, Gateshead
         Corporation, Hitchen and Steane

systems were built. Livett Cartwright
are all in the Leeds area.

Type 3: Framed dwellings with a cladding of
        sheet materials. Examples are:
        Atholl
        Birmingham Corporation
          (House No. 2)
        BISF (first floor)
        Braithwaite
        Cowieson
        Hawthorn Leslie
        Howard
        Keyhouse Unbuilt
        Riley(first floor)
        Weir

Notes: Coweisons and Weirs are all in
         Scotland and both steel sheet clad
         along with Athol, BISF and Rileys, the
         latter being built in small numbers.
         Arrowhead are tile hung.

Type 4: Not framed but comprise steel or cast
        iron panels taking load from roof and
        floors. Examples are:
        Telford
        Thorncliffe

Notes: Telford system dwellings have steel
         panels with wood battens while
         Thorncliffe dwellings have bolted
         cast iron panels which are less
         subject to corrosion than steel.

With the exception of Denis Poulton, Bell
Livett, Dennis, Gateshead Corporation and
Hitchens, all the systems included in the four
lists were the subject of subsequent detailed
reports in the late 1980s or early 1990s by
BRE. Also published by BRE were detailed
reports on 10 other steel framed systems for
dwellings. These were: Roften (1988),
Falkiner-Nuttall (1989), Areal, Arrowhead,
Homevilk Industrialised, British Housing

327 and 328: These two illustrations show two storey semi-detached pairs of BISF steel framed houses. In 327, at the top, the house on the right, probably local authority owned, would appear to be a well maintained example as originally built and with the original windows. The house on the left has a dramatically altered appearance through the simple expedient of painting the ribs of the panelling a contrasting colour and the renewal of the windows in an unusual form. Both have been fitted with a new lightweight profiled steel sheet roofing. On 328, the houses have been refurbished with a new skin of insulating rendering, new roofs and with new windows still within the same original openings even to the extent of the large windows at ground floor level, a particular feature of BISF houses.

329: 'Cussins' steel framed houses and 330 (below) bay windowed 'Cussins' steel framed houses.

(Doxford), 5M, Lowton-Cubitt, Open System Building (all in1991) and Stuart (1992). Open System are of Type 1 classification while both Areal and 5M are of Type 3.

Three of the systems on the BRE list, one from each of types 1, 2 and 3 are illustrated here and are described below:
329 and 330 are photographs of 'Cussins' system steel framed two storey semi-detached

dwellings giving the appearance of conventional brick built dwellings. In fact the facing is of brick slips on concrete panel and on 330 the design is modified to provide bay windows. A feature of this system is that the steel framing extends to the ground floor construction which, prone to rising damp, can become very uneven. A tip for identification from BRE is that door and window heads are not at the same level and that the separating wall in the roof space is made up of a combination of wood wool and plaster over the steel frame.

Illustrations 331 and 332 are of 'Trusteel Mark' 2 steel framed dwellings built in 1953, while 333 is of .Trusteel' 3M system dwellings from the late 1960s the difference being that the former have lightweight 14g or 16g cold rolled latticed steel strip columns and roof components while the latter have pressed steel channels with solid webs containing much

331 and 332: 'Trusteel' Mk2 steel framed houses of 1953 built for the private sector

334 and 335: Original and brick skinned 'Hawthorn Leslie' steel framed houses.

333 : 'Trusteel' 3M steel framed houses of late 1960s.

more metal. The difference extends also to the columns in that the Mark 3 version has lattice steel while the 3M system has two rolled steel channels bolted together back to back, considered much superior. Although much used, BRE list 291 locations, Trusteel Mark 1 and 2 system dwellings are singled out by BRE in the 1987 Report as requiring particularly close attention because of the

extensive use made of highly fabricated steel sheet sections. It was said that some were being taken out of service after a life of a mere 30 years, corrosion being such that it was uneconomic to repair them. Yet 331 and 332 were photographed in the early 2000s, around 50 years after construction. The life of such dwellings is probably very much dependent on the degree of exposure to driving rain, as the steel columns are set in the cavity touching the inside face of the outer leaf. Although the dwellings at 331 and 333 are in an area defined as sheltered, the lightweight structure ought to be opened up for close inspection at the most exposed corner.

Both 334 and 335 show 'Hawthorn Leslie' system steel framed dwellings of Type 3 in the BRE classification. The wall panels on 334 as originally built are recorded as being of compressed asbestos and plasterboard with an applied aggregate finish, while 335 shows

336: These semi-detached BISF steel framed dwellings have been provided with a proprietary insulated overcladding of rendering scored to look like brickwork in stretcher bond. New roofs and windows have also been provided, in the case of the latter, within the original openings. One can only speculate how long the overcladding will remain serviceable and presumably the steel frame, now some 50 years old, will have been closely examined for corrosion.

the framework upgraded with a new brick skin and new pitched roof.

As might be expected, one of the most important factors contributing to the corrosion of the structural components in steel framed houses is exposure to wind driven rain. This is particularly so as in many instances there is only a thickness of material, whatever it might be, of around 100mm between the external wall face and the steel itself. This can allow saturated masonry or rendering to remain in contact with the steel for long periods, or water to be trapped in undrained or rubble filed cavities. Places where this can happen need identifying wherever possible from the results of the BRE investigations based on opening up the structure. Using the earlier subsection here under the heading 'Assessing Exposure', the surveyor will have been able to determine by applying the factors for terrain roughness, topography, orientation and wall shape for the particular dwelling, the spell index and exposure category of the wall likely to be most severely affected by facing into the prevailing wind. BRE found that 55% of

inspections on sites of dwellings in an area where the driving rain exceeds 80 litres per square metre per year, 26% in areas where it was between 40 and 80 litres but only 13% in areas where it was less than 40 litres, suffered from severe corrosion. This BRE defined as the perforation of one or more steel sections. Translated into the more recent

337 and 338: 'Before' and 'after' illustrations of 'Howard' system steel framed two storey semi-detached dwellings, built originally in 1946-1947 and of 'type 3' in the BRE classification. They have a frame of standard rolled steel sections sometimes factory assembled into large units to save site work. Originally they had profiled asbestos cement roofs and were provided with steel angles to form the basis for framing timber soffit and fascia details, which had a tendency to corrode rather rapidly, and an asbestos cement cased steel flue pipe. Walls comprised concrete panels below ground floor window sill height and a cladding of asbestos cement sheets above, all looking rather the worse for wear in 337. A new brick skin was added in the 1970s and further refurbishment of a new tiled roof covering, new windows and a new chimney stack carried out in 1991, to produce a pair of very presentable houses.

parlance of exposure categories as defined in BS8104:1992. 'Code of Practice for Assessing Exposure of Walls to Wind Driven Rain', the 56% would apply to areas categorised as 'Severe' and 'Very Severe', 56.5 litres/m² per spell upwards, the 26% to areas categorised as 'Moderate', 33 to 56.5 litres and the 13% to those categorised as 'Sheltered', less than 33 litres/m² per spell. Accordingly in the broadest terms, surveyors will invariably need to advise opening up for closer inspection where any steel framed dwelling is situated in an area of 'Very Severe' or 'Severe' exposure. As to those dwellings situated in 'Moderate' or 'Sheltered' areas, it will be appreciated that 100mm of even dense concrete on its own is 'not recommended' as suitable construction for a dwelling in the absence of a cavity at least 50mm in width, bridged only by wall ties. While the dwelling taken as a whole may have such a cavity, there are a number of 'types' of steel framed systems where that cavity contains the vital structural elements and in no way can 100mm of any material be considered adequate protection, particularly when, in some types, the steel work is light-weight. Even in areas where exposure is categorised as 'Sheltered' advice therefore, needs to be given, in regard to those particular types of dwelling, for the most exposed parts to be opened up for closer examination.

Of course, a favourable outcome from opening up only applies to the particular section examined and the surveyor needs to stress this aspect and indicate that only the balance of probability suggests that there is no corrosion or only superficial corrosion elsewhere. A purchaser needs to take the risk that may be involved in not exposing the entire frame. Inevitably by 2000 most steel framed dwellings were already around 50 years old and some even more, exceeding what is considered a normal life span of 60 years, and sellers, as well as purchasers, may need reminding of this fact.

Corrosion is a progressive phenomena and this combined with age has to be reflected in any valuation, perhaps more so when compared to some other types of system built dwelling.

Against this rather pessimistic outlook, it is only fair to point out that BRE found and reported in its 1987 publication that about 60% of all steel framed dwellings inspected had revealed either no corrosion or only surface or minor corrosion. Many of those built in the 1920s were, after all, still in use.

339 and 340: This 'Doxford' (British Housing) system two storey steel framed detached house, very much of the 1960s, has a frame comprising standard rolled steel channels and angles bolted together and filled with concrete to form the columns, a roof of profiled asbestos cement sheeting and variable cladding which might consist of tile hanging, cedar boarding or proprietary panels as here on 339. They required a considerable amount of maintenance and a radical solution of providing a new brick envelope has been adopted as at 340 which with its new tiled roof covering and remodelled garage, provides a compact neat looking, if rather basic, house.

Those with rolled steel sections fared better on the inspections than those with the lighter fabricated twisted and bent sheet steel sections. The studies showed that the vast majority of steel framed dwellings provided levels of performance not very different from many traditionally built dwellings of the same age. A small proportion of those in areas of severe exposure have suffered significant corrosion in parts of the structure which are well defined and, once access has been gained, is capable of repair. Provided the repairs are carried out, that raintightness of the envelope is restored and maintained, that there is no increased risk of condensation by inappropriate refurbishment, such as would be the case with the provision of insulation within the cavities, there is no reason why steel framed or steel clad dwellings should not give good performance into the foreseeable future on a par with the life conventionally assumed for rehabilitated dwellings of traditional construction. Bearing in mind that the BRE report is by now not far short of 20 years old and, consequently, the dwellings studied are that much older, it can be concluded that they need each to be treated upon their merits taking into account the findings of that report and the more detailed reports on individual systems subsequently made available.

Timber Framed Walls

Modern

Timber has been used as one of the materials to form the walls of dwellings since time immemorial. We are concerned in this subsection with the softwood variety and its use for that purpose from about 1700. This was roughly when the oak forests in the United Kingdom became depleted and no longer available as a source for the substantial framing which is the characteristic feature of the timber dwellings constructed in the 1300 to 1600s and which will be dealt with in the next subsection.

In the main, the oak timbering was replaced by brick for the outer walls of dwellings in urban locations more so because of the risk of fire than the shortage of the timber. Where this had been used for roofs and floors, it was being replaced in any event by imported softwood from elsewhere in Northern Europe. In rural areas, not subject to the requirement to build in brick or stone, a tradition grew up for building relatively inexpensive dwellings almost entirely of timber. To roofs and floors in softwood would be added softwood framing for the walls, covered externally with weather boarding and internally, first with panelling and later with lath and plaster. 192 is one such example as is 341. As an alternative to weather boarding, lathes could be fixed to the exterior and the upper floors tile or slate hung and the ground floor rendered. Such construction would pass muster as reasonably resistant to the spread of fire for dwellings in the smaller towns where requirements might not be too strict.

Weather boarded exteriors pose no problems in recognising timber framed construction and the hollow sound when the inner face of an external wall is tapped will confirm this. The

341: Weather boarded timber framed house of mid 1700s.

342: In quieter country towns and rural locations where there were no requirements to build in brick or stone to reduce the risk of fire, inexpensive dwellings were often built in the 1700 and 1800s of timber framing with a cladding of weather boarding, as here on the flank elevation of this 1700s house, or, as can also be seen, tile hanging on the front elevation. The weather boarding would either be painted or tarred.

343: This pair of two storey semi-detached cottages from the 1860s have painted weather boarding to the flank elevation and a rendered finish on lath to the front elevation, both on a timber frame.

344: The comparatively light framework of studs and diagonal braces forming the structure of the 1860s two storey semi-detached cottages shown on 343 can be seen from this photograph where internal lath and plaster has been removed but the back of both the weather boarding and the laths for the rendering can be seen. Careful consideration would need to be given on any refurbishment to the positioning of insulation, sheathing perhaps to strengthen the frame and a vapour barrier.

same hollow sound from the inner face of rendered or either tile or slate hung external walls, will also confirm timber framed construction. Such construction happily co-existed in this form outside the major conurbations, along with brick and stone construction, until about the 1920s.

It received boosts at various times. One was the period from 1784 to 1850 when a hefty tax was placed on bricks. Another was the influence of the philanthropists and the Garden Cities Movement, in the late 1800s and early 1900s, to do something about the poor standard of housing in country areas. Various competitions were held to produce designs for 'cheap cottages', many of the entries being based on timber framing.

The timber framed dwellings of the 1700s, 1800s and early 1900s were rather basic, foundations could be rudimentary, the frame lightweight and perhaps inadequately braced leading to racking, weak floors could be over-loaded and there would be an absence of damp proof courses and certainly no insulation. They need the most attentive of visual inspections and may be reluctant to show on the surface

any obvious signs of what could be going on well hidden from view. At the slightest sign of damp, rot or woodworm the surveyor has little option but to advise opening up to expose vulnerable parts of the frame, particularly those near ground level.

Timber framing in softwood was given another boost in the aftermath of the 1914 to 1918 War, when there was a shortage of building

345: An entry to a 'cheap cottages' competition held in 1905. The structure is timber framed with white painted weather boarding. Quite large for a cottage by the standards of the time, it could, conceivably, have been two originally, but refurbished and with a new roof covering is still giving good service after 100 years and with proper maintenance will continue to do so for many more.

346: A pair of two storey semi-detached local authority timber framed and clad dwellings built from prefabricated parts imported from Norway in the late 1920s. Free of any signs of structural distress, minor renewals only were required to some of the more exposed vertical boarding. The lad's shirt gives the game away as to location.

materials and labour costs were very high. This led to encouragement for the development and use of timber frame construction for the provision of local authority dwellings, including an attempt to import the parts to make up complete dwellings from Sweden which was scuppered by opposition from the trade unions. Even so, a number of authorities succeeded in constructing timber framed dwellings using prefabricated parts and components from both Sweden and Norway. Other local authorities developed their own designs and at least two private companies became involved, supplying dwellings for the private sector as well.

The timber framed dwellings produced in the 1920s again pose little problem for recognition, being mainly timber clad, but they are considerably more sturdy in construction. It is with these that the Building Research Establishment first comes to the aid of the surveyor with its studies of types and their characteristics as part of a wider study of most timber framing systems up to 1975, specifically built for local authorities since it is to these that BRE can gain access although, of course, the information gained applies

equally to those built privately for sale.

Three publications by BRE appeared in 1995 and are available as a pack, costing around £80. The first, entitled *Timber Frame Housing 1920-1975: Inspection and Assessment* provides an overall view of the construction characteristics of 11 types built between 1920 and 1944, 9 types built between 1945 and 1965 when the system building programme began to get under way and 34 types constructed between 1965 and 1975 when system building was widely used. The vulnerable parts of timber structures are described in detail over seven pages and a further seven pages are devoted to procedures for carrying out a site inspection to produce a 'structural survey' report, where an initial inspection is followed by opening up for a closer examination of those parts considered vulnerable. There are illustrations and summary information on 55 named systems. Timber framing systems were the only types of system building which continued in favour beyond 1975 and BRE estimates that between half and two thirds of a million such dwellings, between 2% and 3% of the total dwelling stock, were built during the 1900s, the vast majority of them since the mid 1970s, by which time about 100,000 alone had been supplied for the public sector. Of the systems used, 100 have been identified.

The *Inspection and Assessment* publication from BRE is accompanied by two others, *Timber Framed Housing Systems built in the UK 1920-65* which analyses the details of seven of the systems built between 1920 and 1944 and nine of the systems developed between 1945 and 1965. The third publication *Timber Frame Housing Systems built in the UK 1966-75* completes the study by investigating, comprehensively, the details of a further 12 systems.

While the BRE publications of 1995 contain invaluable information which should be available in every surveyor's office, not every known system is described in detail and for those that are, much of the information, as far as identification, details of construction and vulnerable parts are concerned, is derived from inspections involving invasive techniques,

347: A pair of two storey semi-detached local authority dwellings of the late 1940s. These are prefabricated timber frame and clad with vertical timber boarding and were imported from Sweden. They are reported as suffering very little deterioration over the years despite having received little maintenance and they are popular because of generous space standards and good insulation qualities. Some aspects do, however need upgrading.

348: These two storey semi-detached local authority dwellings of the1950s are something of an oddity. They are said to be timber framed but clad with prefabricated reinforced concrete (PRC) panels and hence were designated under the provisions of the Housing Act 1985. However, if the information was accurate and the structural elements were of timber with the PRC panels merely as cladding, then strictly speaking there would have been no reason to designate. They are described as of 'Tarran Dorran' type which would seem to be an amalgam of two somewhat similar systems. There would clearly need to be a very thorough investigation should a surveyor be asked to advise.

not available initially on a purely visual inspection.

BRE give a nod in the direction of a RICS/TRADA publication of 1984 'Structural Surveys of Timber Frame Houses', stating that the BRE inspection procedure augments the RICS/TRADA guidance, particularly where the latter leaves the condition of the dwelling in doubt. BRE points out that while many timber frame systems are essentially similar, each inspection and assessment must be tailored to the specific characteristics of the system being investigated. This seems to be a counsel of perfection in view of the number of systems which were used and is, in effect, only valuable if you know what the system is before you start. However, a surveyor carrying out an inspection envisaged by this series of books will, generally, not know whether he is to inspect even a timber framed dwelling, let alone one constructed on perhaps one of many different possible systems. He probably won't even know by the time he has got through the front door, since the vast majority of timber framed dwellings are clad in brickwork. Most of those that are comparatively easy to recognise as timber framed because they are timber clad were built before 1965 for local authorities, and there were only about 30,000 of those.

As with all inspections, a surveyor has first of all to establish whether the dwelling is of traditional or non-traditional construction. This aspect has been dealt with in detail in the subsection on 'Ascertaining Construction' and the surveyor will have established by assessing age, from the external material used, measuring and tapping that the dwelling is of a non-traditional type. Carrying forward his investigation at the site, there are certain features about timber framed construction which when encountered will lead the surveyor towards the conclusion that that is indeed the form, although it is highly unlikely that he will be able to identify the system at that stage by name. These indications are probably most noticeable in the roof space in the case of terraced and semi-detached dwellings and those where the roofs are gabled as against

hipped. The interior face of the gable wall of brick faced timber framed dwellings will show the stretcher bond of the brickwork, which will need to be tied securely back to either a timber framed spandrel panel or to a properly braced trussed rafter. Separating walls will generally be plasterboard faced on timber framing to produce the required level of fire resistance. The sight of brick or blockwork at this point is no indication either way in terraced or semi-detached dwellings where the roof is hipped, because a separating wall in these materials can be used when the rest of the construction is either conventional or timber framed. While in the roof space, it is worthwhile to try and see the top of the external wall at the eaves. With the aid of a mirror attached to a long handle it is sometimes possible to see the top member of the frame. In contrast to what can be seen in conventionally built dwellings, this is invariably planed as against being left sawn as the wall plate in other cases. The cavity would be closed off usually by means of a timber batten or mineral fibre cavity barrier, in timber framed construction.

Viewed from the exterior, a dwelling with

349: These boxes from the 1970s comprising two storey large double fronted terraced local authority dwellings are timber framed, clad at ground floor level with brickwork and at first floor level with artificial slates and with a flat roof of bituminous felt. The design was developed by a consortium of local authorities.

350: A two storey terraced development from the 1970s, developed by the same authorities as those at 349, but with the provision of pitched roofs, covered with tiles. The artificial slates are still present at first floor level and the arrangement of windows persists, the openings much too near the party wall line.

what seem deeper than usual to the windows and doors, is often found to be timber framed, as both are usually fixed to the timber frame and not the brick skin. With other much thinner forms of cladding, this factor tends to make the windows seem to be set much further forward than usual, while the overall thickness of external walls tends to be much less than would be the case if brick or blockwork had been used and dry lined.

If it is clear that the dwelling being inspected is of fairly recent construction, the sight of soft packing or space left below window sills is a sure indication of timber frame construction. This packing or space left is usually around

351: This terrace of two storey local authority dwellings was built in the 1970s but said to have been designed in the 1950s by Spooners of Hull, who supplied a substantial number of timber stud framed separately clad, in this case with brick, dwellings to local authorities as well as to the private sector. Another later design, known as Spooner Caspon, had large windows and monopitch roofs. The centre dwelling here has been refurbished by the local authority, the dwelling on the left is more or less as original while the one on the right has been refurbished by the 'right to buy' owner. The company features in the BRE lists as suppliers of timber framed dwellings in the period 1945 to 1965 when this terrace was designed. BRE found that the Spooner dwellings inspected, featured a general scarcity of wall ties, allowing the cladding to bulge and lean and in the case of gable walls resulting in collapse at times of gales. Expansive corrosion of poorly protected wall ties and steel lintels has contributed to problems while fire stops were found to missing to the separating walls.

6mm on the ground floor and 12mm on the first floor, and is to allow for differential movement between the frame, and cladding, caused by the initial shrinkage of the timber frame. The shrinkage is usually completed within a short time of construction and, accordingly, if the dwelling is more than say a couple of years old it may be that no gap is visible. The effect of an inadequate allowance for movement being left may be all too evident, however, in disrupted sills either broken, if tiled, or tilted upwards, if solid, and allowing water to flow backwards towards the window.

An unusual relationship between the level of the damp proof course and the weep holes, draining the cavity of brick or blockwork external walls, will alert a surveyor to the probability of timber framed construction. In conventional construction, they are invariably above the level of the damp proof course but with some timber framed systems, they are set below.

Internally, a hollow sound when the inner face of the external walls is tapped might immediately suggest timber frame construction but the surveyor must not forget that brick or block walls can be dry lined on battens, which will also produce a hollow sound when tapped. The part of walls above window and door heads should always be included for tapping where, paradoxically, dry lined walls will continue to sound hollow but timber framed walls will sound less hollow because of the presence of timber lintels.

The presence of all, or some, of the above indications will lead the surveyor to conclude that the dwelling being inspected is timber framed. He may be able to find out which of the many possible systems was used in its construction by enquiry from the owner, the local authority, a neighbour, or, if the dwelling is relatively new, from the builder or developer. If this does not produce the information, BRE suggests consulting Appendix A to its *'Inspection and Assessment'* publication, already referred to, where 51 systems are illustrated and their features summarised and, if that provides an answer, to confirm the system type from the BRE comparison reports covering the details of 28 of those summarised systems. If the system is successfully identified as one of those, then the detailed report should direct the surveyor to the most vulnerable parts of the structure for that particular system. On the other hand, it is debateable whether knowledge of the system used is really essential. BRE acknowledges that many are essentially similar and no particular system is singled out as being structurally worse than any other. Irrespective of whether the name of the system is known and after an inspection of all visible accessible parts any evidence of damp, whatever the cladding, any evidence of a shaky structure, such as popping of the nails of plasterboard or cracks, leans bows or bulges in brick or blockwork cladding, any signs of rot however slight in cladding of timber will leave the surveyor with no alternative but to advise opening up for a closer examination of the structure adjacent to the areas of concern. The same advice may need to be given, depending on circumstances, should there be any indications of alterations which may not have fully taken structural considerations into account.

On the other hand, if there are no visible indications of defects, the surveyor has no grounds for giving advice to open up the structure and it can be left to clients, when appropriately informed and whether they be purchasers or sellers, to consider if they wish to go further with investigations, should it be necessary in the absence of information, to determine the system used. That in itself might involve the necessity for some exposure to establish constructional details, or even doing so on the off chance of finding something wrong. After all the BRE cut off point for its studies was 1975 as far as the systems were concerned and there are plenty of systems which have been developed since then which are said to embody improved techniques, greater accuracy in assembly and a degree of quality control which, previously, had often been lacking.

Medieval

It is always important to consider the structure of a dwelling in its entirety and not just as a diverse collection of parts. This applies irrespective of the materials used in its construction or the way it is put together. Walls depend on floors and roof for restraint and one wall upon another by reason of the bond at corners. Nevertheless, framed as distinct from mass load bearing walls can be looked upon rather differently as if the casing is removed, be it brick or some other material, the frame still remains. This is so for the steel and modern timber framed dwellings already considered but, perhaps, even more so for the medieval timber framed dwelling.

Medieval timber framed walls are part of a series of interconnecting structural components each highly dependent on the other. The framing of the walls is an extension of similar framing making up the floor and roof structure and it is a continuation of those components which takes the load of both downwards to the support provided at or near ground level. The cutting of, or damage to, one component can cause untold damage to the structural integrity of the dwelling as a whole and the full effect can manifest itself quite some distance away. The installation of a brick chimney stack many years after the original construction is a case in point, leaving many a dwelling with a lopsided appearance. However, such is the strength, generally, of the form there is seldom any danger of collapse and careful strapping can do wonders to prevent further movement.

While a medieval timber framed dwelling where most of the panels between the studs are ill fitting or otherwise defective would be costly to deal with, the cost would be nothing like as great as where the frame was badly in disrepair through rot or beetle infestation. The hazards for the frame are, of course, from exposure to driving rain and the attentions of that unwelcome visitor, the death watch beetle.

Repair of damage is costly, not only because the skills to do the work are scarce but also because most medieval timber framed dwellings are either listed or in a conservation area, or both, and will draw the attention of English Heritage, Historic Scotland or CADW when repairs are required. Furthermore, such walls are often at their most vulnerable on account of their construction at ground level. Generally they were provided with a brick or stone plinth on which a sturdy base plate was

laid and to which the vertical studs were dowelled and pinned. There would be little or nothing in the way of a membrane to keep rising damp from the underside of the base plate from day one. Eventually sections attacked by damp from both below and from above, particularly on the elevations most exposed to driving rain, and also probably by death watch beetle preferring as it does damp timber, will rot and need replacement. If replaced with timber, adequate lengthening joints have to be provided to connect new and old sections to maintain the structural integrity of the frame. If they are inadequate or, even worse, brickwork is substituted for timber, as is sometimes found to have been done in the past, the frame at ground level will become weaker and may not be able to resist sideways thrust from the studs. Essential checks need to be made in these circumstances. On other elevations, not so much affected by driving rain, the top surface of the base plate may appear deceptively sound which will necessitate the inclusion of suitable warnings in the report.

On the required basis of inspection, the

352: Jettied construction on both the first and second floors on a dwelling of the mid 1400s. There is fairly close studding but it still requires some wind bracing. When such structures were built on either side of a narrow street, the occupants of the top floors could sometimes shake hands! The state of the street below was probably indescribable.

353: Close studding to the first floor of this dwelling above jettied construction, probably from about the mid 1400s, producing a very sturdy structure and although no wind bracing is provided, there is no movement apparent. The studs at ground floor level which would have been at the same spacing as those above have here been covered with a thick coat of lime plaster and pargetted with geometrical designs.

surveyor will examine all visible and accessible surfaces but there will no doubt be timbers which are visible but not accessible from the surveyor's ladder. Irrespective of whether there are signs, elsewhere, of rot or beetle infestation, the surveyor must advise that those inaccessible timbers be examined, by whatever means necessary, both for rot and beetle infestation and to check, as in other places, that there are no undue gaps between frame and panel which could provide easy access for water penetration. As to internal surfaces where the timber may or may not be visible, the surveyor must advise opening up if he sees any indication whatever that defects may exist in areas hidden from view. With no indications of any possible problems and, as where modern timber framed dwellings are being investigated, the surveyor must include warnings of what might be going on in hidden areas and stress that he is unable to provide insurance against the possibility of defects which may be present but hidden from view.

A line on a graph showing achievement in timber frame building in oak through medieval times would take the form of an arch. From fairly modest beginnings a peak was reached followed by a fall as the availability of the material declined either from shortage or appropriation for other purposes such as ship building. At its peak, the achievement was impressive, finding solutions in the design of layouts and construction and in the design of joints, to leave a legacy of sturdy and

Figure 42: Some idea of the strength and sturdiness of medieval timber framing at the peak of its achievement between about 1450 and 1550 can be gathered from these drawings of a gable elevation and two sections of a two storey dwelling on a corner site and with jettied construction to the first floor on the front and return elevations. It is easy to imagine what the dwelling would look like if all the panels of wattle and daub were removed, as here, leaving just the timber framing. It is not surprising that there are still a considerable number of such dwellings around often showing hardly any signs of structural problems except, perhaps, where some ill considered alterations have been carried out.

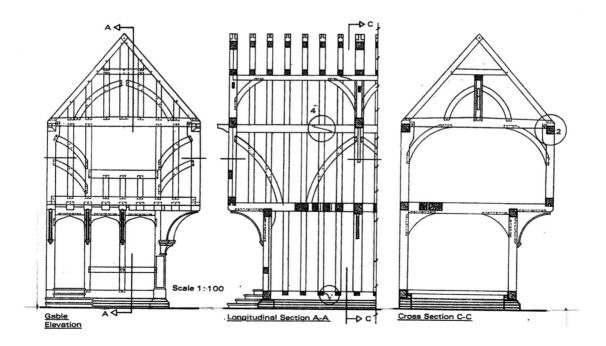

Gable
Elevation

Scale 1:100

Longitudinal Section A-A

Cross Section C-C

354: *The number 2 on the cross section CC on the right hand drawing of Figure 42 encircles the complicated five way joint between the external stud, wind brace, tie beam and rafter and the longitudinal wall plate. The wind brace is as it appears on the elevation of the gable on the left hand drawing but along the side elevation, it would be replaced by an arched wind brace connecting stud and tie beam. When the joint was fitted together there would be very little indication of how it was arranged apart from the position of the pegs or dowels holding the components together. On the longitudinal section shown on Figure 42, the number 4 encircles a scarf joint connecting two lengths of wall plate together and there is another scarf joint of different design connecting two lengths of base plate. Joints such as these, which can often readily be seen, sometimes introduce dampness when the timber shrinks, more likely to be found on elevations with a southerly or south westerly aspect.*

sometimes quite impressive dwellings, 400 to 500 years old. Impressive not only in size but also for construction in an organic material which many would say carried the seeds of its own destruction.

The peak of achievement in the late 1400s and early 1500s is characterised by dwellings made up of substantial storey height closely spaced studs set on a massive base plate and connected together at their top ends by a longitudinal wall plate running both ways in the direction of the wall, with other timbers laid across as floor beams. The floor beams could be jettied outwards if necessary on all four sides of a dwelling to carry a further range of similar studs at first floor level, the

355: *On either side of the peak period of achievement for timber framing in oak, from the mid 1400s to the mid 1500s, with its closely spaced studs and jettied floors, much simpler shapes prevailed. This two storey well preserved dwelling is from the late 1300s, before the peak, and is built on a much simpler box frame system with some wind bracing. Infilling panels are of wattle and daub at this time, before brick making became more general.*

356: *There are contrasting views on the inclusion of brick panels and their effect on the stability of medieval timber frame dwellings. BRE suggest that they provide added strength, whereas others are of the view that they add far too much weight to a frame that can easily be weakened by rot and beetle infestation. The panels of brick, laid in herring bone pattern, do not appear to have had too much effect on the stability of this late 1500s two storey dwelling, built a little after the peak of achievement for timber frame construction. However, the surveyor would need to advise inspection of timber out of his reach from a 4m ladder even if his own examination of timbers lower down revealed little evidence, if any, of rot or beetle infestation which would seem unlikely from the appearance of the timbers here.*

detail to be repeated at the top either as a basis for the roof framing or to carry yet another floor of jettied construction. Jettied construction all round produced a firmer structure and was a useful device in the narrow confines of medieval streets, enabling encroachment into the air space above the footpath so as to gain the maximum usable floor space from the site.

Both the earlier and later periods of timber framed construction depended on the box frame system. This was inherently weaker than the close studding of the middle period. Many of the dwellings of this form will exhibit the appearance of settlement, leans and bulges yet even when looking as though they would fall down at any time they often have a reserve of strength which should make the surveyor cautious of coming to hasty conclusions. A factual report on condition without speculation is required although if unduly worried, a recommendation to either buyer or seller to consult a structural engineer specialising in the analysis of timber framed structures should be made.

357: (below left) The flank elevation of this 1600s dwelling, built well after the peak of timber frame construction, is of very basic square panels with an infilling of brick in a chevron pattern and a mere two wind braces. Sturdy timbers were employed which is more than can be said for many of the timber structures of the time. The arrangement of the timbers in the Queen Post style roof can be clearly seen.

358: (below centre) Mock timber framing on this early 1900s two storey dwelling with brick ground floor. Again the surveyor must advise inspection of the inaccessible timbers by whatever appropriate means, for rot and beetle infestation even if none is found elsewhere and even though the cost of renewing a few planks would be relatively small. A timber treatment firm would immediately advise that the whole dwelling be sprayed with insecticide should a painter subsequently find a few holes on an external redecoration, whether it was justified or not.

359: (below right) Mock half timbering is still to be found of quite recent origin. This is on a 'right to buy' privately owned 1920s ex local authority dwelling of cavity wall brick construction. With new windows provided it is not in itself unattractive but fits rather uniquely into the locality, which was no doubt the intention.

360: Modern timber framed construction is mostly brick cased to look like other conventionally brick built dwellings. This could also be the fate of some medieval timber framed dwellings with claddings of either brickwork, tiles including the mathematical variety to look like bricks, weather boarding or lath and plaster. It is believed that this pair of 1400s dwellings were given a different appearance in the late 1600s, still in evidence on the one on the right and then the one on the left an entirely different make-over in the 1700s to the front elevation with a small moulded cornice, rendering in the shape of panels, a pedimented porch with fanlight and double hung sash windows.
The side elevation and no doubt the interior will give the game away as to the real form of construction.

361: This window opening in a multi-storey late 1800s dwelling is spanned partly by dual timber lintels, the outer face of which is positioned a mere 100-120mm from the exterior. If the wall was in an exposed position, driving rain could easily penetrate to the timber and induce rot. In such a confined and unventilated space it would almost certainly be dry rot. Most of the load from the 1½ brick wall above the opening is carried by the lintels but the outer half brick thickness is supported by the top section of an artificial stone window surround. This is made up from partly hollow clay units placed in a mould and cast in sections with a stucco type of cement mix, part of which can be seen on the right of the photograph before replacement.

Wall Details

Arches and Lintels

Openings for windows and doors in mass load bearing walls of both solid and cavity construction are a source of weakness. If they are numerous and too close together, thin, weak, narrow piers may be formed and if the openings are not adequately spanned across at the top by an arch or lintel of brick, stone, timber or steel may give rise to settlement and undue thrust on the supports on either side. This is not quite the case where framed walls are involved since at the construction stage, openings are normally provided in the panels. If additional openings were found to be necessary later, it would be foolish to cut and remove part of the frame, although it is

not unknown for this to be done. An error such as this is more likely to have been made internally to form an opening in a partition where it was not realised that the partition was framed in such a way as to be virtually self supporting. Sagging in overloaded floors can be the result.

Brick built dwellings are usually provided with brick arches which can take various forms, some more effective than others. The misnomer of a brick lintel being called a 'soldier arch' is unfortunate and the difference between a brick arch and a brick lintel should be recognised by surveyors and the correct descriptive term used. Both have near flat horizontal soffites, but the true arch has wedge shaped voussoirs while with a brick lintel the 'voussoirs' are set truly vertical, the bricks generally laid unshaped on end but, occasionally, on edge. The span of a brick

362: Settlement in the flat brick arches of the ground floor windows of this early 1700s dwelling, probably due to rotted timber lintels, has led to the brick courses above becoming out of level and the sills of the windows in the floor above, to fracture.

363: The rebuilding of brickwork to the left of the lower window on this illustration suggests that major problems have been affecting stability. These have disrupted the setting of the arch leading to fractured brickwork above, the crack extending through all brick courses and fracturing the sill of the window on the upper floor.

lintel should not normally exceed about 900mm unless provided additionally with a steel bar or angle. The flat brick arch is not greatly stronger than a brick lintel but will safely span an opening roughly half as much again which can also be increased by the use of a steel bar or angle.

Both flat arches, in gauged and rubbed brick-work, and brick lintels will be encountered frequently as will segmental and semi-circular arches. Less frequently encountered will be pointed arches in, among others, their Venetian and Lancet forms and the weaker, four centred, Tudor form. Semi-elliptical arches are considered the weakest for taking the load from above while those that take on more of an equilateral shape, where the rise approaches the span, are considered the strongest.

For dwellings built up to around 1920, one of the most frequently used forms of construction employed either brick arches or lintels in combination with timber lintels. The arch or brick lintel, for the sake of appearance, would carry the first outer half brick thickness of walling with either one, two, or more timber lintels to take the remaining load, depending

on the total thickness of the wall. This brought the outer face of the timber to within 100-120mm of the exterior. Exposure to driving

364: The very severe movement, perhaps due to differential settlement, causing the long vertical crack, which is wider at the top and which on the way passes round the left hand windows at both levels, has so disrupted one of the arches that a voussoir has dropped at least 75mm with consequential disturbance to the brickwork above. There is a possibility that the dislodged shaped brick could fall out to the considerable danger of passers by if not dealt with urgently, as a matter of serious concern, before due consideration is given to other repairs.

rain would often mean that the lintel would be affected by damp penetration through the joints of the brickwork and, with little ventilation in such a confined space, rot would eventually set in. The timber lintel would originally be sharing the load from above with the brick arch or lintel but when weakened by rot the bulk of the load would be transferred to the brickwork upsetting the stability. In the likely event of the arch settling, or being pushed outwards, brick courses above will settle and cracks may be induced.

Very popular in the latter part of the period up to 1920 were both real and artificial stone mouldings used for decorative purposes around windows and doors. If expected to span too great a distance because the supplier's advice was not followed, there will be a tendency to fracture and they will almost certainly do so, if there is settlement or subsidence in the piers between openings, or there is an inadequate bearing at each end.

Band Courses, Cornices and Window Surrounds

A dwelling with little or no decoration at all in the form of projections of any sort can look somewhat dreary, 385 is a case in point. However it has to be admitted, on the other hand, such projections can sometimes be the cause of problems. Ideally they should all be covered with a flashing if damp is to be prevented from entering the main structure and appearing internally. Another factor is that they can be a danger, if neglected, to the extent of falling off and causing injury.

To a degree, inspecting the exterior of a dwelling in the pouring rain, while unpleasant, can help to identify the places where damp penetration could be occurring. If inspecting on a dry day and there are damp patches on

365: Ideally the tops of all features projecting from the face of walls should be protected by a flashing, tucked into the main wall at or above the splashing level, normally considered to be about 150mm, dressed over the feature concerned and finished with a drip on the outer edge to throw water clear of the wall below. Such flashings are seldom found to the degree of those subsequently provided to this late 1800s five storey block of flats, suggesting, perhaps, that a lesson has been well learnt from past experience.

walls, the cause of which is not readily apparent, then a suggestion to this effect can sometimes yield good results.

Cornices in themselves, by their sheer size and weight can be a source of problems. In the past, all sorts of materials and methods were used in their provision among them stone, corbelled brickwork and building up with tiles and cement, the two latter usually being stucco rendered. Often some iron would be used for tying back and additional support and to cramp sections together. This can eventually rust and expand, causing severe cracks. Very few were provided with proper flashings and the weighing down of the back edge by the parapet would be inadequate to prevent the outer edge sloping down and pulling the top section of the front wall and the parapet forward.

Where maintenance of the dwelling has been neglected and access for inspection is limited the surveyor must advise that a close inspection be made.

Damp Proof Courses, Plinths and Air Bricks

Somewhere near the base of the external walls is where the surveyor will hope to find a sign of a dwelling's damp proof course (dpc). The incorporation of a dpc has been a requirement for all new dwellings since the passing of the Public Health Act 1875. It is no earthly use, however, for the surveyor to assume that because the dwelling he is inspecting is either known, or assessed, to have been built since that date, it necessarily has a dpc. What the law required is not always carried out and the surveyor needs to follow a routine, which may involve a bit of dashing in and out, to establish if he can whether there is a dpc and, if so, whether it is effective or not, irrespective of the age of the property.

Old dwellings can have a dpc installed. Damp proof courses installed are not always effective, some can wear out and others can be rendered ineffective by subsequent action. Effectiveness is a matter of demonstration with the moisture meter, but it is salutary to be reminded that the presence of damp can render a dwelling at law 'unfit for human habitation'. Since it is the external walls which are being dealt with in this section, it is what can, or cannot, be seen in the way of a dpc from the exterior which will, initially, be considered. Effectiveness is, strictly speaking, best dealt with from the internal viewpoint but as presence and effectiveness are inexorably linked, it is appropriate to deal with the latter aspect at the same time.

Damp proof courses were by no means a novelty in 1875, so that some dwellings would have already been provided with one from the time of their original construction before that date. They would probably have been mainly those dwellings not provided

with basements. With principal rooms on the ground floor, the marks of rising damp would have been all too evident. For dwellings with basements the provision of a dpc would be considered not all that important. The servants worked there, and some might even sleep there, but they would be expected to put up with any rising damp, while its effects would be unlikely to reach the ground floor.

Some of the earliest types of damp proof course are usually the easiest to spot from the exterior. Two courses of either blue or red engineering bricks laid in strong cement mortar are easy to spot, as even the red have a different texture and are probably a different shade of colour than other red bricks used for general walling. Two courses of slates laid to break joint in cement mortar are equally easy to spot. Neither are ideal for the purpose of providing an adequate barrier against the rise of dampness from the ground. The bricks and slates themselves are fine and impervious to moisture. It is the mortar in which they are laid which transmits the damp, more so if the perpends between the bricks are filled with mortar. Both brick and slate dpcs can be rendered ineffective, should there be any subsidence causing them to fracture.

Another early type of dpc which is easy to spot is of asphalt, laid on a level bed of mortar. This will produce a joint thicker then the remaining brick joints and, even if some of the asphalt has not been squeezed out to be highly visible as often happens, the thick joint in that position is a good indication of the presence of an asphalt dpc. However, the provision of an asphalt dpc near the base of all the walls of a dwelling does provide a plane of possible slippage and movement of the structure horizontally above the level of the dpc has been known to occur, particularly on sloping sites. This can show up as a slight overhang on one side of the house while the opposite effect will be apparent on the other side.

A strip of lead makes an excellent dpc and has, of course, been available for the purpose over a very long period indeed. It is thin enough, however, to be embedded in a brick joint without anyone being aware that it is there at all. It is only when it is arranged so that it projects slightly beyond the wall face that it can be seen. However builders seem reluctant to allow this and prefer, for the sake of appearance, to leave it so that the joint can be pointed in the same way and at the same time as all the other brick joints. Lead used for this purpose in new or near new dwellings needs to be coated in bitumen or tar to protect it against the action of the lime in the mortar and, of course, should be of a thickness appropriate for its purpose as set out in BS8215:1991:Code of Practice for Design and Installation of damp proof courses in masonry construction.

The thin nature of many of the other materials used for dpcs, both past and present, such as those of bitumen with a hessian, fibre, felt or asbestos base, whether incorporating a core of lead sheet, along with those comprising plastic membranes makes it equally difficult to detect whether they have been used or not. There is a further drawback to not leaving any indication of their presence. This is because pointing the outer edge of the joint can undo the protection which a dpc installation provides because of the bridging effect. It will allow rising damp to by-pass the dpc and appear higher up the wall, albeit perhaps to a limited extent.

At the height in the external walls where the surveyor would hope to see signs of a dpc, he might see instead a line of holes, either well or badly made good, or in other instances, a copper band secured to the walls and extending right round the house. These would be signs that something has been done to deal with damp stains internally. In the first instance this would be by means of insertion of a chemical dpc, in the second case, by a system based on electro-osmosis to prevent dampness rising in the walls above a certain height.

If the surveyor is reasonably certain that he has located the position in the wall of a possible dpc, even if he cannot determine the type with any degree of certainty, he should then proceed to trace that position round the house and note its relationship, at all points, to ground or paving level and to all other features that could be a source of bridging, steps, a patio, and banked earth by the keen gardener, all possible sources of damp.

Strictly speaking, evidence of damp caused by bridging of the dpc is in the category of penetrating damp, not rising damp, and the surveyor should distinguish between the two. It is usually a great deal cheaper to eliminate the causes of penetrating damp than rising damp due to a lack of, or a defective, dpc. BRE are of the view that the incidence of true rising damp is less than is generally thought and that most damp in the lower part of the walls of dwellings at ground floor or basement level is due to other causes. Bridging has already been mentioned but another cause is the presence of hygroscopic salts in the plaster. These could have been deposited in the wall originally by rising damp which has subsequently been eradicated but the affected plaster has not been renewed. On damp days the salts take up moisture from the atmosphere and the plaster looks, feels and registers as damp on the meter. Renewal of the plaster is usually necessary for a minimum of a metre above floor level and is a vital part of the process of installing a dpc in an existing wall. The situation internally will not necessarily be improved until this has been done.

Irrespective of whether the surveyor finds evidence of the presence of a dpc in the external walls, he is required to test with his moisture meter for any evidence of damp within the dwelling. If there is a dpc this use of the meter will demonstrate whether it is effective or not. If there is no external evidence of a dpc, the surveyor might well consider pursuing this aspect internally at this stage. In many dwellings the lack of a dpc will reveal itself fairly readily by high readings on the meter but, in others, it may not be so clear cut. Steps to hide the rising damp or lessen its incidence may have been taken. Lining the walls with metallic foil backed paper is one method which nullifies the use of the moisture meter by short circuiting the probes. Securing bitumen impregnated lathing to the walls, plastering and decorating is another although tapping will reveal a sound less solid than when the inside of the external walls are tapped elsewhere. A method claimed to lessen the effect of rising damp consisted of installing absorbent hollow tubes at low level which ventilate to the outside air. There are methods of installing chemical dpcs which when carefully carried out and all areas disturbed are made good, are virtually undetectable. Clearly there is a need to question the seller if there is no sign of damp and no evidence of a dpc after exhaustive use of the damp meter. Of course, it is highly probable that the seller having seen all this going on, will be only too happy not to wait to be asked but will be trumpeting the success of whatever damp proofing system the dwelling possesses. Details of the system and particulars of any guarantee which accompanied the installation will need to be obtained so that the surveyor can provide his opinion on its likely long term effectiveness.

Dampness, or firm evidence that it is merely being hidden is a category of defect requiring urgent attention and leaves the inspector with

no option but to advise further investigation into the most appropriate form of eradication. Discovery at the inspection stage before the preparation of a Home Condition Report, gives the inspector an opportunity to discuss with the seller or his agent the implications. Is the inspector to proceed and the report lodged containing the finding that there is damp and advice that further investigation is required? Is the dwelling to be marketed on that basis? Alternatively, is that further investigation to be carried out and although the recommended work may not be undertaken, an estimate be attached to the report? Yet again, the seller may have the work done to the satisfaction of the inspector and the logical sequence of events could be set out in the Home Condition Report. Obviously following the latter alternative would be likely to secure a sale on the best terms but not all sellers would necessarily have the capital to have the work undertaken or would wish the disturbance involved with replastering. The middle road would provide a reasonable solution in these circumstances with the seller's asking price adjusted to take account of the estimate. If the seller was not prepared to proceed on

366: Evidence of the probable insertion of a chemical dpc in a house built some 20 years or so after legislation requiring the provision of a dpc came into force but where there is no evidence that one was ever provided. Even now the effectiveness of the comparatively new installation is a bit hit and miss and, ideally, something better should be installed and the exterior making good far more effectively carried out.

either basis and wished the surveyor to proceed regardless, it would need to be explained to him that complications and delays would be likely to arise on the sale.

Remembering that it is the presence of damp and not the absence of a damp proof course that can render a dwelling unfit for human habitation, failure, if that is the right word, by the surveyor after exhaustive searching and use of the moisture meter to find any evidence of the presence of either should make him very cautious of the words he uses in his report. He cannot be expected to test for dampness every few millimetres near the base of all walls in a dwelling, but he should know, having regard to the age, type and general condition of the dwelling coupled with his inspection, where the most likely places for rising damp could be. He needs to keep a careful note of all the positions where he has tested for damp and, inevitably, he will need a diagram for this purpose. If the terms of engagement require testing at certain intervals then he must be sure to have at last fulfilled that requirement. Unlikely though it may seem, there are dwellings with no dpc and no evidence of damp at or near the base of the walls but, fortunately perhaps for the surveyor's appointment book, they are rare. What could be lengthy procedures are more often short circuited by the finding of damp.

The relationship of four of the items at or near the base of the external walls of most dwellings, dpc, ground level, plinth and air bricks is important and requires some consideration by the surveyor. Already clear is that any plinth should be below the level of the dpc to avoid bridging. In turn the damp proof course should be at least 150mm above the level of any paving, the generally recognised limit of splashing from rainfall. While it should also be below the level of any timber built into or attached to external walls

it should also be below any wall plates on top of sleeper walls used to support a joisted hollow timber ground floor. This puts the ideal level for the top surface of a hollow timber joisted ground floor at about 350mm above external paving level, roughly two steps up to the front door. These are ideal for positioning a long narrow air brick, or a series of smaller ones, to introduce a through current of air below the entrance hall, provided always, of course, that there is a corresponding air brick at the rear. To ventilate the general area of hollow timber flooring satisfactorily, other air bricks should be positioned immediately above the damp proof course in sufficient numbers for an adequate amount of through ventilation. Testing whether there is a through current of air is necessary and can be done with the aid of long matches when on all except the very stillest of days, there should be some indication of air movement. Departure

367: The relationship of items to each other at the base of the external walls, such as damp proof course to ground level, to the level of a hollow timber ground floor and its essential ventilating air bricks and to any plinth is vital to the satisfactory performance of walls and floors at this point. The surveyor needs to give them his full consideration and departures from the ideal should provide him with the spur to further investigation into any observed defects. For example there is no sign of a dpc in this square bay window, there is a plinth extending above the level of air bricks which appears damp ridden. It would be very surprising if a clean bill of health could be given.

from the ideal relationship at the base of the external walls can introduce the risk of problems and the surveyor need to weigh up their significance and assess their effect.

Of course, all the foregoing presupposes that there is, indeed, a hollow suspended timber ground floor but there are quite a number of dwellings, particularly some of those built in the late 1940s and 1950s, which were provided with solid floors. Here the ideal arrangement on a level site is a damp proof membrane in the floor lining up and lapped with the dpc in the external walls, bringing the top of the floor to a little more than 150mm above outside paving level. On a sloping site, matters become more complicated and as the dpc changes level so there should be a vertical dpc to link floor membrane and wall dpc. There is more opportunity then for miscalculation, errors and subsequent changes in ground level by the owner, all of which can lead to bridging and be a possible cause of damp patches appearing at low level on external walls.

Applied Surfaces

Tile, Slate and Shingle Hanging

It may come as a surprise to some that the only forms of construction capable of resisting 'Very Severe' exposure conditions, as set out in BS5628:Part 3: 2001, are two types of what are termed 'external cladding'. One of these is 'weather resisting' backed by a 'moisture resisting layer' and comprising tile or slate hanging, natural stone, cement based products or wood which has overlapping dry joints. Recognition indeed that it is the joints in masonry construction that mainly provide the path for damp penetration.

Tile, slate and shingle hanging will be

368: Tile hanging to the first floor of a two storey terrace house of the late 1800s with brickwork to the ground floor in stretcher bond. This suggests timber frame construction which might be confirmed by a sight of the inside face of the gable wall. The damaged area of tile hanging on the corner needs repair with angle tiles to match the original.

369: An attractive arrangement of tile hanging to a late 1800s dwelling comprising alternately two courses of Arrowhead to three courses of Club shaped tiles, the variation in colour adding to the effect along with the ornamental carved bargeboard. The junction of the tiling with the double hung sash windows is crisp as the tiling is carried across the face of the box frame but provides a somewhat insubstantial appearance. Although it could be effective it would need to be checked since it is the junctions with other features which are very often a weak spot.

encountered frequently as the external cladding of dwellings particularly above ground floor level and provided they are kept in good condition should provide satisfactory service for many years. For new or near new construction information on the materials has

370: The opportunity for attractive patterning is not so great with slates as it is with tiles and special slates for turning corners are obviously not possible. Damage here needs urgent repair to restore protection against damp penetration.

371: Old centre nailed slates in process of being rehung on new battens correctly set out to provide three thicknesses of slate at the lap, which would appear to be about the minimum expected to be satisfactory when hung vertically in a location bordering on 'Severe' as far as exposure is concerned. Are the battens treated, what is the underlay and what type of nails are being used are some of the questions to be asked.

372: Shingles hung vertically at first floor level on a pair of 1930s semi-detached houses probably as originally provided, the timber used, now weathered to a silver grey shade. When backed by a 'moisture resisting layer' shingles, vertically hung, are a form of wall construction considered capable of coping with exposure conditions categorised as 'Very Severe', though not required to do so in this location.

373: Red cedar shingles hung vertically on the first floor of a detached dwelling built in the 1930s in an 'olde world style' but somewhat let down by the rather mean overhang to the thatched roof and the characterless windows. The remainder of the elevations are covered with overlapping horizontally hung waney edged boarding and as long as all the walls had been provided with a moisture resisting layer beneath the cladding would, perhaps, surprisingly be classed as satisfactory construction to cope with exposure conditions in the 'Very Severe' category according to BS5628:Part 3: 2001.

already been provided in section 3 for where they are employed for roofs.

Not so frequently encountered and indeed probably rarely so on dwellings will be the alternative form of construction to cope with 'Very Severe' exposure conditions. This is 'external cladding' which is 'impervious' rather than just 'weather resisting' and can comprise metal, plastic, glass or bituminous products which are jointless or have sealed joints.

One of the main problems with the vertical hanging of tiles, slates and shingles is that they can easily be the subject of impact or vandalism at ground floor level. For this reason the form of construction is usually accompanied by a ground floor of masonry and which would only be considered capable of coping with the lower levels of exposure category depending on the materials used and the thickness of masonry employed if solid. However, if of cavity construction 'Severe' or 'Very Severe' could be accommodated provided the cavity width was increased.

Weather-boarding

A weather-board, or clapboard, cladding to a softwood timber frame, in white pine, hemlock or spruce has been a favoured method of weather proofing since around the end of the 1600s. In the early 1900s Western Red Cedar became available for the purpose and has largely superseded the use of other woods. It has also been used when fixed to battens as a form of weatherproofing to walls of inferior brickwork which in itself has proved to be unreliable as a means of keeping out the weather.

375: Weather-boarding covering all the elevations of a two storey dwelling of the 1920s unlikely to have an underlay of a type which would bring it up to the standard of a dwelling able to cope with 'Very Severe' exposure conditions but nevertheless still able to provide weather proofing superior to that of inferior brickwork.

376: Shiplap boarding, stained black, on a detached two storey house of the 1980s. Contrasting panels of boarding were a fashionable design feature from the 1960s onwards, often providing problems at the junctions with other materials unless carefully detailed. Here however the boarding covers the bulk of the elevations above a partial brick ground floor, rather more than a plinth and therefore probably precluding the dwelling's construction as being suitable for an area where exposure conditions are 'Very Severe'.

374: Not usually suffering from the disadvantage of impact or vandalism at ground floor level, as is the case with tile, slate and shingles hung vertically, weather-boarding over a 'moisture resisting layer' will also cope with exposure conditions in the 'Very Severe' category. This two storey dwelling from the late 1700s, reclad with cedar boarding in the early 1970s could well have the appropriate underlay but unless it is possible to check the dwelling's capability in this regard would have to remain as no more than a reasonable speculation.

Examples of weather-boarding on timber frame dwellings have already been illustrated at 192, 342, 343 and 345 but other examples are included here since when combined with a moisture resisting layer it is a type of construction appropriate for resisting the weather in areas of 'Very Severe' exposure. It has the advantage over the likes of tile, slate and shingle in similar exposure conditions of not being prone to impact or vandalism at ground floor level.

Painting or the application of tar or other forms of preservative to the bare wood is an essential measure to protect the timber from the ravages of the weather. Flaking paint and bare patches of timber are a source of damp penetration and will eventually set up wet rot in the timber.

Self coloured PVC-U plastic boarding can be manufactured which when fixed, can provide the same effect as white painted timber weather-boarding. Trumpeted as maintenance free it has to be considered a fairly short term product and will lose its sheen, discolour and become brittle in a far quicker length of time than well maintained timber weatherboarding. Plastics are not, however, included as one of the materials which with overlapping joints and an appropriate underlay provide construction

377: Ceramic tiling to three of the six floors of a block of flats built in the 1990s. The tiled first, second and third floors are devoid of projections, even to the extent of the sills for the windows, so that a continual stream of water will run down the face of the impervious surface during rainfall seeking a path through the joints, which, because of the size of the tiles selected, are even more numerous than would be the case if impervious glazed bricks had been used. The surveyor would need to give warnings, not only of possible damp penetration but also of tiles being loosened because of frost action and becoming a danger.

able to cope with 'Very Severe' exposure conditions.

Ceramic Tiling

It is a puzzle why some designers persist in cladding their buildings with ceramic tiles after the experience of the 1960s and 1970s when such cladding, more or less, routinely fell off after a very short time. The expanses of cement and sand substrate which remain with the pattern of where the tiles were once pressed into place before grouting are a constant reminder because nobody has bothered to replace them.

If a surveyor has the task of inspecting a dwelling the majority of which is encased in ceramic tile he will need to be very wary indeed. Water streams down the impervious surface, soon to find a way in through the joints to the back of the tiles where it freezes, expands and forces off the tile. One tile off and the process becomes unstoppable. Often it is started by the bedding down of a newly completed structure. The slightest movement is usually sufficient to cause fracturing in the tiles. The surveyor might be able to do some judicious tapping to see if sections of tiling have become loosened but, as with some of the other applied finishes, it may be a case of warning of what might be happening at inaccessible points and advising a closer inspection.

Rendering

Rendering has been used over the centuries mainly for two purposes, first, as an aid to keeping the weather at bay and, second, as a way of improving appearance.

Probably all medieval timber framed dwellings

started life with the panels filled but the frame left exposed. If the frame was of storey height closely spaced studs where there was an ample supply of oak, the basis for the filling would be on split oak laths sprung and wedged into grooves left in the sides of the studs. If the supply of oak was running out or in areas where it was sparse, the panels were more square in shape and oak staves would be sprung into holes left in the horizontal members to provide a framework for hazel boughs or laths to be woven through and interlaced to form a close basketwork pattern of wattle. Panels would then be daubed with a mixture of wet clay with added flax, straw or cow hair thrown on from both sides in layers and, when dried, completed with a coating of smooth lime and hair plaster. The panels did not always end up draught proof and as a result over the years many were subsequently plastered all over, including the frame, to improve comfort. The coats of plaster were much thicker than is common today and lent themselves in some areas to quite fanciful decoration known as 'pargetting'.

379: Pargetting on this two storey timber framed but wholly plastered dwelling was probably carried out in the 1700s when the exterior was remodelled to include double hung sash replacement windows. The panels are mainly of geometrical patterns in basketweave and sunrise designs with a more elaborate panel of foliage between the windows.

Other than on cottages, rendering went completely out of fashion for the brick and stone terraced dwellings of the late 1600 and 1700s but came back into favour in a different form for about 50 years into the first half of the 1800s. Initially smooth it would be incised to imitate the joints of stone ashlaring to cover the front elevations, but sometimes the

378: In some parts of the country the thick coating of lime plaster on the panels, or even covering the whole of many timber framed dwellings, was often modelled and relief designs formed of foliage, heraldic figures, animals and the like, a technique known as 'pargetting'. Here the pargetting and the adjacent shell door canopy is dated 1690, presumably to celebrate the arrival of William and Mary on the throne, but is worked on a dwelling of a much earlier date.

380: In the first half of the 1800s there was a strong demand for smooth rendering incised to imitate the joints in stonework to cover over cheap brickwork or, as here, unfashionable timber framing.

381: The full effect of painted stuccowork is reflected in the long lines of four, five and six storey terrace dwellings typical of the mid 1800s of which this is a well maintained five storey example. The full array of cornice, band course, window surrounds, full width first floor balcony, reticulated ground floor and columned portico is present. Compared to some, however, it is comparatively light on decoration, no ornamental balustrades for example but, even so, expensive to maintain.

382: Elaborate stuccowork when neglected and damp penetrates to force off sections can be a danger to passers by, as here. Until assurance can be given that there is no danger, the surveyor should advise that scaffolding and fans be put up.

383: A good period for the quality of rendering, both as to materials and workmanship, came in the early 1900s, before the general practice came about of incorporating Portland cement in the mix. The Arts and Crafts movement favoured cottage style dwellings with a coating of a thick rough textured rendering, as here.

whole of, the terraced dwellings of the period, some of which were popularly known as 'plaster palaces' on account of their overall smooth stucco rendering. Some of these dwellings were painted early on in their lives and this practice became common. The upkeep of the rendering and the cost of the regular painting required can be a hefty burden on owners but if neglected and damp gets behind the rendering, large hefty chunks can become loose and fall to the danger of owners and passers by alike. The surveyor needs to be liberal with his warnings as inevitably there will be substantial areas which are inaccessible without long ladders, scaffolding or hoists.

Later the 1800s saw stucco rendering in the main confined to the lower part of the front elevation of the larger terrace dwellings at ground floor and basement levels. At the same time the Arts and Crafts movement of the late 1800 and early 1900s favoured a return to cottage styles of building and roughcast, still based on lime but with a coarser aggregate to produce a rougher texture became fashionable again though it had continued in use as a protection of the structure in areas of 'Severe' and 'Very Severe' exposure. Harling in Scotland, for example.

The inclusion of too much Portland cement in the mix for renderings was not a good idea

384: The considerable expansion of the suburbs in the period 1920 to 1940 with the typical three up, two down semi-detached dwelling and the continued use by most house builders of one brick solid walling for the external walls saw a considerable number finished at first floor level with pebble dash rendering. With a mix containing a high proportion of Portland cement there was usually a considerable amount of shrinkage which caused cracks and allowed damp to penetrate and assist any tendency to sulphate attack. Once this got under way, large areas could become defective. On this pair of semi-detached dwellings, not entirely typical of the period, the rendering has had to be totally renewed.

385: While beauty, as they say, is in the eye of the beholder and, strictly speaking, appearance does not form part of a report on condition, it is an undoubted fact that a bright, cheerful looking dwelling will sell quicker and may even achieve a higher price. It is difficult to say precisely what makes this 1920s detached dwelling look quite so gloomy. It is doubtful that repainting the rendering, obviously needed, or even restoring some of the vegetation which appears to have been carefully removed will do the trick. Perhaps the early metal windows are so rusty as to need renewal and selection of a different type might help. Everybody probably has their own idea. Maybe it is the width of that gable?

and the period 1920 to 1940, extending also into the 1940 to 1960 period, saw poor performance from the mainly pebble dash and smooth coating of the time. Applied in much thinner coats than hitherto, the Portland cement often made it too strong for the substrate and possessing a stronger shrinkage element on cracking and falling off it would often bring with it parts of the brickwork to which it had been applied. The situation in regard to renderings has not changed a great deal since except that there does appear to be a great understanding of the types of rendering most suitable for the different categories of substrate.

Mock Stone

The idea of aspiring to be able to afford a stone dwelling as distinct from one of brick or some other material which was rife in the latter

half of the 1700s and the first half of the 1800s rose again to prominence in the second half of the 1900s. Whereas the earlier manifestation took the form of rendering the brick and then making it look like stone by incising a pattern of joints on the surface, the modern incarnation takes the form of applying thin panels of a material to imitate stone to brick or other surfaces. The panels are moulded to look like squared rubble stonework with individual 'stones' sometimes given a contrasting colour.

Some panels are more convincing than others in what they purport to represent but it is unlikely that any would fool a surveyor into thinking the mock stone was real. That might just be possible by applying the panels to a totally detached house but generally when applied to a dwelling either semi-detached or in a terrace they stand out like a sore thumb. That probably is the intention so that when

386: Mock stone applied by the purchaser, under the right to buy provisions, to the end house of a two storey terrace of Hawthorn-Leslie steel framed system built dwellings. The veneer does not seem to be wearing well on the single storey entrance addition, perhaps because of a lack of heating internally leaving it much colder than elsewhere. The adjoining house, clearly now also privately owned with its small pane replacement windows, also has a small panel of mock stone below the new bow fronted projecting ground floor window. The roof covering to the new window looks somewhat unkempt. Problems perhaps from damp penetration?

388: Mock stone applied by the right to buy new owner of a two storey attached house built in the early 1950s in a pedestrian enclave. Because of the stepped layout the 'improvements', which also include the addition of an asymmetrical roof porch and the removal of the shiplap boarding below the left hand window do not look anything like as incongruous as they could do in other circumstances. The removal of the projecting sill to the left hand window could however pose problems for the future.

387: Mock stone applied to a brick built two storey terrace house of the late 1800s, looking out of place compared with its neighbour on the left which is pretty well as originally built and even when compared with its neighbour on the right where the original brickwork has been painted pale blue. The routine for relating details at, or near, the base of the external wall should reveal whether in applying the veneer, air bricks have been covered over or the damp proof course bridged, faults reported from other locations. It is curious to see that a representation of an arch with stone voussoirs the size of bricks is provided over the windows, highly unlikely to be encountered with genuine stone construction. Even stranger are the 'stone' lintels over the windows in 386.

389: Perhaps almost the ultimate of inappropriateness is the addition of mock stone to the first and second floors of this four storey brick built end of terrace dwelling of the early 1800s which along with other alterations completely changes its character. Apart from introducing the possibility of problems for the occupants, one wonders how the local authority is getting on with its enforcement procedures and, for that matter, what will the dwelling be like when it succeeds.

speaking to the seller, the surveyor perhaps ought to be careful of what he says!

While some of the mock stone panels seen have been of fibreglass, the sheets merely stuck on with adhesive, others are in the form of a mix of sand, cement and pulverised fuel ash with a colouring agent pressed on to a cement and sand coating. Whenever materials for walling are used which have a comparatively smooth face, there is the same likelihood as there is with shiny bricks or ceramic tiles of water running down the face, finding its way through the joints to affect the interior or freeze, expand and blow off the applied material. This has happened at a number of locations and leaves the surveyor with no option but to advise inspection of those areas of applied mock stone which are out of his reach.

Painting

Painting the outside of the external walls of a dwelling, as distinct from the woodwork, is fairly easy to do but rather difficult to reverse. There is a difference however, in that once done the burden of repainting at regular intervals becomes a fact of life.

It is the renewal in a time of change in the formulation of paints that sometimes causes problems. When there was a long local tradition handed down of what to do and whoever did it before would probably do it again, the right sort of paint would almost automatically be used. Limewash would be applied over limewash, oil paint on to oil paint. Now many of the constituents of the older types of paint are frowned upon, lead and solvents for example. The disastrous results of painting a cottage built of cob with an impervious coating of plastic paint have already been touched upon and stuccowork painted at regular intervals over the last 150 years or so will not take kindly to being repainted with a water based paint currently being strenuously pushed as being safer to use. Old established professional decorating firms will know these 'wrinkles' but much repainting is a DIY activity or, at best, carried out by small firms who quoted the lowest price and who may not be

aware of the pitfalls.

Surveyors should be able to put their finger on what has gone wrong when they see the effects of ill-considered decisions on what paint to use on both old and newly painted surfaces. They should also be able to identify another effect of automatically accepting the lowest estimate, that of failing to check the specification, which hopefully accompanied it and, if satisfied, seeing that it was carried out. Generally even on the smallest of jobs, the painter will provide a little more information than just the estimated price and, of course, in most cases this will cover the woodwork of windows and doors and the metal of pipes and gutters as well as masonry, if the latter is indeed included which is not necessarily always the case. The owner may not give this specification the attention it deserves but, at the very least, it should include the cleaning of surfaces, their preparation, types of paint and number of coats.

Defects in the form of flaking, bare patches or the separation of coats are easy to recognise and if accompanied by a fading of colour or dulling of the sheen where gloss paint has been used would indicate that the interval for renewal has been exceeded and additional costs for preparation will be involved when that renewal is carried out. BRE puts renewal for solvent borne paints at around four to six years and for water borne paints at slightly longer five to eight years, including the water borne cement paints normally applied to masonry. If paintwork is in good condition but not looking as though it has been applied recently, the surveyor needs to make an assessment of when the next maintenance repainting should be carried out, differentiating if necessary between different areas.

Should flaking and separation of coats of paint be evident on surfaces which appear to have been comparatively recently redecorated indicated by the colour being still fresh and an absence of dulling to the sheen, the surveyor should enquire from the owner when the work was carried out and ask for a sight of the specification and the estimate given at the time. He might then be able to identify the cause of failure.

Where painting has been newly carried out and no flaws are apparent, the surveyor can do no more than report the fact. The basic problem in these circumstances is that however ill-considered the decisions which may have been taken about what paint to apply and the standard of workmanship, the surfaces will probably not show any effect if they were wrong or inadequate for about a year or 18 months. It is essential therefore for the surveyor to ask for the specification and estimate covering the work to be produced if possible and for him to comment on the likely performance and durability. If he is dubious about the contents of the documentation he can include his reservations in the report. Even if he is satisfied on both specification and estimate he was still not present to supervise the work as to preparation number of coats etc. and needs to make this point clear in the report

Vegetation

There is quite a bit to be said for the odd climbing plant loosely wired, but not entwined, to a sturdy trellis attached to a batten fixed to the external wall of a dwelling. The colour of foliage and contrasting blooms, some with a heavy scent on a warm summer's evening, can be captivating. The advantage of a trellis screwed to a batten is that it can be taken down when required for the repainting of external walls and then put back. Even some of those plants which can run riot

without any help by way of support can look attractive but they do have a cardinal advantage and not just to a surveyor endeavouring to carry out an inspection, see 1 on p37. They all help to retain moisture against the wall reducing the opportunity for evaporation and they can clog gutters and rainwater pipes. Some are even destructive. While Virginia Creeper attaches itself to masonry by suckers, turns attractive shades in autumn and dies down in winter, ivy is a great deal tougher with roots that burrow into the joints of masonry, growing all the time and can eventually dislodge whatever gets in its path. The surveyor's advice must be to get rid of it if it is already causing damage and even if it is only likely to do so in the future.

As for Virginia Creeper and other plants such as Wisteria which climb and can attach themselves by entwining around anything that is available, pipes for example, they need managing and it is better to control their activity than allow them to run riot.

Windows

It is appropriate to include a subsection on windows in this section on External Walls since they are another part of the structure which helps to keep out the wind and the rain, the heat in and cold out. Windows tend to be dealt with in all the forms of report under consideration in a separate paragraph as a summary pulling together the results of an inspection of each window both from the exterior, physically at ground level or by binoculars and from the interior where each should, and hopefully can, be opened. This is a necessary part of every inspection. The window can be inspected from the aspects of safety, security and performance while the surrounding area, the sides and above and below the window can also be examined. Furthermore, a view can often be obtained of features not visible from elsewhere, the roofs of extensions for example.

For the making of windows, wood has reigned supreme throughout and it is estimated that 60% of the total stock of dwellings have windows of wood construction, roughly 15% of the double hung sash variety produced mainly between 1700 and 1920, with the remainder in casement form. The joinery manufacturers produce something like 1.5 million windows a year to this day. Other materials, of course, have also been used, iron for a while in the 1600s and again after the Industrial Revolution on a small scale but it was not until the 1920s that windows made from rolled steel sections began to encroach on the dominance of wood. Now metal windows, mainly of steel but also of aluminium are estimated to be installed in about 15% of the total stock of dwellings. The other material which has come strongly into the reckoning since the 1970s is PVC-U and windows of this material now grace about 20% of the total. As to glazing the vast majority are still single glazed but it is estimated that about 30% of all windows are now double glazed, a proportion likely to increase fairly rapidly in the future.

Since the early 1980s, it has been possible to purchase windows in all the above materials in five different categories according to their performance for weather tightness. This, for the first time, enables a surveyor inspecting a new or near new dwelling, or one where there is evidence of replacement windows being installed, to question the owner and preferably, examine the specification for the windows supplied. The surveyor will then be able to include in his report, information as to whether the windows comply with the exposure category appropriate to the location, as defined in BS6375-1, which sets out the

performance requirements for windows when tested in accordance with BS5368-1, -2 and -3 Methods of Testing Windows for -1:1976 Air Permeability, -2:1980 Watertightness and -3:1978 Wind Resistance.

The categories for windows are related to wind speed during the most severe storms expressed as a design wind pressure in Pascals (Pa). The lowest two categories, less than 1200 Pa and 1200 Pa cover windows which are suitable for installation in low rise dwellings up to three storeys in height, situated in built up areas or sheltered countryside. Windows in the next category up, 1600, are also appropriate for low rise dwellings but situated on exposed hillsides and high ground in open country. The two highest categories, 2000 Pa and over 2000 Pa will only be of interest to surveyors in the most exposed locations likely to suffer the severest storms, for example north west Scotland and some south west facing coastal areas in England and where for new construction only windows in these categories should be installed.

Relating a dwelling's overall form of construction to exposure categories has already been dealt with and where it presupposed long period 'spells' of wind and driving rain likely to saturate walls and lead to damp penetration. The testing of windows, on the other hand, is concerned with the gusting of wind and rain over quite short periods. The testing of the types of window submitted is carried out on special test rigs and, accordingly, there is no way of in situ testing of existing windows in the same way. They could be removed, of course for the purpose but even if this were possible, it would be rather pointless since an existing window is no doubt already providing abundant evidence of its performance and it has to be admitted unlikely to survive the stringency of the British Standard tests.

A concomitant part of new replacement windows and even with some improved and refurbished older windows, is the double glazing unit, which began to be used in the 1960s. The units should be provided to BS5713:1979 Specification for Hermetically Sealed Flat Double Glazing Units. The Standard includes tests, which are likely to be made more stringent under EU proposals, and there is provision for kite marking on the spacer bar to include the name of the maker and the year of manufacture. Of interest to the surveyor is another British Standard, BS7543:1992 Guide to Durability of Buildings and Building Elements, Products and Components which states that the design life of a double glazed sealed unit is intended to be 30 years but notes that in fact service life can vary from 5 to 35 years. BRE on a survey found that 30% failed prematurely, with misting being the most common problem.

Manufacturers of double glazing units usually provide warranties of 5 or 15 years, so that the surveyor needs to ascertain from the unit, or units, the date of manufacture and examine the warranty for himself. These are very often conditional upon installation being carried out in accordance with the maker's specification and subsequent maintenance again carried out in accordance with the maker's requirements. Failure to follow what the manufacturer requires can vitiate the warranty. Fixing can be difficult in older windows because of the 16mm and 20mm gaps which are industry standard and which are often too wide for existing rebates. Furthermore there are preferred glazing methods. For example double glazing units in wood windows often fail because of the absence of drainage and ventilation provision. For these various reasons it is sometimes better that secondary glazing with a minimum gap of 100mm is given consideration which in any event can provide the advantages of both heat and noise insulation.

Adequate windows, to provide both light and ventilation in most instances, are, of course, a requirement of the Building Regulations and have been since their introduction in 1963 and before then under bylaws in most areas. There will be times, however, when the surveyor may consider it necessary to calculate whether the correct proportion of window to floor area has been provided in the case of new or near new dwellings or whether it is reasonable in the case of older dwellings. His report should reflect his findings and express his view as to whether both light and ventilation are adequate. BRE contrasts the efficiency of some older windows which those of modern times in this regard. The 1800s double hung sash window with frame takes up about 15% of the opening while with many windows nowadays the framing takes up around 35%. Lowering the top sash a little provides adequate ventilation above head height in most circumstances as do the small opening top hung square fanlights in metal windows whereas a horizontally pivoted window lacks such fine adjustment, providing a draught rather than room ventilation. The surveyor should comment in appropriate cases.

The Building Regulations also embody requirements for safety glass to be positioned in windows and doors and adjacent to circulation areas where there is a risk of injury. This is a field which has been neglected in the past and a surveyor needs to keep the requirements in mind. There may be no requirements to bring older dwellings up to standard but there are many windows and doors, particularly from the period 1960 to 1980, glazed down almost to floor level which can pose a hazard. The surveyor must draw the attention of both buyers and sellers to these and the implications if there is no evidence of safety glass positioned at crucial points.

Another field which has received comparatively

little attention in the past is security which has grown in recent years to assume much more importance. While BS5950:1997 Specification for Enhanced Security Performance of Casement and Tilt/Turn Windows for Domestic Application provides much useful information on what can be done in regard to windows, aspects of security are very much related to the occupier, his possessions and the location of the dwelling. There is also the question of insurance which no doubt takes the same factors into account. The surveyor will, no doubt, take note, even if only mentally, of the closing and locking devices as a part of his inspection of every window and external door as much because of safety as security. Inadequacies as to safety he must mention but it could be considered reasonable if he mentions no more than a woeful lack of security items. Every window should have a catch to prevent opportunist entry, every external door should have, in addition to a latch, a dead lock operable by a key together with a bolt. A lack of any of these items should be mentioned but not of the electronic sensors which might be considered essential to an occupier with a valuable jewel collection.

The surveyor might have two thoughts on the provision of internal child proof safety locks for windows. If present fine, if not should their absence be mentioned? On balance the answer must be yes.

Apart from checking that the windows are of the appropriate exposure category according to their location, windows in new or near new dwellings or replacement windows in older dwellings need also to be checked for manufacture according to the material used. For wood windows the current Standard is BS644-1:1989 Wood Windows: Specification for factory assembled Windows of Various Types. The steel windows of the 1920s and 1930s were standardised in BS990:1945 but

the current Standard is BS6510:1984 Specification for Steel Windows, Sills, Window Boards and Doors which is the one for when replacements to match the old, with their slim frames are needed. Aluminium windows are covered by BS4873:1986 Specification for Aluminium Alloy Windows which deals with frames as well. A great advantage now is the availability of both steel and aluminium windows in a polyester powder factory applied coating to BS:6496:1984 which avoids the need for painting on site and repainting for maintenance purposes, washing only being required.

The lack of any need for maintenance is also one of the main advantages favouring the use of PVC-U for the windows of new dwellings and as replacement windows for older dwellings. A whole raft of British Standards now covers PVC-U, the main one being BS7412:1991 Specification for Plastic Windows made from PVC-U extruded hollow profiles. Others are BS7413 and 7414, both of 1991 for White Type A and B material respectively and BS7722:1994 for surface covered PVC-U extruded hollow profiles.

These Standards for windows of different materials might also be relevant where replacement windows are in evidence but where they were not supplied and certified as being in one or other of the exposure categories or, indeed, were supplied before such categorisations were available. For other existing windows not covered by British Standards, it is better to consider their characteristics and durability potential from the point of view of the material from which they are made.

Wood

The durability of wood windows like all others, is very much related to their degree of exposure to the elements. This is mainly dependent on their positioning in the building and within the opening in which they are situated. In other words, windows at high level, facing the direction of the prevailing wind and set well forward in the opening in line with the face of the wall are more likely to need repair sooner than those at low level, facing the opposite direction and set well back.

Because the principal use of the double hung sash occurred between 1700 and 1920 when the vast majority of the dwellings built during that time which have survived were terraced and consisted of basement, ground and two or more upper floors it follows that a fair proportion of the surveyors work will consist of inspecting that type of window many in quite exposed positions. They present quite a lot of surface visible and accessible for inspection and it can be time consuming to give each the attention they require. For such windows in high exposed positions in dwellings built in the very early 1700s when frames were set flush with the wall face the bottom of the sash boxes and the sills need judicious probing

for rot along with the bottom rail of the lower sash. Probing may not be necessary if the appearance is such that rot can safely be considered present, not only because of flaking paint and bare patches but also because of the presence of cracks, opening joints putty cracked and part missing from the glazing and discolouration of the wood.

Double hung sash windows obtained more protection from the weather when, in London, they were required to be set back 100mm from the face of the wall, the length of a half brick, in 1709 and even more so in the 1770s when the sash box not only has to be set back that amount but also accommodated in a recess, leaving only a thin

391: Softwood double hung sash windows in the very early 1700s would be set flush with the face of the wall in a plain reveal below a wide rubbed brick flat arch, four courses high, as here, or a segmental arch. When positioned in upper floors facing the prevailing wind and rain regular maintenance would be essential to keep the paint film intact and the junction between the brickwork and the wood moulding well pointed. Here the sashes have been renewed but out of keeping with the original, the glazing bars being too thin for the period and horns, more appropriate to sashes with larger single or twin squares of glass, not likely to be generally available for at least another 150 years or so.

390: Windows in medieval timber framed dwellings would be provided in the panels between the oak frame, and themselves might comprise wrought iron frames containing leaded lights sometimes as here in complicated patterns. Some would be arranged to open.

392: Incredibly no proper sill has been provided for this badly neglected double hung sash window which must mean that it is highly likely that damp has penetrated to the interior plaster below the window. The neglect of maintenance has left the glazing putty cracked and broken away, bare patches to the framing and undoubtedly wet rot in the sash boxes and elsewhere.

strip of wood visible from the outside. This made the necessity for the junction of masonry and wood being kept watertight even more essential as if damp got round the back of the sash box, it would remain in contact with the wood for very much longer than if surfaces were exposed and there would be a risk of dry rot. Legislative changes affecting windows imposed initially in London, were soon followed elsewhere through the publication and availability of pattern books although local custom also took precedence on occasions. For example, the use of horizontal sliding sashes in Yorkshire and the West Country and, in Scotland, it was always the custom to set windows well back in the opening. Furthermore, in softwood timber framed dwellings it was usually necessary to set double hung sashes well forward to give them support from the frame.

The condition of windows will depend so much on regular maintenance of the paint film to both wood and glazing putty. Undue neglect has been known to lead to dangerous conditions. Both glazing and whole sashes have been known to fall out when they have become jammed and force has been used to attempt to open them or make them close

properly. Another aspect of safety with double hung sash windows concerns the cords. If one breaks all four should be renewed as a matter of urgency with the appropriate strength of cord. If this is not done, fingers could be trapped and injured when the next one snaps. Brittleness in the cords indicates that they are old and nearing the time for renewal. It might be better to replace with chains if the sashes are very heavy.

Wood casement windows, as with double hung sashes, when well maintained and of reasonable quality will last a long time. Neglect will lead to warping, twisting, open joints, cracked and loose glazing putty, rot and rusty hinges and fastenings. There was a good period for joinery in the late 1800 and early 1900s, when casements began to oust the supremacy of the double hung sash. The quality of both wood and workmanship at this time was high. The period of 1920 to 1940 saw a reduction in the size of members and a weakening of strength, not so able to withstand the unfortunate neglect of the 1940s, when resources were otherwise engaged, and which has resulted in the replacement of so many of the wood casements of the period. The 1960s saw a paring down of sizes to an even greater

393: Evident wet rot at the base of the sash box of this window indicating that a great deal of renewal will be required. The prodding, with what looks like a screw-driver, is a bit heavy handed and is more appropriate when the defects in the timber are not so evident.

extent in the rush to build more dwellings, so that for example, components such as sills were built up from small sections. Combined with poor quality wood with too much sapwood present, the slightest fall off in maintenance saw the need for much renewal. If the original windows are still present in the typical semi-detached dwelling from the1920s onwards and they have recently been repainted, sills and bottom rails, particularly at the joints need testing for rot. This can often be done best from the underside, frequently left unpainted but still visible and accessible, but not quite so visible as to raise the ire of the owner.

Much in the way of design and construction of both wood casement and double hung sashes of the past seems to anticipate the need for maintenance and repair. Most can be maintained in a straightforward way, which

395: Wood casement windows set in square bays to a low rise three storey block of flats of the late 1980s. The windows are double glazed and most fitted with a trickle ventilator. Were the windows specified and supplied as satisfying the appropriate exposure category for the location? The glazing is secured by beads. Is the glazing system satisfactory for the type of unit and could the beads be prized off the facilitate unauthorised entry? Is there a warranty available for the double glazing units?

394: Casement windows of very high quality in a two storey square bay window to a stone fronted semi-detached dwelling of the late 1800s. The bow fronted extensions to the square bay have curved glass and ornamental stained glass leaded light fanlights. The stone front of the dwelling is of composite construction, the flank wall wholly of brick, the upper floor rendered. Brick courses line up with the stone, two brick courses to one of stone.

does not just consist of slapping on a coat of paint but involves making sure they are eased regularly and work efficiently. If repairs are needed, they can be taken apart, new sections inserted and then put back together again. There is ample evidence that this has been common practice with the result that many old windows are providing good service to this day. Admittedly the work is labour intensive but much can be done by the home owner himself exerting a modicum of skill and this is more than can be said for windows in some other materials.

Metal

The rolled steel window with its bronze fittings and slim Z and T shaped sections, sometimes set in a wood frame but appearing even slimmer when set in a frame of the same shaped sections, was dominant for many years. It was the archetypal window of the latter half of the 1920 to 1940 period and again much used in the following 20 year period. Projecting hinges for easy cleaning from the inside came to be fitted in the 1930s, enabling steel windows to be used in flats and continual developments have produced ranges on different modules, top and bottom hung windows, pivot hung both vertical and horizontal, folding and sliding windows along with doors. Aluminium on a smaller scale came to be used for similar types but because aluminium is an even poorer insulator than steel, a plastic thermal break needs to be incorporated within the frames to prevent undue heat loss to the outside.

Before the advent of hot dip zinc galvanising, the main problem with steel windows was rust which could rapidly develop if the paint film was damaged, the expansion and the build up of corrosion leading to broken glass and fittings. Galvanising largely dealt with this problem but neglect and rough treatment can nevertheless introduce the possibility of damage to the zinc coating and rusting to the steel below.

396: Steel windows are prone to rust when the paint film, as here, is neglected. A little judicious scraping with the owner's permission, should determine whether rubbing down, treating with rust inhibiting primer and repainting is all that is needed to restore the window to near new condition, which would seem here more than likely to be achieved.

Similarly, the experience with aluminium and anodising led to dissatisfaction with performance and it is only with the introduction of factory applied white polyester powder coatings for both steel and aluminium windows that the situation has improved when maintenance is not always carried out when it should be.

One of the reasons why metal windows tended to suffer unduly from corrosion lay in the fact that to derive full cost benefit from the use of metal frames and sills, windows were set well forward in the openings, a mere 50mm in from the face of the wall. While this was popular with occupiers, since it provided a more generous internal sill, it subjected the putty used for glazing, which was not always the correct type suitable for use with metal windows, to the full force of the elements. This lead to shrinkage, gaps forming and rainwater penetrating to lie in the angle of the sections for lengthy periods against unprotected metal.

There are degrees of corrosion, of course, and the surveyor should avoid blanket condemnation of metal at the slightest sign. The preservation of metal windows is perfectly valid in the context of dwellings from the periods when they were much used. This is particularly so of those of the white walled, sometimes flat roofed or tiled in bright colours, variety of the so called 'modern movement', a part of which consisted very much of the slim lines of the frames and glazing bars of steel windows. Light rust can be rubbed down to firm smooth metal which can then be treated with a rust inhibiting primer and repainted. Even where the metal is pitted, it can be repaired with an epoxy based car body filler and repainted. It is only when there is hardly any metal left on some sections that it may be necessary to take out the window and send it to a specialist workshop for repair. The repairer will no doubt possess a number of the many tossed out metal windows from dwellings now

397: The period around the 1960s saw the adoption of enormous single glazed windows with hardly a thought given to heat loss. Successive hikes to oil prices brought home the folly. The centre horizontal pivot hung windows, as here, provided to these five storey terrace houses, are inefficient as providers of ventilation as they are too large. There is either too much or too little.

398: Metal replacement windows for wood double hung sashes in two storey late 1800s terraced dwellings. They represent a reasonable compromise in that the overall proportions are not greatly different from those they replace. This is particularly so of those in the left hand dwelling with thicker sections, the upper 'sashes' almost giving the impression of being set slightly forward.

with PVC-U replacement windows and from which he can cannibalise appropriate sections for welding in as replacements. Even bent or twisted sections can sometimes be straightened out in a workshop.

If the surveyor finds corrosion beyond reasonable possibility of repair and it is not possible to renew sections, then he should be aware that it is still possible to purchase rolled steel

windows in the shapes and sizes used in the 1920s and 1930s. Now they have just those slight variations to the sections to provide for double glazing, draught proofing and if polyester powder coating is not required, then stove enamelling can be provided for a glossy everlasting, maintenance free finish. Double glazing should cope with the condensation which always seem to have been a problem with metal windows and where there was seldom any provision for draining it away, leading to the conclusion, sometimes, that it was a lack of weather tightness.

Plastic

Plastics of various types have been the favoured choice of the homeowner since the 1970s for replacing whatever type of window previously graced his dwelling. Until recently, there has been no control, except in conservation areas or in respect of listed buildings, on what that replacement consisted of, no regard was had to whether it matched in style what was there before or, even, whether it went some way towards complementing what already existed in the locality. This was curious as in plastics any type of window shape is possible, since the hollow extruded sections are cut to size and the corners heat welded to form any desired shape. It was the government's concern over energy consumption that prompted the introduction of control.

For the homeowner, good overall performance has been achieved with the added advantage that no maintenance is required. The quality of the material has been improved over the years with additives to reduce brittleness and the effect of ultra violet sunlight. Nevertheless, problems can arise and the welded corners can be a source of weakness if quality is poor and if insufficient allowance has been made for the expansion of PVC-U in sunlight.

399: As BRE points out, the proportion of framing to the size of the opening is so much greater for windows of the present time than it was in the 1800s, something like 35% as against 15%. This cut away section of the fixed light of a PVC-U window shows why. To obtain sufficient strength, the extruded sections have to be substantial in size, particularly when incorporating a double glazing unit, as here, utilising the drained and ventilated method of glazing. Where there are opening lights it will be appreciated the width of framing is about double that shown here.

PVC-U is basically white and the introduction of wood grain effects in the 1980s, which involved dyeing the material, has meant that the darker shades heat up to a greater extent in the sun. This can lead to deformation if additional reinforcement has not been provided in the hollow sections and the increased movement not allowed for in the design. Dyes, of course, should last the lifetime of the window but, on poor quality material, they have been known to fade fairly rapidly.

Doors

Doors, like windows, form part of the external envelope of a dwelling with the purpose of keeping the elements at bay and, accordingly, are appropriate also to be included in a sub-section on external walls. Houses have a front door as a means of entrance and final exit, designed often to impress visitors, and a rear or side door. Many also have separate doors to the garden, up to the 1960s often in the form of French windows and since then as large glass patio doors. Older terrace dwellings could also have doors in the form of long windows giving access to a balcony at first floor level. Flats may have an entrance door leading to the exterior, if provided with balcony access, or to an internal staircase or lobby. There may also be a door to a private balcony.

Security for external doors was touched upon when locks and bolts were discussed for windows, but what of the construction of doors? It is not possible to purchase doors to cope with specific degrees of exposure in the same way as it is possible to do for windows. Indeed, a door opening inwards from outside is difficult to make weather tight to anything like the same degree, although a great deal can be done in this respect. This is one reason why a porch or canopy with side cheeks extending to a minimum projection of 750mm is recommended, certainly for the front entrance door. The other reason is, of course, that it gives some protection from the pouring rain to visitors waiting for the door to be opened and to occupants fumbling for their keys.

The dominant material for the construction of external doors has always been wood, totally so up to the Industrial Revolution, but with comparatively small numbers of metal framed doors added from the 1920s, usually in combination with metal windows and employing a considerable proportion of glass. As far as the aspects of corrosion and repair of metal doors is concerned the same factors as for metal windows need to be taken into account.

400: (above left) This six panelled softwood entrance door to a 1700s terrace house faces south west and has no protection from hot summer sunshine. As a result the solid panels each made up of three sections, tend to shrink and then swell in damper conditions, disrupting the paint film in the process. It may be that joints of framing and, in this case, panels will eventually need regluing but, in the meantime, a sun blind would help. The kicking plate suggests there may be rough usage.

401: (above centre) A much neglected six panel entrance door to an early 1800s house. If not rubbed down and repainted soon, probably much more radical repairs will be required.

402: (above right) An attractive panelled entrance door with stained glass leaded lights to two large and three small upper panels and similar treatment to the sidelights for this early 1900s house, now divided into flats. The door is sturdy despite its size and the presence of glazed panels, so that a well carried out repainting, after adequate preparation, should restore it to near its original condition.

Although some wood battened doors, usually of the framed, braced and ledged variety, will be found in cottage style dwellings by far the majority of external doors are framed and panelled. Some will be of hardwood in sturdy form but, unfortunately, as the years have progressed so the thickness of the framing of softwood doors has tended to lessen, leaving many installed since the 1950s insufficiently robust to withstand even ordinary use, let alone any assault on them by intruders. A kick will often break the door itself even if the fastenings are able to hold secure. With the use of too much sapwood at the time and a lack of preservative treatment, many external softwood doors succumbed to wet rot in as little as 10 to 15 years.

BRE recommend that both main and subsidiary doors are of solid wood to a minimum thickness of 44mm. This applies to the panels as well as the framing. Should a flush door be present, it should be of solid core construction, as was common in the 1920s and 1930s, not the hollow type which continuing the desire into the 1950s to avoid ledges, which harboured dust, tended to used later. Tapping will distinguish. Glazing should be in such sizes and positions that if broken it should not be possible to reach latch type locking devices.

Many external softwood doors provided for dwellings, built up to about the 1940s, will be found to be of adequate thickness so far as

404: Warnings need to be given by the surveyor in regard to detailing and possible effects of damp penetration around this rather flimsy extension, providing access to a flat roof. Older flat roofs usually lack any ventilation and damp penetration can lead to outbreaks of dry rot, very expensive to tackle and the extent cannot be gauged before opening up.

403: Doors from the 1950s onwards were made insufficiently robust in many cases. This flush door is very flimsy not helped by the large glazed panel. It needs attention, but does at least have a weatherboard.

405: Over enthusiastic and rather unnecessary prodding with a bradawl to what is clearly a defective bottom rail and stile to an external door. Depending on the condition elsewhere, repairs could be possible, although a new door might be cheaper.

the framing is concerned even though some of the panels might be a bit short of the full thickness. Although as with wood windows regular maintenance of the paint film is essential, sometimes however warping will occur making doors difficult to secure and allowing draughts to enter. Doors can be flattened in the workshop, but moving the door stops and draught proofing can often be a ready solution. Aspect plays a part and panelled doors facing south west and due west are particularly prone to shrinkage in hot summer sunshine and swelling in damper weather. This eventually causes the glue to fail and gaps to open along the shoulder of the joints, often within a year or so of repainting. The only real solution usually is to take down the door, reglue and clamp the joints, but it is a pity that the simpler solution of providing a sun blind for such doors seems totally to have gone out of fashion.that the simpler solution of providing a sun blind for such doors seems totally to have gone out of fashion.

The arrangement at the threshold is important to the performance of a door, particularly those giving access to balconies above ground floor level. Even when a canopy is fitted driving rain can often reach doors in these positions and rum down the face. A weatherboard near the foot of the door is the first essential to throw clear as much water as possible. Rebates, drips and water bars formed and fitted correctly in the right places are the last line of defence, but not always provided, to prevent rain water getting under the door and penetrating to the interior by capillary action. It is not unknown for this to happen over a period of years and set up wet rot and even dry rot in adjacent areas of flooring. Depending on the evidence from the door and adjacent areas, it may be necessary to advise opening up for closer inspection if, for example, fitted carpets are damp stained.

Foundations

It may seem perverse to include a subsection on foundations in a book dealing with visual inspections of dwellings as they are never seen. Flaws, however, in their design or construction or subsequent changes in the soil below can sometimes have striking visible effects on the structure above ground level. Nevertheless, before any blame can be attributed to foundations or the subsoil it is necessary to be reasonably sure that any defects seen are not due to some other cause. If other causes can be shown to be exclusively involved then there is little need to give the foundations of any dwelling much further thought as long as the dwelling is more than 10 years old. This is because by this time, and very often sooner, any flaws in design or construction will, it is generally thought, have become apparent. What does need further thought, however, is the subsoil on which the dwelling is founded and the surrounding topography and not that just within the boundary.

Of course, for a dwelling within that ten year time scale a surveyor will hope to see details of the site investigation if at all possible

and examine how the results have been incorporated in the design adopted. This can be considered necessary whether there are any signs of foundation problems or not since a prospective purchaser will hope to have some assurance of likely future performance.

The nature of the ground in which foundations are placed obviously varies considerably across the United Kingdom, some more suitable than others for building and some made more or less so by mans' activities, ballast or brick earth extraction, underground mining, industrial processes and the waste tipping which produces land fill sites. It is an essential part of a surveyor's knowledge to be aware of the paramount type of soil for the locality in which he practices as well as the area's history. Accordingly it is necessary for his office to have a set of the local 1:50,000 scale Geological Maps published for England, Wales and Scotland by the British Geological Survey and available from the Survey's offices in London, Nottingham and Edinburgh. The information on the Geological maps is a coloured overlay to the detail of roads, railways, rivers etc. provided by the Ordnance Survey on its topographical maps to the same scale, so that it is possible to pinpoint with a reasonable degree of accuracy the location of any dwelling being inspected.

To a surveyor unfamiliar with Geological maps it might be appropriate to arrange for a discussion with a local practicing structural or civil engineer so that he can explain the meaning of the symbols used and how to determine from the key provided on the maps the type of soil on which properties are likely to bear and whether such soils are of types likely to cause problems. This information can be coupled with valuable local knowledge and hopefully a working relationship with the local Building Inspector since Geological maps do not always accurately record the shallow deposits within the top 4-5m of the surface where most foundation problems occur. Whether mention should be made in a report on the type of subsoil must be a matter of individual circumstances relating to each inspection but certainly in a Court of Appeal case, *Cormack* v *Washbourne* 1996, it was held that the surveyor was negligent in not consulting the geological map where there were diagonal cracks in the walls of a 1969 house and trees within 7-8m.

Methods of site investigation prior to building have developed very considerably over recent years and where a large estate is proposed it can be carried out comprehensively by taking deep samples from boreholes at many points and building up profiles. The end result should be a foundation design appropriate to the subsoil and the development. Special provision might need to be made where old features are to be removed, areas filled in or tress cut down. Notwithstanding, a sample cannot be taken at every point and in many cases it still boils down to a decision by the site team as to the appropriate level to place the foundation. While such level for each dwelling should be approved by the Building or NHBC Inspector, it may be that assumptions are made and that what is considered suitable for one dwelling will be suitable for others. It is not surprising therefore that on occasions cracks appear fairly soon after completion should a pocket of poor soil be missed.

It is generally the level at which foundations are placed that is the problem. Up to around 1900 most dwellings were provided either with cellars or substantial room height base-ments. Accordingly the foundations would be much lower than the normal ground level and the removal of the volume of soil to form the basement and its replacement by the loading from the dwelling balanced the equation. It will be found that such dwellings experience

few problems from subsidence,

While the loads in domestic construction are light, it is nevertheless necessary where foundations are placed at around ground level that they are at least below the level where the ground is influenced by climatic changes. Where areas are only grass covered this is normally between 1.0 and 1.5m, but where vegetation is present can be much greater, perhaps up to 5.0m. This necessity has not always been followed with two to three storey suburban properties built since the early 1900s without cellars or basements. Even if an appropriate level is provided at the time of construction, the foundations can be upset by subsequent planting, or its removal, or other form of disturbance.

Where foundations are set around ground level there is nearly always an initial overall settlement as the ground below is consolidated. It is usually quite small and as long as it is of a uniform nature often passes un-noticed.

If the dwelling comprises a single structural unit but the initial settlement not be uniform, portions can settle at a different rate or by differing amounts and cracks can appear. These will almost certainly be of a diagonal nature and will taper. If the settlement is at the extremities of the dwelling, they will be at their maximum width near roof level but if the

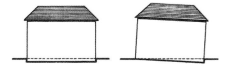

Figure 43: What happens when a dwelling is provided with foundations at around ground level is that there is a uniform settlement, often quite slight, as the ground below consolidates. Even if the whole dwelling tilts, as in the diagram on the right it often passes unnoticed.

settlement, less likely, is at the centre and the ends are stable, the maximum width will be at ground level passing through the damp proof course and probably extending down through the foundation itself, though not of course seen without opening up. The cracks will also appear in roughly the same place on both sides of a solid wall, but where there is a cavity there could be a marked difference in where they appear, bearing in mind that it is not possible for a subsidence crack to be restricted solely to one leaf. One instance where 'settlement' could appear to be taking place at the centre is where trees have been removed which formerly extracted moisture from a shrinkable clay soil. On removal of the trees the clay soil recovers the moisture formerly lost, swelling in the process. This is a phenomena known as 'heave' where the extremities are forced up, giving the impression that the centre has settled.

Of course, the walls affected by 'settlement' or heave are very often not blank walls but contain window and door openings introducing

406: Subsidence has severely affected the corner of this two storey double fronted house of the early 1900s. Window surrounds, the band course and arches are distorted and the balcony fractured. The cause could have been the roots of a tree near the corner, although there is no trace of this now, or, perhaps, leaking drains. Substantial expenditure will be required to restore the dwelling to a presentable appearance, once it is determined that movement has ceased, although value will always be affected.

407: Subsidence in the pier between the two front entrances which is also, of course, the front end of the party wall, of these three storey, non-basement houses of the mid 1800s, converted two at a time horizontally into flats, hence the blocking off of one entrance. The effect is dramatic in that brick courses have dropped by as much as 75mm, cracks from ground level extend round the door arches to the sills of the windows above disrupting in the process the adjacent windows and their arches and fracturing the artificial stone mullion necessitating emergency restraint.

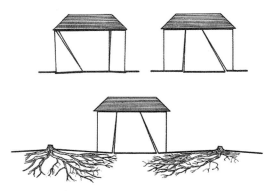

Figure 44: Differential settlement in a single structural unit invariably causes cracks, usually diagonal and tapering. Should the extremities settle and the centre remain stable, as in the top left diagram, crack widths are at their maximum at roof level. However, if the centre settles and the extremities remain stable, the maximum width of cracks is at ground level, continuing below ground through the foundations as in the case of the top right diagram. In the example at the bottom of soil recovery, known as 'heave', where trees have been cut down before or subsequent to construction, the soil swells as moisture is recovered forcing the extremities upwards. Heave cracks do not always display the diagonal tendency of subsidence cracks to the same extent but are always similarly tapered.

planes of weakness. The cracks accordingly go round these, appearing as gaps between frame and wall.

Movement causing cracks involving foundations and appearing within 10 years of construction is not covered by normal subsidence insurance. Insurers refer to such movement as settlement restricting the term subsidence to movement involving foundations occurring after the 10 year period. It would however be covered by the NHBC, or any other similar insurance warranty, provided the damage was notified in time. If the dwelling had no cover of this nature a remedy might be sought from a builder or developer vendor under the Defective Premises Act 1972 but action would have to be taken within six years. If outside either period and a subsequent sale was involved the surveyor could find himself in difficulty if tell-tale signs had been missed.

If the dwelling comprises a number of different structural units, it is not uncommon for these to settle or be subject to the effects of ground movement in contrasting ways. Much will depend on how closely related are their forms of construction. More pronounced effects are often apparent where foundations are at a different level, for example where a back addition, founded approximately at ground level is attached to a main building with a basement. Another is where there is a garage, sometimes with a room above, at the side of a dwelling but most often where there is a bay window with minimal foundations, set out from a wall extending lower. Where the structures part company is often not where the crack occurs. This will often appear a little way along the wall of the lighter structure at a plane of weakness such as a window or door opening.

The foregoing descriptions and diagrams illustrate in outline the main types of settlement, subsidence and heave and their effect on

408: Close up of the crack starting at ground level in the pier of brickwork between the door and windows of 407, fracturing the bricks and increasing in width with height. Underpinning may be necessary in this instance if the movement is progressive.

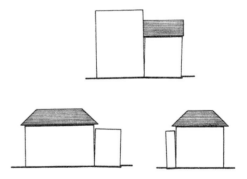

Figure 45: Attached different structural units can behave in contrasting ways and have a tendency to part company unless of very similar construction and with foundations at the same level. The top diagram shows a back addition attached to a taller main structure, very typical of dwellings up to the 1920s, with or without basements. The diagram, lower left, shows a garage alongside a two storey dwelling with a room above, while the lower diagram on the right shows a bay window again attached to a two storey dwelling. All are parting company from their host structures.

masonry structures, brick, stone or concrete block, by far the majority forms of construction

409: Indications of movement to the two storey bay window of an early 1900s terrace dwelling. The crack extends to ground level and increases in width with height providing strong evidence that the cause is subsidence. The pattern is repeated on other dwellings in the terrace but where carefully repaired is difficult to detect. As plaster cracks have been repaired internally and show no sign of re-opening the movement is probably not progressive and careful repair here could restore appearance. It is regrettable that the original double hung sash windows have been replaced by an out of style type. Surveyors should warn that this type of window is vulnerable in that the louvres can easily be removed from the outside.

for dwellings in the United Kingdom. Of course, the evidence on site is seldom as straightforward as the descriptions and outline diagrams suggest and requires, at times, careful analysis to determine with any reasonable degree of certainty whether defects originate from below ground level or above. Movement in the structure above ground level from causes such as thermal effects and shrinkage produce crack patterns different to those caused by subsidence but other causes, such as failure in a lintel or beam, can produce diagonal cracks which also taper. However, they seldom extend and are traceable, even if interrupted along the way, throughout a wall. If such cracks do and, in particular, if they extend below the lower windows down to ground level, they are highly likely to be caused by subsidence or heave. Careful viewing of whole elevations from the exterior will sometimes help the inspector to trace the entire length of a crack which may have been noted previously in sections internally from plaster cracks in rooms or the staircase.

A comparatively small proportion of dwellings are of framed construction, timber, steel or concrete and while many of these are clad in brickwork, others are clad in timber or PVC boarding, tiles, steel or other forms of sheeting. If affected by subsidence they will develop a different pattern of cracking, more likely related to the separation of cladding from frame or the crushing of infilling panels and require a specialist engineer's investigation. Timber cladding, PVC boarding or slate or tile hanging applied to masonry construction does not display the external evidence of subsidence so readily and to the same extent, the components adjusting their position to accommodate the movement, and evidence may be restricted mainly to internal plaster cracks, although gaps in soffit or fascia boards are often an indication that movement has taken place. Fortunately slate or tile hanging is seldom taken down to ground level because of the risk of physical damage, so that cracks below the lower windows indicating likely subsidence or heave may still be visible.

Having decided that the visible defects are most likely to have been caused by foundation movement the advice to be given to a client can only be formulated after a number of other aspects have been considered relative, mainly, to the possible cause but also to the size of the cracks, whether they are likely to be progressive and the age of the dwelling.

Foundation movement occurs when there is a volumetric change in the ground supporting the foundation, causing the dwelling to move downward. Engineers and the insurance industry define this as subsidence. It can be due to a shrinking or softening of clay soils, washing away of fine material in granular soils, further compaction of fill or compression of peat other than that due to the load of the dwelling itself. This latter cause, along with any movement within a structure due to the redistribution of loads and stresses within the various elements of its construction within the first 10 years, would be known as settlement within the insurance industry and is usually excluded from domestic policies covering subsidence, landslip and heave, being considered the responsibility of the builder to put right.

Volumetric changes to the ground only occur as a direct result of outside external influences such as changes in ground water levels which may be caused by leaking drains, the extraction of water by trees or other vegetation, extremes of climate for example prolonged rain or persistent drought, decomposition of organic material and landslip, mining or other excavation. Mention has already been made of the need for surveyors to be aware of the subsoil in their area of practice. A very considerable proportion of dwellings have foundations placed in clay soil, some more prone to shrinkage than others, but all with a propensity for volumetric change if the naturally occurring balance of composition is altered by external influences. Clays of highly shrinkable potential cover large areas and a surveyor's local knowledge should extend to an awareness of specific areas where such clay is likely to be present. Sometimes a quick look around the garden of the dwelling being inspected, along with the grassed areas around trees in the locality, during dry weather will reveal quite substantial fissures in the ground, as much as 20 to 25mm wide and up to half a metre deep, all features of highly shrinkable clay.

Trees

Of the various external influences which can cause volumetric change in clay soils, by far the most commonly occurring, giving rise to

not far short of three quarters of all claims under subsidence insurance in dry years, is the extraction of moisture by the roots of trees or other vegetation. This causes the clay to shrink depriving the foundations of support. Knowledge that the dwelling is founded on clay necessitates the surveyor first considering its proximity to trees and, if trees are present nearby, their type. The presence of such trees is not, of course restricted to the gardens of the dwelling being inspected but may be situated in neighbouring gardens or the public footpath.

Determining whether the trees are likely to be responsible for the damage can be a fairly complex matter and depends on a number of factors such as the species of the trees, the characteristics of their growing patterns and their ages, their proximity to the dwelling, the depth of the dwelling's foundations and recent climatic conditions. Each factor can be variable and the interaction between factors even more variable, resulting in the anomaly that damage can be caused in some cases by a tree at some distance away from a dwelling, whereas in other different but similar cases no damage is caused by a tree of the same species and size at a much closer distance.

While it is probably true to say that any tree or large shrub with its roots in highly shrinkable clay growing close to a dwelling can, in certain circumstances cause damage, it is also true to say that many trees are unjustly blamed for damage which is really attributable to other causes. Trees, however, vary considerably in their size and characteristics and some are more prone to causing damage than others. The trees which are known to cause damage are more likely to do so when long hot summers are separated by relatively dry winters, 1975-1976 and 1989-1990 being prime examples with perhaps 2003-2004 being added to the list.

new growth

COMMON OAK 25m

Common Oak. Twisted, heavy branches. Often with burrs on the short bole

TURKEY OAK   30m

Turkey Oak. Branches swollen where they leave the bole, *p.141*

Figure 46: Oaks

Fast and vigorous growing broad leafed trees with shallow roots reaching maturity and producing large crowns and a substantial area of leaf surface are the most likely to cause damage. Such trees spread a myriad of fine roots over a considerable area of ground, much nearer to the surface than is commonly believed and extract moisture at a rate often well in excess of what can be recovered in the winter months. As such, there is a continual deficit of water in the ground when compared with its natural state. This deficit of moisture can only be made up if the tree dies or is cut down but this can produce a substantial problem from heave as the soil can take many years to recover all the moisture that has been lost, continuing to swell as it does so. Rather than cut the tree down reducing the size of the crown to lessen the summer requirement for moisture could lead to a less damaging gradual recovery instead.

It is not an easy matter to tell one type of tree from another and the expert advice is that for identification the tree should match with its description in all respects, not just the leaf shape, since different families of tree have leaves of comparatively similar shape. For surveyors the Judge in the case of *Daisley v BS Hall & Co* 1972 said: 'it really is essential to make sure of your tree recognition because Poplars are not the only trees of which there is more than one species'. The case concerned an eight year old house where the surveyor had attributed cracks to 'minor settlement and bedding down' failing to mention either the subsoil or a line of trees 8-9m away which were a close relation of the Black Poplar.

To assist towards the identification of the trees most likely, and indeed known to have caused damage to dwellings, the following pages contain illustrations of 14 varieties of deciduous, broadleafed trees. Each illustration shows the general appearance when in leaf,

the silhouette when the leaves have fallen in winter, the bark, the leaf shape and colour and where appropriate buds, flowers and fruits. Below the general appearance, the likely mature height is given in metres, although these heights are a little more than those indicated elsewhere, for example, on a list produced by BRE for the same trees when in an urban situation and on clay soil. Once mature, trees can live a remarkably long time. Of those illustrated, Oaks can live 500 years, Common Lime, London Plane and Sycamore trees 400 years and Sycamores up to 350 years.

The illustrations are compiled from the pages of *The Pocket Guide to the Trees of Britain and Northern Europe* by Alan Mitchell and John Wilkinson, an inexpensive and invaluable publication from A and C Black of London which should be part of the on-site armoury of every surveyor inspecting dwellings. The guide contains illustrations and descriptions of over 600 species and varieties to be found in the gardens, parks, hedges, town streets and woods of the United Kingdom and reference to its contents when necessary should help to distinguish those trees unlikely to cause damage from those illustrated here.

As shown in succession on these pages the 14 trees follow a descending order of threat as set out in a 1999 BRE list of 21 varieties, but no illustration of a Hawthorn is provided which appears seventh in the list between Horse Chestnut and Lime, since this can hardly be mistaken for anything else because of its long hard spikes, comparatively low 10m mature height and red or white flowers. There are also no illustrations of a Rowan/Service tree, last in the BRE list but for this is substituted a Whitebeam which is of the same family, or the fruit trees, Apple, Pear, Plum and Damson which are also of relatively low mature height and in the presence of which owners are invariably only too happy to gush.

Also not illustrated at this particular point are Cypresses, which appear next to last on the BRE list as known to cause damage, but are featured later, see Figure 63 on p425, in the Leylandi variety, as a nuisance, the planting of which has had to be controlled by legislation.

The Institution of Structural Engineers (ISE) also publishes a list of trees thought to be implicated in damage. All 14 varieties illustrated here appear on that list as well as on the BRE list but the ISE suggest a further 13 possibilities. Many of these including Holly, Laburnum, Magnolia, Pine, Spruce, Walnut and Yew are noted as rarely implicated and the remainder include the four fruit trees and Hawthorn and Cypresses already mentioned above. The ISE list is mentioned here since comments on each tree are also included on its ability to accept an element of management by way of pruning to reduce the leaf area and correspondingly its ability to transpire and extract moisture from the ground. This is strongly suggested as an alternative to felling or the underpinning of a damaged dwelling, both of which engineers and arboriculturists consider to have been too readily recommended and implemented as solutions in the past, often without prior appropriate investigation which they now believe ought to have been carried out.

As to the trees illustrated here, Oaks Figure 46, comprise one of the biggest of all groups and according to BRE are way out in front for causing damage. They all produce recognisable acorns which could ease recognition but also a wide variety of leaf shapes, not all of which would be commonly thought of as oak, some quite plain with saw tooth edges for example. Oaks are likely to cause a persistent deficit of moisture in clay soil which can result in long term problems from heave if they are cut down.

It is not generally realised that Poplars, Figure 47 and Willows, Figure 53, are of the same family, the contrast between the tall, graceful Lombardy variety of Poplar and the Weeping Willow, both providing the popular conception of their names, being quite striking. They are all fast growing with shallow roots but with a number of different crown shapes. The Black Poplar is fairly common in the Northern towns, being resistant to smoky air. Aspens, not illustrated, with their fluttering leaves of the same shape as those of a Grey Poplar grow to about only half the height. Not all Willows have the long slender leaves of those shown, some are broader, but all varieties produce silky catkins, 'pussy willows'. The Bat Willow, straight and more upright than others, grows extremely fast mainly in East Anglia and South East England and, as its name implies, provides the wood for cricket bats. Both Poplars and Willows tolerate heavy pruning, which might prove an option where damage has been caused but, as during growth they can cause a persistent deficit in a clay soil's moisture content, care needs to be taken to cope with subsequent heave should it be decided to cut down.

The Common Ash, Figure 48, often very shallow rooted grows well in cities and is the only variety of ash producing coal black buds. Unlike the example illustrated, many tend to fork much lower, nearer the ground. An Ash tree will tolerate the crown being pruned and this can be done but still retain a reasonable shape.

The harsh upper surface of the leaves of the English Elm, Figure 49, differentiate it from other varieties and it is said to have a unique outline to its crown. The leaves of all Elms are to an extent asymmetric at the base. A virulent form of the Elm Disease fungus arrived around 1960, carried by the Elm-bark beetle and to which English Elms readily succumbed, necessitating their removal when affected.

Figure 47: Poplars

25m
COMMON ASH

Common Ash. Some branches can carry fruit in winter

bud

fruits

Figure 48: Common Ash

The Robinia, Figure 50 False Accia or Locust Tree to give it its various names, grows well in towns and cities but is not common north or west of the Midlands. A smaller golden coloured variety called 'Frisia' is now a commonly planted tree for small gardens. Both varieties are known for dropping branches unexpectedly.

Familiar to all schoolboys for its conkers in the autumn with their spiky green casings, the Horse Chestnut, Figure 51, is easy to recognise by its large leaves and long tresses of white flowers in Spring. It has an unfortunate tendency to shed heavy branches in summer but will tolerate heavy pruning.

Described in the *The Pocket Guide* as the worst tree known for lining streets and avenues, the Common Lime, Figure 52, often achieves a great height and can live a long time, but its roots are known to be invasive, it sprouts around the base and is usually infested with greenfly. As anyone who has parked a car near a Lime tree will know, the sticky droplets are a pain to remove. It will, however, tolerate heavy pruning.

The Beech family of trees, Figure 54, which as a family includes Sweet Chestnuts and, incidentally, Oaks, are large and bear nuts which are partially or wholly enclosed in a cup or husk. The common Beech prefers freely draining soil, is not often encountered on clay and accordingly is only rarely implicated in damage. In the United Kingdom, it is confined mainly to England. Old trees have a reputation of being unstable, a point to bear in mind if fully grown mature specimens are encountered in gardens.

The London Plane, Figure 55, is a hybrid and some confusion continues to arise over the name, the Sycamore being known as a Plane in Scotland. It is common in towns. In London

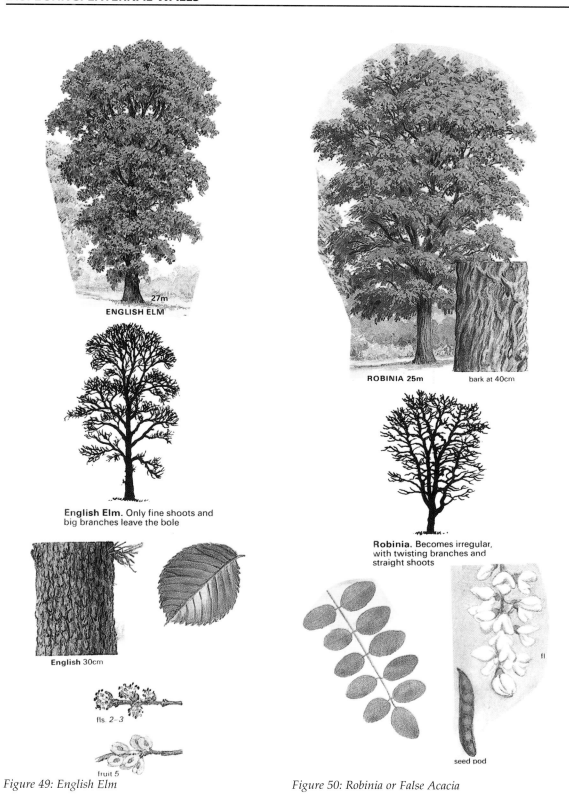

27m
ENGLISH ELM

ROBINIA 25m   bark at 40cm

English Elm. Only fine shoots and
big branches leave the bole

Robinia. Becomes irregular,
with twisting branches and
straight shoots

English 30cm

fls. 2–3

fruit 5

seed pod

fl

Figure 49: English Elm   *Figure 50: Robinia or False Acacia*

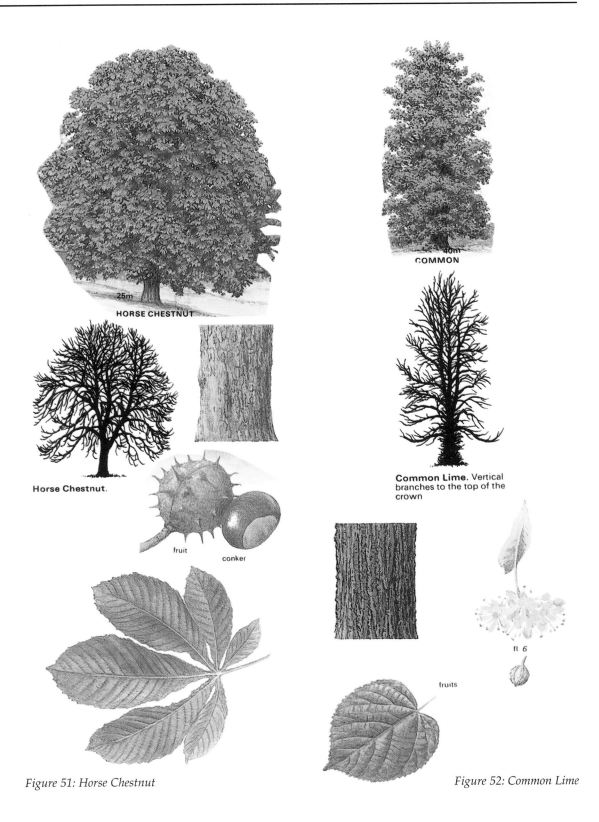

Figure 51: Horse Chestnut

Figure 52: Common Lime

CRACK WILLOW

20m

BAT WILLOW

25m

Crack Willow. Heavy, low branches make a broad crown,

Weeping Willow. Shoots *yellow*, brightest after mid-winter

WEEPING WILLOW 18m

Figure 53: Willows

Figure 54: Beech

the distinctive form is Common and known as Pyramidalis, has a burred trunk which copes well with the atmosphere by regularly shedding its surface layer, a rich glossy green three lobed leaf and only one or two large fruits on each stalk which can persist on the tree throughout the winter. It is regularly subjected to heavy pruning, successfully to control height and size of crown.

Described as a dull tree in the *The Pocket Guide* the Sycamore, Figure 56, is part of the Maple family, the Acer species, but nevertheless is very accommodating, thriving better than most other trees in coastal areas, smoky city air or the high hills. It shades out other trees, while its heavy leaf falls blanket out anything growing underneath. Its little 'whirly bird' seeds get carried everywhere and many a surveyor will have found little trees growing happily in the silt of out of the way gutters. It is tolerant of heavy pruning.

The Rose family embraces a wide variety of trees, including Apples, Pears and Cherries, Figure 57, Hawthorns Cotoneasters, Rowans and Whitebeams, Figure 59, and the curiously named True Services Tree (Sorbus). The difference between Rowans and Whitebeams is that the former have 'pinnate' leaves, a number of leaves on a single stem, rather like those of the Ash or Robinia, False Acacia, while the latter have the simple leaves shown. Commonly planted in streets because of its low mature height, not often implicated in damage and, if necessary, will tolerate heavy pruning. The Rose family also includes the Prunus species of Plum, Almond and Peach trees as well as Cherries. All can have very shallow roots with a risk of damage if planted by the unsuspecting owner in shrinkable clay very close to a dwelling. The Wild Cherry, Figure 56, is also much planted in gardens, parks and streets and is considered a first class amenity tree. The crown is whorled, i.e. the

PLANE

'Pyramidalis'

♂

♀

5

London Plane. Twisting branches. Many hanging fruits persist through the winter

30m

fruit

Figure 55: London Plane

WILD CHERRY   18m

bark at 80cm
SYCAMORE

Sycamore. Short-jointed, narrow shoot systems; in woods the bole can be long

27m

new growth

bark at 20cm

Figure 56: Sycamore   *Figure 57: Wild Cherry*

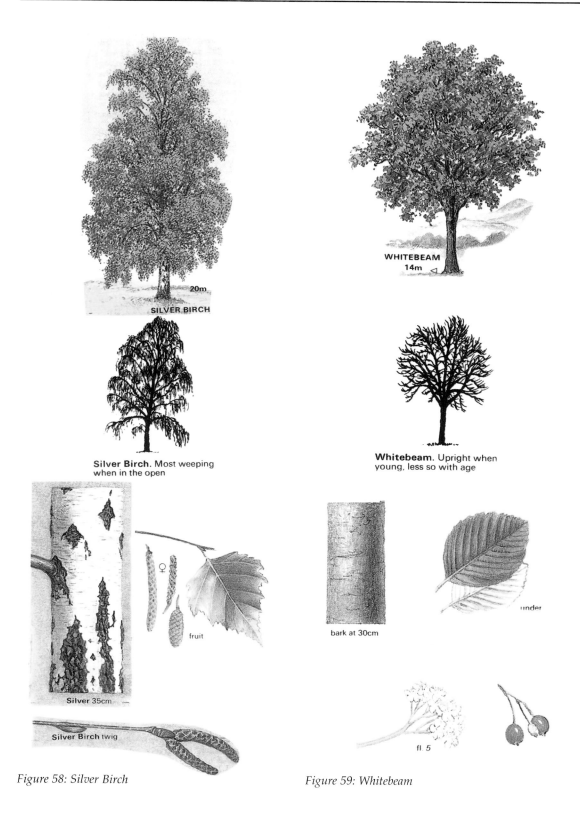

WHITEBEAM
14m

20m
SILVER BIRCH

Silver Birch. Most weeping when in the open

Whitebeam. Upright when young, less so with age

fruit

Silver 35cm

Silver Birch twig

bark at 30cm

under

fl. 5

Figure 58: Silver Birch

Figure 59: Whitebeam

smaller branches and leaves ring round the centre stem, unusual in a broadleafed tree but like many conifers. It does not take kindly to pruning, as is also the case with other examples of the species.

The timber of the Birch family, the trees of which grow further north than any other, is much used for making plywood and all are fast growing. The Silver Birch, Figure 58, is easy to recognise, particularly from its bark and 'weeping' silhouette in winter but, favouring sandy soils, is rarely implicated in damage. It is a tree that requires careful pruning if that should prove to be necessary. Alders are of the same family.

Trees near dwellings but beyond the distance of the mature height shown on the illustration, can be ignored generally, although in the case of Poplars and Willows it is suggested that the distances for these trees should be 35m rather than 26m for the former and 40m instead of 25m for the latter. Risk of damage increases substantially the closer the tree to the dwelling and those dwellings with shallow foundations will always be more vulnerable, two storey dwellings from the 1920s onwards, without basements, being most likely to be affected while extensions, porches, garages, conservatories and bay windows of any era, even more likely.

Ground Movement

In the absence of trees, but still with the evidence of subsidence damage to a dwelling, other possible causes may need to be considered. These can fall into two categories, those which can be said to be within sight and those which can only be a matter of pure speculation but, nevertheless, put forward with a distinct degree of possibility, perhaps reinforced if possible by the surveyor's local knowledge.

The most obvious possible cause within sight could be the affect, if any, on the dwelling of its proximity to sloping ground. Surveyors should be aware of the tendency of sloping ground to slip. Dwellings on or at the top of slopes accordingly can be affected by subsidence. Steep slopes are more inclined to move than shallow and looser soils more so than those which are stiffer or more cohesive, although even rock is not immune if cracks or fissures are present. Clay interspersed with layers of sand is much more sensitive to ground movement. Indications of unstable slopes include out of plumb fence posts and trees, fissures on the surface of the ground and local 'slumped' areas of ground. Where clay is known by local knowledge to be present, a slope of 8° is the maximum that can be thought of as free from risk. The surveyor needs to note where there is evidence of sloping ground and any signs of downhill movement in the dwelling. If excavations have been carried out nearby their influence or possible effect needs to be considered.

Leaking drains can be the cause of subsidence. The external area in the region of cracks thought to have been caused by subsidence should be inspected for the evidence of drainage runs, inspection chambers, gullies, soil and rainwater pipes. Leaking drains can cause a softening of ground below the foundations rendering it incapable of supporting the load of the dwelling or they can cause ground to be washed away depriving the foundation of support. This can be a smaller scale version of what can happen when there is movement of ground water caused by, for example, a fractured culvert, pumping of ground water out of excavations, discharge of rainwater pipes on to the ground or leaching of water into backfilled material in excavations. The likelihood of a dwelling being founded on subsidence sensitive soil other than shrinkable clay needs to be borne in mind although probably only a surveyor's local knowledge will draw him towards such a conclusion. The action of both ground and surface water on some soluble rocks such as chalk and limestone is a cause of cavities forming in the ground which may be infilled with soil, water or just air. If the infill material destabilises, subsidence can occur. Applied Geology Ltd on behalf of the government has prepared a comprehensive review of instability due to natural underground cavities and maintains a database of over 20,000 listed sites.

Un-natural cavities can be produced by the mining of below surface deposits, the collapse of the roofs of which can cause subsidence. Those mined in the past include coal, slate, limestone, tin and iron among others and local archives may need to be drawn upon if such a cause is suspected. Longwall and pillar and stall methods, along with other methods, for metal extraction are described in the Institution of Structural Engineer's booklet mentioned on p 320. Deterioration of the pillars can cause collapse and subsidence many years after mining has ceased. Reinstated open cast mining becomes another back-filled site.

Back filled sites can continue consolidating for some time depending on the material, the placement and subsequent compaction. Decomposition can continue for many years as well and particular dangers from subsidence occur where the development straddles a boundary between a filled site and virgin land. Vibration due to pile driving on nearby sites can cause yet further consolidation. Peat is described in the ISE booklet as 'reinforced water', highly compressible and liable to be displaced under load. Non-uniformity is a problem and significant variations in depth can occur. Lowering ground water in peats, silts and loose granular soils can reduce volume levels and cause subsidence. Freezing of the water

410: *A clear example of subsidence affecting this terrace of late 1800s two storey stone fronted cottages, situated in a mining area. The whole of the terrace beyond the rainwater pipe some 10 dwelling is tilted outwards in relation to the section on the left. It is just possible to see, by the rainwater pipe, the brickwork which has had to be used to make up the gap in the party wall created by the subsidence.*

content of soils and its consequential expansion can cause uplift, 'frost heave', but is an unlikely source of damage to heated dwellings, but could affect unheated garages or outbuildings if the foundations are less than half a metre deep.

To conclude the description of less likely, but possible, causes of subsidence, mention should be made of chemical attack on foundations by sulphates in the soil but which is only feasible in the presence of water. Sulphates occur naturally in some soils, for example Kimmeridge, Oxford and London clays, Keuper marl and Lower Lias and solutions of the sulphates in such soils attack the hardened concrete causing it to expand. Sulphate resisting cement should, of course, have been used for the foundations if the presence of sulphates in the soil had been known about or ascertained, but tests were not always carried out in the past and may not always be on smaller developments today. Even if sulphates were not present in the soil, the use of sulphate laden backfill materials such as colliery shale, brick rubble with adhering plaster and industrial wastes has caused problems in this regard, lifting both ground floor concrete slabs and the walls they support. Trench foundations, piers and piles are, however, less liable to damage from chemical attack than slabs.

As to the investigation into the causes of subsidence and the remedies available the Institution of Structural Engineers has produced a booklet of 174 pages entitled *Subsidence of Low-Rise Buildings*, ie those up to four storeys in this instance, which is now in its Second Edition of 2000, having first been published in 1994. It is the work of a task force brought together by the Institution representing owners, insurers, loss adjusters, lenders, contractors, local authorities, surveyors, arboriculturists and, of course, engineers.

It is, however, perhaps regrettable that the Task Force did not appoint an author to write the booklet, reserving to itself the function of editorial control. The booklet, as produced, unfortunately bears all the hallmarks of authorship by a committee in that, while no doubt containing all the necessary knowledge, it is diffuse, repetitious and contradictory at times. The stated excuse for the repetitions, that it saves time referring backwards and forwards, is rather undone by the fact that what appears to be a repetition can be either larger or smaller in scope than what came before. Nevertheless, the booklet in its entirety contains much of value and in particular the first six chapters up to where 'Further Investigations' begin to be described, following an Initial Appraisal, are of considerable

interest to surveyors, who are on the receiving end of a certain amount of implied, although not direct, criticism in their capacity as advisers to owners, buyers and lenders and sometimes insurance companies.

The Institution had fully recognised that a combination of increasing home ownership, changes in building design and construction over the years which seemed to make dwellings more brittle, two hot summers separated by a dry winter and the perceptions of occupiers had led to a vast increase in the number of claims against insurance companies. They were made under the insured risk of subsidence and landslip, first introduced generally in the early 1970s at the request of the Building Societies and extended to include heave in the 1980s, following publicity being given to a number of cases concerning the removal of large mature trees from public footpaths.

Whereas in rented accommodation occupants would tolerate cracks as not being their concern, homeowners could see the value of their investment and lenders the value of their security being put at risk. In particular after the very dry summers of 1975 and 1976 a substantial industry had grown up devoted to underpinning dwellings at insurers' expense. Further dry summers in the 1980s and 1990s have led to annual expenditure on such work often amounting to between £300 million and £500 million with an average of 40,000 claims a year. Insurance premiums had increased considerably while the number of disputes had grown, inducing an air of bad feeling all round.

Much of the expenditure on underpinning came about because the prospective buyer of a dwelling up for sale could not obtain finance due to the presence of cracks thought to be due to subsidence. If confirmed as subsidence and the insurer accepted the claim, pressure would be brought to bear for something to be done quickly and whereas in other circumstances a much fuller investigation over time as to causes and possibly less drastic remedial measures might have been acceptable, underpinning would be carried out with a view to getting the sale back on track. Even so, some insurers and lenders continued to be sniffy about dwellings which had been underpinned. Since, as has been seen, some 65%, more in years of drought, of all subsidence damage to dwellings can be attributed to trees and large shrubs on clay subsoils, it is not surprising that if a tree was nearby it became the culprit and the object of removal, with other possible causes given scant attention.

It was perhaps the concern of arboriculturists, as well as some engineers, which led to the preparation of the second edition of the booklet, a mere six years after the first, containing as it does much more information on tree management and recommending the felling of trees only in exceptional circumstances. The task force's hopes are that with a wider and better understanding of the genuine concerns of all parties at times when subsidence is suspected but, particularly, when dwellings are in the process of being sold and bought, better knowledge of causes and a greater understanding of the variety of remedial measures available, there can be more co-operation and less confrontation between the parties involved. This is perhaps a pious hope but it is, of course, the surveyor inspecting a dwelling for a seller, lender or buyer and owing a legal duty to all three who brings such questions into the open. Too much caution on his part can perhaps hinder a sale but too casual an attitude, or recklessness, can lay him open to a charge of negligence.

Categorising cracks by their size and dealing

with their structural significance is dealt with in the ISE booklet following on from the issue in the 1980s of a BRE Digest with a table dealing with 'The Assessment of Damage in Low-Rise Buildings with Particular Reference to Progressive Foundation Movement'. The first edition of this Digest included the subjective description of cracks in walls up to 5mm wide as 'slight' and between 5mm and up to 15mm as 'moderate'. The ISE while no doubt sharing the general view that such descriptions would never be accepted by the average home owner, who is inclined to view as unacceptable, if not alarming, any crack however small, seems to have regretted the dropping of the adjectives in the revised BRE Digest issued in 1990 and now includes its own adjectives in two tables in its booklet, cracks up to 2mm being defined as 'fine' and up to 5mm as 'moderate' reserving 'serious' for those from 5mm to 15mm in width.

While the ISE's adjectives are a little more acceptable perhaps than those used when the BRE Digest was originally first published they, none the less, are still very wide of the mark as to what would be thought appropriate descriptions by home owners and, accordingly, are best avoided. At the beginning of this subsection, under the heading of The Inspection, on p 161 it is suggested that for descriptive purposes the width of cracks alone is used and if they are tapering the width at each end given. It is not that there need be any objection to a general collective adjective being used to describe cracks in a room or dwelling for example 'the cracks are serious for a house of this age' or 'considering the age of this house the cracks are slight', although even here there is an element of subjectiveness.

Describing and classifying cracks by their width and giving a number to a category of damage as do both the BRE Digest and the

ISE booklet goes only a small way towards assessing their real significance. Their location, the pattern they form and if they taper whether the widest part is at the top or the bottom will indicate whether they are likely to be attributable to subsidence but what clients as sellers, buyers and lenders want to know is whether they are progressive or not. If a reasonably positive answer to that question can be given a much better picture of the significance of the cracks will be obtained and an even greater significance when they are related to other aspects of the dwelling, its age, its site characteristics and surrounding features.

All the aspects of the cracks as listed in the previous paragraph, bar one, are matters of fact which can be recorded on a single inspection. What cannot be recorded on one visit is whether they are progressive or not. The only way to be able to express a positive view on that is to monitor their width over a period of time sufficient to determine whether movement is in excess of that which can be considered normal. An opinion, but only an opinion, might be expressed as to whether cracks are likely to be progressive or not but that opinion would be mainly based on evidence other than the cracks themselves.

As already discussed, dwellings when first built will probably suffer an initial settlement, which may or may not be noticeable by way of cracks. Because most new dwellings involve a mixture of wet and dry trades in their construction, they may also exhibit cracks from shrinkage as the water used in those constructional processes dries out. These cracks will show up on the initial decoration provided by the builder. However, since they are normal to the process of new building and not caused by any failure by the builder to follow the NHBC Standards and therefore outside the terms of the warranty, they will

fall to be made good either by the builder. However, there is provision in the after sales service of the contract or by the first owner if and when he chooses. It is generally considered that any cracks arising from these two causes should never exceed 2mm in width and will generally be very much less. Repair accordingly will be comparatively simple involving filling in at the time of the first proper redecoration

In addition to initial settlement and shrinkage, the surfaces of a near new dwelling and indeed a dwelling of any age have to cope with seasonal variations between hot and cold and wet and dry periods. These variations can cause slight flexing of the fabric of a structure, particularly when it is founded in a clay soil, which can manifest itself at a point where a previous crack had developed. When a wall has cracked for whatever reason filling in and redecorating does not mechanically rejoin the two parts together, it merely fills the gap with a material which in itself may be subject to shrinkage as it dries out and hardens. This shrinkage and any slight flexing between seasons can be sufficient to re-open cracks to an extent of 1mm or so. The traditional way to combat this is to line the repaired area but the slight movement may still be sufficient to tear the lining material.

It is not the purpose here to deal with any concerns which the purchaser of a new dwelling might reasonably have about the scope of the NHBC Buildmark Warranty and Insurance Scheme. A surveyor's advice on that and on the new dwelling itself would take on a greatly different form to that derived solely from a visual inspection. However, in the majority of cases where a surveyor is instructed to carry out an inspection on a near new dwelling being sold to a second or subsequent purchaser within 10 years of the issue of the NHBC Certificate, he could be faced with either cracks or the evidence of making good to cracks of no more than 1 to 2mm in width. This degree of disturbance would be consistent with the effects of initial settlement and shrinkage coupled with any subsequent seasonal movement in a dwelling on a reasonably flat site, with foundations in a clay soil which is not unduly shrinkable and where the dwelling is well away from any trees. Provided the cracks or making good did not form an array, i.e. a series of cracks fanning out from a single location, the sum of the width of which should be taken as the equivalent of the width of a single crack for assessment purposes, or a pattern which could be attributable to subsidence and furthermore a period of five years had elapsed since construction had been completed it is not considered that there would need to be any cause for alarm. Nevertheless in the apt phraseology of the ISE booklet, relatively new dwellings are 'untested' and it is therefore necessary to adopt a particularly cautious atti-tude to such dwellings if cracks or other evidence of structural movement are found to exist or there are indications that such cracks have been made good. For this reason any inspection of such a dwelling needs to be even more exhaustive than usual to ensure that differentiation is made as far as is reasonably possible between cracks which may have different causes. Having done this it is important to demonstrate in the report how conclusions have been formed and to set out the evidence to show that they were not reached lightly but only after due diligence and consideration of all the factors.

This care and diligence is particularly important in the circumstances set out above in view of the different considerations which apply to a second or subsequent purchaser in relation to the NHBC Warranty. While the warranty will be carried forward it will only be honoured in respect of defects which develop after completion of the sale and not to the repair of those which existed, or which could have been ascertained to exist by inspection, at the time the sale took place. For this reason if a favourable view is to be given in a report it may be that to obtain an adequate sight of all parts of the dwelling, furniture may need to be temporarily removed.

It is unlikely with a near new dwelling that such a favourable view could be expressed in any other circumstances than those described above. Cracks, or evidence of making good to cracks larger than 2mm, a shorter period of time, a pattern of cracks pointing to the view that subsidence could be the cause, a steeply sloping site, clay soil with trees nearby are all possible reasons, either singly or collectively for the surveyor indicating a distinct risk of proceeding, with the reasons for such advice being fully set out in the report. It is salutary to be reminded that over half of all the claims made under the NHBC Warranty scheme relate to the foundations and that claims are made on something like 10% of all dwellings completed in a typical year. There is, there-fore, all possibility that a surveyor will be inspecting a dwelling where it is purported that not only initial settlement and shrinkage have been dealt with but also a degree of subsidence where work to foundations not originally carried out in accordance with the NHBC Standards has been stabilised. If such stabilisation has been carried out within a period of five years before a sale to a second or subsequent purchaser, then the only way to demonstrate that the work was successful is for it to be monitored professionally for at least that period, irrespective of how well it might be claimed that the earlier investigation had been carried out and the quality of the work of stabilisation. Even if the situation had been monitored for at least a period of five years and there was no sign of further movement it would still be necessary for the surveyor to consider critically what had been done, or, if unable to do so himself, advise the employment of an engineer for the purpose.

Such considerations apply equally to any dwelling of an age beyond where it can be considered new or near new although in the absence of a warranty there will be less temptation to a prospective buyer to proceed since he will have to bear the cost of repairs if the surveyor has been sufficiently careful to stress the risk. The most dangerous situation from the surveyor's point of view is where a dwelling has been completely and newly redecorated and there is evidence of cracks having been made good. Here again the surveyor is in no position on a single inspection to say whether the cracks are likely to re-open or not and if they did whether they would be progressive. The redecoration may have innocently been carried out to improve the marketability, as is sometimes advised, but as the Judge said in the case of *Morgan* v *Perry* 1973, human nature being what it is from time to time vendors will take steps to cover up defects. The surveyor is not to know which of these circumstances apply and questioning the owner will not always elicit a satisfactory reply. The case mentioned concerned the inspection of a four year old house built on sloping ground, the surveyor mentioning only 'hairline' cracking but no location given. He made no comment on making good to a cornice or the redecoration which had been carried out, the sloping floors or cupboard doors which were sticking. More cracks appeared after purchase and damages of £11,400 were awarded. Cracks could possibly re-open if the making good is in places where subsidence cracks could occur and there are present any of the situations, clay soil, presence of trees, sloping ground, for example, likely to induce it. There would be no alternative but for the surveyor to advise that proceeding to purchase would involve risk,

more perhaps in some cases than others but that would only be a matter of degree. The risk could only be discounted by a period of monitoring to ensure that cracks did not re-open to an extent of more than 0.5mm over at least two years. If evidence, photographs etc., could be produced to show that such monitoring had been carried out professionally for that period at least before the entire redecoration, then the surveyor would have no valid reason to stress the risk of proceeding unduly but he should be very careful to explain his reasons for being less concerned so that he could, if challenged at a later date, show that he took all the reasonable care that any prudent surveyor would take in similar circumstances.

Where the decorations in a dwelling beyond the near new stage are clearly of some age with wall covering faded, the contrast between that behind pictures and elsewhere rather striking, paintwork yellowing, worn and chipped and ceilings and walls dis-coloured by the occupants habit of smoking or marked by centrally heated air currents, cracks can usually be discerned as either old or fairly new if they themselves are either as grubby as the rest, and contain debris, or look sharp and have a fresh appearance. Such indications are, however, not wholly reliable. The cracks seen on a single inspection show the total amount of movement since the decoration was carried out but not conclusively whether it was comparatively sudden or whether it has developed slowly over the whole time. Advice here will depend very much on individual circumstances, the type of soil, certainly local knowledge, the presence of trees and sloping ground will all need to be taken into account for an opinion to be formed. Just occasionally when either none of the circumstances likely to induce subsidence are present or if they are, they are too remote and can be discounted, it may be possible to say that any movement in the past at such stated locations in the dwelling has ceased, is of long standing, is non-progressive and is acceptable. It could be that the older the dwelling the more likely is this to be the case.

Radon

The National Radiological Protection Board (NRPB) is a statutory body set up following the passing of the Radiological Protection Act of 1970 and charged with the duty of advising the government on the protection of the community from radiation hazards. Its advice that there was a danger to the health of occupants of dwellings where the concentration of radon, a radioactive gas, was above a certain level while disputed by others was, nevertheless, accepted by the government. It concluded that the advice was in conformity with the international scientific consensus. Accordingly, owners of dwellings together with prospective buyers and their lenders can now find out from local authority searches and by means of logging on to the NRPB website www.nrpb.org whether the dwelling in which they are interested is in a 'radon affected area'. Owners up to the early 1990s had been going around blissfully unaware that such a danger existed, but now are told that radon in their homes could be causing some 2,500 deaths a year from cancer, mainly lung but also possibly skin and prostate and also leukaemia, although the latter, due to exposure to radon, is considered far more speculative.

What this may mean to the average person is not really known but many will have heard of the horrific injuries sustained by the survivors of the atomic bombing of Hiroshima and Nagasaki and the devastating effect from the spread of radiation following the explosion at the nuclear power plant in Chernobyl. Alarm,

accordingly, would not be surprising. Besides confirming that the dwelling, by enquiry, is in a radon affected area, it has now become the surveyor's duty to explain the significance of the information. First of all he should ask the owner if he was aware of the fact and, if so, has he done anything about it. If he has, he will no doubt be able to inform the surveyor of the radon measurement result. This is obtainable through the post from NRPB, for a fee, by use of the detector which the board supplies and which records the level over a three month period, the results being analysed to produce the average radon concentration over that period. More than likely it will be the first time he has heard anything about it. If he is proposing to sell then, clearly, he needs to put obtaining such a figure high on his list of priorities. Fortunately, however, though in a 'radon affected area', many dwellings have a low average radon concentration, but until that average over a three month period is known, nobody can say whether any action need be taken.

Radon being a colourless, odourless and tasteless radioactive gas, discovered in 1901, comes from the decay of uranium which is found in all soils and rocks, varying in amount from place to place. The gas rises from the ground and is diluted in the atmosphere so that the concentration in the open air is very low, indeed negligible. However, it gets into dwellings through cracks in the floors and the walls at low level, where the concentration increases as a result of the pressure being lower inside than out. It is thought that the more prevalent adoption of central heating and the improvement in insulation levels in many instances has reduced air movement in dwellings to the extent that levels of concentration have increased over the years.

The Board has estimated that the average indoor level of concentration for radon gas in the 24 million dwellings of the United Kingdom housing stock is 20 becquerels per cubic metre, Bq/m^3, a becquerel being a unit named after the Nobel prize winner Henri Antoine Becquerel. NRPB points out that the figure for dwellings includes both low and high rise units, so that it could be expected that average figures for houses would be higher. The government has set the 'action level' at an annual average radon gas concentration of 200 Bq/m^3 and indicates that if the concentration level is either at or above that level, action should be taken to reduce it.

It will be noted that the 'action level' is set at 10 times the estimated average level. The 'radon affected' areas are those where it is expected that 1% of the dwellings have radon levels in excess of 200 Bq/m^3 and exposure at that level for a whole lifetime will cause about three extra deaths from lung cancer in a group of 100 people, i.e. a 3% risk. This is in addition to the normal incidence of six deaths from lung cancer, mainly due to smoking, which would occur in the absence of radon exposure. The risk would, of course, be compounded and therefore higher for smokers, 10% as against 1% for non-smokers.

Initially, areas thought to be most 'radon affected' were identified as the granite regions of Cornwall and Devon, limestone subsoil regions of Somerset, Derbyshire and Northamptonshire and less severely the granite areas of Grampian, Highland, Dumphries and Galloway and parts of Wales and the Midlands. The reason why these other granite regions are less severely affected is because the rocks are far less fissured than those of the South West of England. Further areas are constantly being added to NRPB's atlas.

Whereas in the case of new dwellings the Building Regulations 1991 Part C2 ('dangerous' and 'offensive substances found on or in the ground') in Approved Document C, 'Site Preparation and Resistance to Moisture', cites a report by BRE which gives detailed guidance on the localities where protection against radon in new buildings is required, together with suitable construction details, there are no requirements for existing buildings found to be above the 'action level' in radon affected areas. For one thing, sources acknowledge the undoubted fact that radon levels in a dwelling can be up to ten times different from those in a similar dwelling next door, they vary from season to season, hour by hour, day by day and according to the weather. Nevertheless, if readings over three months produce an average radon concentration in excess of the 'action level' of 200Bq/m^3, owners, prospective owners and lenders must be informed of the fact and be advised to take, or require action to be taken, to reduce the level. If no work has been carried out, then a recommendation for further investigation is needed to ascertain what form, having regard to the level of concentration and the type of existing construction, of protective measures are required. Until these are known and their cost, a handsome allowance should be made in any valuation as the range of costs for protective measures is considerable.

It is the fact that readings over three months are required by NRPB to assess the average concentration and the acknowledgement that the actual level can, seemingly, vary by the minute that begins to sow a degree of scepticism over the whole situation. For example would another three month's data collection produce an entirely different figure? It certainly would if taken at a different time of year. Eminent scientists in the field have cast doubt on NRPB's estimate of 2,500 deaths a year attributed to radon, pointing out that there is not a shred of evidence to support the figure. Is this just another imagined scare like so many others, based on no real evidence, man's influence on global warming for example?

Everyone agrees that massive doses of radiation do harm, as discovered in the case of uranium and other miners, but rather like small doses of the poisons arsenic and strychnine which act as a tonic, so small doses of radiation can help to boost the immune system, just as immunity to disease can be induced by small doses of the appropriate virus. This is borne out by evidence that there is less incidence of lung cancer in areas where radon levels are highest, Cornwall for example, backed up by similar findings in the United States, Sweden and Finland, to the extent that some consider the small dosage of radon to be actually helping smokers suffering from lung cancer and those in urban areas affected by atmospheric pollution to live longer. The Radiobiology Unit of the Medical Research Council concluded that little is known about the risk of exposure to radon in the average domestic environment of the United Kingdom and considered that over estimation may have resulted from assessments being based on the incidence of lung cancer among miners, which had been extrapolated downwards. The House of Commons Environment Committee acknowledged the differences of opinion in 1991 but the government, no doubt fearful of being caught out by saying that there was no danger, came down on the side of caution, setting the action level at 200 Bq/m^3 as against the 150 Bq/m^3 adopted in the United States.

For inspections and reports on existing dwellings for buyers and lenders, the surveyor, for all reasons, has to set out the advice of government, the NRPB and BRE. Clients, however, before expending, or being required to spend, perhaps considerable sums of money which they are not required to do by law, should be given the opportunity to

review both sides of the argument and make up their own minds. Sellers, on the other hand, in 'radon affected' areas would be well advised to establish the average level of concentration for the dwelling and then decide if it is at or above the 'action level', whether steps should be taken to reduce it, bearing in mind that a prospective buyer may take the government's view that there is a danger to health, rendering the dwelling unfit for human habitation and either be put off entirely, require a reduction in the asking price or even protective measures to his approval being carried out before completion of the sale. Whereas the protective measures which would be necessary would depend upon the outcome of a further investigation, the surveyor needs to be aware of the various methods which are available. Briefly these are as follows: simple sealing of any openings allowing entry of the gas from the soil: diversion of the soil gas through extra ventilation to underfloor space or the building of a sump below a solid floor with an outlet pipe connected to a fan, which depressurises the soil and draws the gas away. Whereas the methods sound fairly simple it is difficult to achieve a substantial reduction down to the recommended maximum concentration of 150 Bq/m^3 with any degree of certainty and costs can be into the thousands of pounds if unpleasant conditions and disturbance from mechanical noise is to be reduced to a minimum.

Section 5 Interiors

Introduction

It is important that the structure of a dwelling is considered as a whole and not just as a series of individual components. A sound and satisfactory roof and well constructed walls are, obviously, vital elements but, more often than not, they in turn are dependent on the internal structural features of partitions and floors to provide the necessary support and restraint. Without these, external walls can develop leans, bows and bulges and roofs can sag. Even when the necessary support and restraint are provided, inadequacies of design and construction can undermine what would otherwise be satisfactory construction elsewhere. For example, an internal partition wall could well be carrying a substantial load from floors and roof. Failure of its foundation, causing it to settle, can lead to sloping floors, a distorted roof and possibly disturb the stability of the external walls. It is vitally necessary therefore to establish, on the internal inspection, which partitions provide support to the structure above. It is often overlooked that internal partitions can also be crucial in providing vertical restraint to assist the prevention of outward movement in the external walls. As to floors at first storey and above, examination should determine the nature of the bearings providing support to ensure that intended lateral restraint to the external walls has not been altered or rendered ineffective by decay. The internal inspection needs also therefore to take the aspects of construction, support and restraint of floors and walls, both external and internal, into account in every room on every floor.

Partitions

Apart from those bungalows built from around the 1950s when lightweight timber roof trusses began to become available to span large areas of floor space, most dwellings will have at least one internal structural partition to support the floor of the first and any further upper storeys. However, even bungalows built before then may have required an internal load bearing partition to support the struts forming an essential part of the on site cut and fitted close couple roofs of the time.

Whether the dwelling is of the two storey three up, two down semi-detached or detached type of the periods 1920 to 1940 and 1940 to 1960 or the average sized three to four storey terrace dwelling of the 1700s, 1800s or early 1900s, the main internal load bearing partition will generally be found to be that separating the front and rear principal rooms on the ground floor. This partition not only supports the floor joists of the first storey but also those of any floors above and furthermore provided the most convenient position from which to support the struts forming part of the roof structure, being roughly half the distance of the total roof span.

It will be appreciated, therefore, that the internal load bearing partition of the typical average sized dwelling built over a period of some 250 years will be supporting a load in excess of twice that of any of the external walls. While it is true that for domestic construction, most of the external walls, because of the need for weather proofing, are built far stronger than technically necessary, it has not always been the case that the requirement for this extra degree of strength by the internal load bearing partition has in all cases been fully appreciated and accommodated. Thus, it will sometimes be found that while the external walls may be substantial, the partition can be of comparatively flimsy construction. Recognition where this is found to be the case often provides a simple explanation for many of the internal defects found in dwellings.

Of course, typical average sized dwellings are being considered here but this is not to overlook the fact that it has always been possible to arrange the structural elements of dwellings to provide for the types of accommodation required. For the more expensive dwelling of the early to middle 1900s with wider frontage, where large areas were needed at ground floor level, structural partitions could run from front to rear, dividing a basically rectangular dwelling into separate though integrated structural units. In the 1700 and 1800s, on the other hand, large spaces were more often needed at first storey level, seemingly interrupting a transverse structural partition, which is apparent at basement, ground and second storey levels in a large terrace dwelling, but not in the first storey. This could cause the surveyor a degree of puzzlement and, accordingly, he needs to be aware that trussed partitions, based on similar joints with the metal straps and tension rods used for the elaborate roof trusses of the period, could be arranged so that a floor could be suspended below the partition which would also, if required, be supporting a floor above.

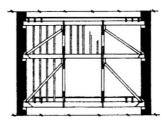

Figure 60 : An example of a trussed partition used in the 1700 and 1800s. These partitions could be used at an upper storey level when large through rooms were required in the storey below. They were braced against the walls on either side and supported a floor above and carried the floor below, though in this case with the lower tie showing as a 'beam' on the ceiling underneath. If no opening in the partition was required there would be no need for the 'beam' to be seen as it could be positioned above the floor level, holding up the joists.

Partitions are often considered a source of annoyance to owners who wish to make improvements. Throwing the front and rear rooms into one is a popular measure. It can be done at ground storey level, always assuming that an adequate beam is provided *in lieu* of the structural partition. A smaller opening in the form of an arch or a door still needs care and the adequate spread of the load from above on either side is essential. Where the partition is of timber and perhaps to the uninitiated thought less important, unwise cutting of sill plates, struts or braces can lead to a serious loss of support. A case in 1996, *Gardner* v *Marsh and Parsons*, concerned openings formed by a developer in the central spine partition during the conversion to flats of a listed five storey mid 1800s terrace dwelling. The surveyor missed the significance of rucking wallpaper, indicating developing cracks which later necessitated the insertion of steel beams. The Court of Appeal upheld the view of the lower court that the surveyor had been negligent.

For ascertaining the construction and significance of internal partitions the age of the dwelling is an important factor, as always, towards providing clues. Partitions invariably present a face of plaster on both sides so that in addition to a knowledge or an assessment of age, reliance needs to be placed on the partitions thickness, its sound when tapped plus information derived from surrounding features. The adjacent floors either side may reveal floorboards running parallel, a good indication that the partition is supporting floors joists. Boards running at right angles suggest joists are running parallel to the partition. The layout of rooms above and below may need consideration such as whether the partition under review continues in the same form above and below and in the same position.

As to measurements of the thickness of partitions these are often most conveniently taken on leaving a room. This should be part of a routine sequence in combination with an examination of the door and frame. If the latter are out of square this can be an indication of movement in the partition itself. Solid sounding partitions in dwellings built before 1920 will almost certainly be based on brick ie 260-270mm if of one brick thickness and around 130-140mm for a half brick thickness, with plaster both sides. Comparatively rarely, solid load bearing partitions in large houses of the 1700, 1800 and early 1900s will be found to be of 1½ and 2 brick thickness, 370-380mm and 480-490mm thick and if the dwelling is of stone construction, upwards of 450mm. For houses built after 1920, the load bearing partitions are more likely to be of cheaper concrete blocks or fired clay units and thicknesses would be usually some 50-60mm less than would be the case if they were of brick. Unless a sight of a partition is obtained where it is not plastered, only reasonable assumptions can be made based on the evidence obtained and put forward as such.

It should not be assumed that a hollow sounding partition is of no structural significance, even if it is of comparatively slender thickness. With good quality timber, careful design and construction with adequate ties and bracing, softwood timber is perfectly capable of supporting floors and a degree of roof loading in two storey domestic construction. With lath and plaster both sides, the thickness would unlikely normally, to be less than 140-150mm and could possibly be much more. Once again only assumptions can be made of construction, but information derived from adjacent floors and from rooms above and below should enable the status of the partition to be clarified as structural or non-structural.

The important factor about a timber and plaster load bearing partition is the treatment at the base. In dwellings built before 1900, very often the partition would be built off a plate fixed to the top of a sleeper wall at the same height as other plates supporting the timber joists of a hollow suspended ground floor. Such plates should, of course, be laid on a damp proof membrane to help prevent, along with the ventilation provision through the sleeper walls, any rising damp causing an outbreak of rot. The damp proof membrane however, was often omitted and even though the ventilation may be adequate to prevent an outbreak of dry rot, over time wet rot in the plate can eventually affect the effectiveness of the partition to sustain the load without settling.

A cautionary note needs to be introduced here in regard to the transverse load bearing partition in the speculative housing of the 1700 and 1800s. This will often sound solid and provide a thickness measurement indicating brick construction. A common practice of the time, however, was not to build the partition with sound bricks, proper bonding and adequate mortar but to use overburnt, inferior misshapen or underburnt soft bricks along with broken bats and inferior mortar, all put together in a rough and ready framework of timber. Ground leases in the early 1700s were short, as little as 33 years at times, and the builder's intention was that the construction should last very little longer than the length of the lease. Even when leases became more standardised at 66 or 99 years, the practice continued and was common until the late 1800s. As with the hollow timber and plaster partitions noted above, the important factor was the performance of the base plate in relation to rot and degradation.

Partitions are not necessarily of the same construction or of the same thickness throughout their height. If the floor for the topmost storey has been provided with adequate support and there is no load to be carried from the roof, the uppermost portion of the partition can be constructed to a lesser standard, sufficient merely to separate rooms.

Generally, the foundations of structural partitions should have followed the practice at the time for the construction of the foundations for the external walls and where this has been carried out there is perhaps less likelihood of problems than there could be with the external walls. The dwelling itself protects the ground in which they are placed from the effects of undue weather changes. However, short cuts and economies may be taken leaving the partition less able to perform its function than would otherwise be the case. The inspector will not be able to see the form of the original construction but sloping floors, binding doors and cracks in the plasterwork will, where they occur and are serious, need to be reported and coupled with advice that further investigation will need to be undertaken.

In particular, the founding of internal structural partitions on the oversite construction below hollow timber joisted ground floors can lead to problems of settlement where no allowance for the loading on the partition has been taken. Even founding on the solid concrete ground floor slabs which were frequently used because of timber shortages in the 1940s and 1950s produced an array of problems due to insufficient preparation of the site, inadequate compaction and often the omission of appropriate thickening of the slab where the partition occurred.

The quality of the dwelling may well have been a determining factor in the construction of non load bearing partitions. Expensive dwellings are likely to have brick and plaster partitions in anticipation of the need to support large glazed pictures and other heavy objects or fittings on the wall, for superior sound insulation and for a more impressive effect. The occupants can find the thin timber and lath and plaster or plasterboard partitions found frequently in cheaper dwellings less than ideal in these regards. Even more inadequate are those proprietary partitions consisting of two layers of plasterboard separated by an egg-crate shaped carboard interior, since plasterboard needs a patent type of hook for hanging even the lightest object.

Non structural partitions at any storey level which are not a continuation of a partition in the storey below need some form of support unless they are of very lightweight construction and the partition is running at right angles to the floor joists below. If comparatively heavy, it should have been provided with support from a beam. Otherwise support is normally obtained where the partition is running parallel to the floor joists from a doubled up joist. In the conversion of larger dwellings into flats, support for the new partitions, which are normally needed, does not always receive the attention it deserves with the result that cracks can appear in the ceiling below as deflection in the floor occurs.

The ubiquitous breeze blocks with which solid partitions were often formed in houses built between 1918 and 1960 were of dismal quality and difficult to cut to form angles or openings. If intact they are best left as they are but cracks and or bulges suggest instability and this may entail renewal on a more extensive scale than might, at first, be thought. The introduction of a range of concrete and lightweight blocks has made a vast improvement but examination of cracks and distortion is just as necessary in both structural and detailing terms as with timber partitions, to determine whether there are major or local faults.

Floors

Hollow Timber

In domestic construction, the vast majority of floors are of wood, almost universally so above the lowest storey. Along with partitions, they have a contribution to make to the overall structure of a dwelling above ground level by reason of their self weight and imposed loads from furniture and occupants, being supported on some of the external walls and which, in turn, benefit from the very necessary restraint they provide.

Little attention, however, was ever paid to the aspect of providing restraint to the other walls of a dwelling until the 1960s. Accordingly it was only those walls which supported the ends of floor joists which received the benefit of restraint. Walls parallel to the run of floor joists in many dwellings built before that time were left without any restraint at all. This often resulted in those walls developing the leans and bulges, even in two storey dwellings, described in Section 4.

Similarly, there was little attention by way of regulation paid to the requirements of strength in the construction of upper suspended floors until the introduction of Bylaws in the late 1800s. In earlier times, but particularly in the 1700 and 1800s, reliance would be placed on rules of thumb, or pattern books, to determine the depth required for floor joists to span the largest room in the proposed dwelling. The rule would assume usage of the common width of 50mm for softwood structural timber available from the merchants at the generally adopted spacing for joists of 350mm. Once selected, this would set the depth of the floor at that level throughout the dwelling. The margin of safety provided by the rule of thumb and the quality of timber obtainable at the time meant that, in the main, a satisfactory result was produced. Difficulties could arise, however, when to save money, a lesser standard was adopted, 50mm less, say, for the depth than the answer the rule of thumb produced. Furthermore, in practical terms all the timber floor joists of the 1700 and 1800s have subsequently been notched and drilled for the installation of the service pipes to supply water, gas and then electricity and the improvements found to be necessary to the systems when new bathrooms and kitchens have been installed during the 1900s.

For new dwellings, the size of floor timbers are set out for England and Wales in the Building Regulations 1991 in Appendix A to Approved Document A or for Scotland in section 5.10 of the Small Buildings Guide produced in conjunction with the Building Standards (Scotland) Regulations 1990. It is to one or other of these documents that the surveyor needs to turn when he has the opportunity to examine the specification for a new or near new dwelling for to confirm that the specified sizes are in compliance with the regulations or standards. Should the specification not be available then the surveyor is in no better position for checking the adequacy of an upper storey floor than he would be in respect of any other dwelling.

In the absence of permission to open up, the information for checking the adequacy of floors could be limited though it can be greater in some circumstances than others. Fitted carpets which cannot be lifted at any point and prevent a sight of the floorboards and therefore the direction of run, deprive the surveyor of even information on the span distance. In these circumstances, where there is doubt about the ability of the floor to support domestic loading because of undue deflection, the only advice which can be given is to expose the structure for further investigation.

411: *In process of refurbishment, the ceiling of the front ground floor room with cornice, seen on the left, is higher than that of the rear room beyond, providing a sight of the inner end of the joists of the floor above and the timbers of the braced stud partition carrying the ends of the joists and separating the rooms. The removal of a stud to form a later door opening has caused some deflection in the top plate.*

412: *The underside of the landing to a staircase where the ceiling has been removed providing a sight of the floorboards, the substantial joists and, beyond, the exposed underside of a flight of stairs. The carriage is clearly visible as is the bottom of the top landing newel post. The curved dado rail on the wall beyond follows the curve of the winders making up the intermediate treads.*

In other cases, where floorboards are visible, the run of joists will be evident, enabling the span to be measured. The frequency of nailing will probably also be apparent giving the spacing of the joists. It should be possible to measure the thickness of the floor at some point on the staircase which, with a reduction of 40mm for ceiling and floor finishes in pre 1940 dwellings, provides the depth of the joists. The only item on which information is missing is the width of the floor joists. For a dwelling built up to 1940 a reasonable assumption for the width would be the 50mm already mentioned and, with this amount of information, it is possible to compare the size as assessed with the tables in the regulations and standards and check whether the correct size for the span was originally used. A short-fall in the assessed size in comparison with the correct size from the tables using the lowest strength of timber would indicate that any undue deflection is due, at least partly, to an inherently wrong choice of joist size. If the comparison suggests, however, that the joists ought to be satisfactory, then the cause of the inadequacy could lie in the floor's subsequent history, for example excessive notching and drilling for the installation of services as described previously.

For houses built after 1940, however, it would not be reasonable to make the assumptions suggested because of the much wider possibility for the use of different sizes of timber and different forms of construction. Advice on opening up for a close examination would then need to be given.

Some of the information which it is necessary to assemble before an opinion can be given on the inherent strength of an upper floor is that which should have been collected on the room by room inspection of the interior, particularly the run of floor joists. Other

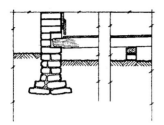 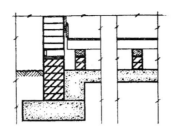 

Figure 61 Many early hollow suspended timber floors, as on the left, were deficient in underfloor ventilation. Furthermore, there was no site concrete, no damp proof course and the ends of joists were built in to external walls. Rot was the almost inevitable result. In the centre is a diagram of what might have been done to improve the situation to bring the floor up to the standards of the 1900s. Underpinning, site concrete and a new sleeper wall to carry the ends of the joists and air bricks for through underfloor ventilation would be provided. The drawing on the right shows the typical 1900s floor structure for a dwelling requiring $1\frac{1}{2}$ brick thick walls.

information, more related to the development of subsequent faults, is whether the floor is sloping, whether as a whole or in part and in which direction, whether it sags towards the centre, whether there is a gap between the floorboards and the skirting or whether a moulding has been fixed at that point to disguise a gap which exists. It is, of course, in connection with ascertaining these faults that the surveyor's spirit level is put to good use.

Not quite as obvious as the above defects which are visible are those which may be hidden but which are possibly contributing to the overall appearance of a defective floor. Among these could be decay and rot in the ends of joists where they are built into external walls for support and rot in the timber plate on which they bear. Building in to external walls was the practice until the 1940s and as no ventilation was ever provided to the timbers in upper floor construction, persistent driving rain could penetrate through what was, in the case of walls of one brick thickness, little

more than 120mm or so of brickwork. Floors of dwellings built after 1940 can also have problems where joists instead of being built in, are supported on joist hangers. It is important that joists are properly seated, fitted tightly to the back of the hanger and that the hangers are securely built in to the brickwork and prevented from twisting by the fixing of a batten between them, attached firmly to the adjacent wall.

It is unlikely that solid construction will be found for the floors at upper levels except in the case of purpose built blocks of flats. Such floors are likely to be of reinforced concrete construction if built from the 1920s onwards or of rather specialised types of construction in the mansion blocks of flats built from the 1880s to around 1920. This is not so, however, in the case of floors constructed at ground level and it may need a sharp stamp on the floor or a 'heel drop' with the feet together to distinguish solid from hollow construction. The sound of a hollow suspended timber floor can be produced at various degrees of intensity, bright and clear to signify a tight, sound, well constructed floor or a deep rumbly, dead sound with a distinct degree of vibration to indicate a rather shaky floor, suggesting less than adequate construction or the development of faults.

Generally, suspended timber floors at ground storey level are formed on sleeper walls built off such site ground cover as there may be, rammed earth, ashes or weak concrete depending on the era of construction and the rigorousness, or otherwise, of the enforcement of the regulations in existence at the time. Early versions in the 1700s could be built off the compacted earth after the top soil had been removed with a baulk of timber or a plate on a course of bricks, the ends of the joists built into the external walls and with minimal underfloor through ventilation. In the absence of any damp proof membranes, the ends of the joists would probably degrade first from wet rot and then dry rot. Later versions would, *in lieu* of building in, provide an additional sleeper wall close to the external wall and a greater depth above an ash covered site with more air bricks for under floor through ventilation. At the end of the 1800s and into the early 1900s, site concrete and damp proof courses would become common

413: *A section of suspended hollow timber floor construction at ground level has been taken up here for renewal. A new sleeper wall of three courses of brick with wood plate has been provided to carry the new joists. The surveyor will need to check that a damp proof membrane has been provided below the plate and that there is adequate underfloor through ventilation. None is to be seen on the section visible here. The old joists, one can be seen on the right, are carried on the partition the bottom of which can be viewed beyond the new sleeper wall.*

to bring the standard up to that which generally pertained up to the 1940s and, indeed, in many instances continues to this day.

The trouble was that until the period 1920-1940 when the typical house became more or less square in shape, those of the later 1800s and early 1900s were very far from square. With their back additions or 'offshots' usually provided with solid floors, the suspended hollow timber floors of the main part would be left with quite large areas of subfloor space inadequately ventilated, particularly if there were comparatively few air bricks positioned in the external walls. On a damp site or where dpcs had been omitted or had failed there could be the ideal conditions for the development of dry rot.

Many of the typical terrace houses of the 1700s, 1800s and early 1900s were, of course provided with cellars, either over the whole of the ground floor areas of the dwelling or limited to that area beneath the entrance hall. These were primarily for the storage of coal which provided the only form of heating up to around 1900 and even up to the 1940s was still the cheapest. Cellars, now often full of old or unused household appliances, provide inspectors with useful access from which to view the underside of the floor at ground level. If a cellar is provided over the whole site area then, of course, the floor above will as likely as not, be totally suspended in the same way as the floors to the storeys above but enabling, assuming there is no ceiling, the surveyor in these circumstances to see the size and spacing of the joists. He will probably also be able to identify and see where the main internal structural partition is placed, establish its construction and measure its thickness along with any other partition that may, for example, be supporting the staircase and landings. Entrance hall floors are often tiled on top of joists and floorboards and the

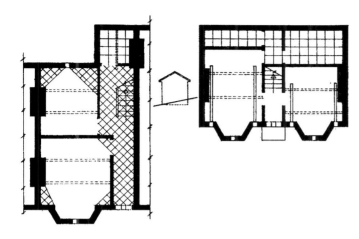

Figure 62: The suspended hollow timber floor at ground level in the average terrace dwelling built in the latter part of the 1800s and early part of the 1900s often, as originally built, has limited ventilation to the underfloor space. With only one air brick in the front bay, another in the front threshold and only one in the rear wall pockets of damp air would accumulate in the cross hatched areas shown on the left hand drawing. If the back addition or offshoot had a solid floor, as usually the case, the whole of the space below the entrance hall would be unventilated. An even worse situation would arise in a detached dwelling built on the side of a hill where rooms at the rear had solid floors as shown on the drawing on the right. The situation might be improved, as shown, by the provision of new air bricks on both sides of the bay windows and truncated ventilation to the rear of the main rooms.

membrane below the wood plates and also the adequacy of the openings left for through ventilation in the sleeper walls. These partial cellars in many ways can be considered the saving grace of such houses since not only do they provide useful space and access for inspection but the extra degree of ventilation can keep the area free of the problems which may arise if the space below the hallway flooring was more confined and deficient in ventilation. The not uncommon burst or pin hole leak on an old lead mains water pipe is a case in point. It will usually be spotted early on and dealt with if there is a cellar. If not and it occurs below a floor with tiling or fitted carpet it may go unnoticed for a long time until an outbreak of dry rot causes the floor to collapse.

surveyor will be able to see how this form of construction is arranged. Without a cellar this would be impossible to ascertain without opening up and, in the presence of tiling, permission to do this is unlikely ever to be forthcoming unless some serious defect, obvious to all eyes, is present.

Even though the cellar may only extend to the area below the entrance hall, a sight of the construction of the sleeper walls supporting the bulk of the floor at ground level is usually obtainable. This enables a check to be made on the position, the thickness and the presence, or otherwise, of the damp proof

414: Generally hollow timber suspended floors at approximately ground level are single, i.e. the joists are supported directly off the plates on the sleeper walls. In a somewhat unorthodox fashion this floor is double. Substantial cross beams support the floor joists which in turn are carried on brick piers, approximately 850mm high.

Higher up the scale, so to speak, from 'cellars' are 'basements' with, usually, their better light, ventilation, a proper ceiling and plastered walls. The durability of any hollow suspended timber floor at this level, well below the level of the surrounding ground, will be very much dependent on the degree of dampness present and any provisions for its prevention in the enclosing walls or amelioration but, in particular, on the availability of through ventilation to the underside and the avoidance of pockets of damp air which are liable to accumulate. This can be achieved through the provision of external areas front and back for the full width of the dwelling extending to a depth of around 400mm below the internal finished floor level and the installation of air bricks at each extremity and at intervals in accordance with current requirements, 1500mm² for every metre run of wall.

The external areas should extend, front to back, from the walls enclosing the basement, free of obstruction, to at least the minimum current requirements wherever possible. These ideal requirements will be found to be present in very few old dwellings even apart from the factors of damp proofing, often difficult to achieve in the basements of the older terrace dwelling limiting the use of such accommodation to utility purposes. The surveyor's routine and the requirements of his Conditions of Engagement, Practice and Guidance Notes will necessitate the use of the damp meter at frequent intervals and more than likely the finding of considerable areas of dampness. The shortcomings of a basement need identifying and describing in the report.

Solid Construction

It was not until the period 1940-1960 that solid construction for floors at ground level

began to take on the structural significance that they were to assume for some years, particularly in many local authority and New Town Corporation housing schemes and those schemes at the lower end of the private sector price range. For example in one English New Town BRE estimated that 97% of houses had solid floors at ground level.

It was not a good period for the propagation of such construction on a major scale even though the reasons of urgent need for new housing, shortage of timber and economy seemed sound at the time. The policy of retaining green belts around the major cities meant that sites not previously considered suitable for house construction were being brought into use, some steeply sloping, with drainage problems, where household waste had previously been dumped or minerals extracted. Skills were lacking and less experienced developers were in the market. It was thought that sites could be quickly cleared and levelled, a layer of hardcore put down, a slab of concrete laid and the floor was ready for its finish. As a bonus the slab provided support for partitions, structural or otherwise, together with all the fittings in the kitchen and support for the staircase.

Poor sites, poor preparation of the site, the wrong materials in the hardcore, collery shale and pyrites for example, indifferent compaction of the concrete, resulted in overall and differential settlement in many cases and in others substantial heave, as much as 150mm in some, due to sulphate attack on the concrete. BRE record 600 houses in one New Town with major problems of heave in the ground floor concrete slabs in the late 1970s and a bill of £2 million at 1990 prices to put them right. NHBC was paying out over £1 million a year in the 1970s and 1980s on claims in respect of defective floors slabs, in some cases at the rate of £7,500 per house at prices then ruling. It had to raise its standards in 1974 to require a suspended floor at ground level where the depth of fill exceeded 600mm. Much of the subsequent advice and the requirements for the construction of floors at ground level tends to reinforce the long held traditional practice in Scotland where for house construction, the site was cleared, covered with pitch and the lowest floor suspended from the walls, in the same way as upper floors, with a suitable space beneath for ventilation but with no reliance placed on oversite concrete.

When the surveyor is advising on a new or near newly constructed dwelling with a floor at ground level of solid construction he needs to ensure that the requirements produced in the 1970s have been followed both as a consequence on the outcome of the site investigation and in relation to the topography of the site. Of course, a suspended floor does not necessarily have to be of timber. There are a number of pre-cast concrete beam and block systems which have been increasingly used in housing developments since the 1980s. The surveyor may need to do some research and make enquiry from the developer or the local authority to find out the type used and information on possible problems should he suspect that all is not well with what he finds.

Despite the popularity at present of hardwood flooring in both professional and DIY forms, this is still the age of the fitted carpet and in scanning the walls and ceilings and in examining windows, doors and other fittings the surveyor is in danger of skimming over the carpeting or applied hardwood and giving the floors less than adequate attention. A few cases from the 1980s will stand as a reminder of what needs consideration. In one case the surveyor on a mortgage valuation missed a crack in a floor. Damages of £29,000 were awarded

although the house only cost £14,850, *London and South of England Building Society* v *Stone* (1983). In another case, damages of £5,000 were awarded where a surveyor missed a 20mm gap below the skirting to rooms on the ground floor, *Westlake* v *Bracknell District Council* 1987. In the third case, *Cross* v *David Martin and Mortimer* 1989, the surveyor failed to relate the evidence of sloping floors and doors out of alignment to the degree of settlement in a house built on a sloping site. The award in this case included amounts in respect of other items where the surveyor was also found negligent.

Should solid construction for the floor at ground level be encountered in a dwelling built within the last 40 to 50 years and the floor is covered with fitted carpet, it will not be possible to see whether there are cracks in the surface and even, perhaps, whether there is a gap below the skirting. Nevertheless the surveyor's spirit level will show whether the floor is sloping and, if so, should there be displaced framing and door openings out of true alignment in partitions he should carefully consider whether the two defects are related. A conclusion that they are, could well provide the reason for other defects such as roof and staircase distortions.

As already mentioned, solid construction for floors at upper levels in domestic buildings are unlikely to be found except in the case of flats. Although multi-storey blocks of flats were common on the continent and in Scotland, producing apartments and tenements, they were comparatively rare in England and Wales until the charitable trusts began to build them from the 1860s onwards. These were followed from the late 1800s and early 1900s by private developers of mansion blocks and local authorities, all producing flats to let, now of course, mainly sold on long lease. The flats in the mansion blocks could

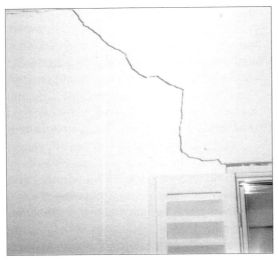

415: A diagonal crack in a partition separating these two rooms from the top corner of the connecting door coupled with distortion in the frame is a sign of subsidence in the solid floor on which it is supported.

be quite large, often built to a high standard which could extend to an expensive finish to the floor structure such as superior hardwood boarding, parquet or wood block. There could also be accommodation for servants with separate access and the provision of services to include centralised heating and hot water systems, lifts and porterage.

The form of construction for the external walls and internal structural partitions would traditionally be of brick or stone, as although a patent for reinforced concrete had been taken out here in the 1850s, development and use was mainly on the continent and its employment for the walls of commercial buildings in the United Kingdom did not commence until the early 1900s, examples being the Royal Liver building in Liverpool and Kings College Hospital in London, both of 1909. For domestic construction it was even later, the 1920s and 1930s, for use in the structural frames of blocks of flats. It had not been until 1915 that the London County

Council introduced regulations governing the use of reinforced concrete and the first British Standard Code of Practice did not appear until 1934. This lapse of time, however, did not occur in the case of floors which as long as they had continued to be formed of timber represented a hazard, allowing a fairly rapid spread of fire and resulting in substantial conflagrations and loss of life. Such fires led to the earlier development of floors with a degree of fire resistance.

Fire resisting floors became possible following the Industrial Revolution from the late 1700s onwards. Cast iron had been an early outcome and was used as beams of various shapes from inverted T-section to those not too dissimilar from the familiar rolled steel joist of today. From the lower flange on either side would be sprung a shallow brick 'jack' arch, the top being brought level with an infilling of mass concrete, rediscovered in the late 1700s after a lapse of over 1000 years from the time the Romans departed. Early rediscovered mass concrete was first based on lime, then on Parker's Roman cement from 1796 until the patenting of Portland cement in 1824. Cast iron was brittle but strong in compression and therefore satisfactory for columns but weaker in tension and following the collapse of a number of bridges in the mid 1800s was replaced for beams by wrought iron which was much stronger in tension and had been used since medieval times, in small quantities, for cramps and tie bars.

Combinations of steel beams and mass concrete were known as 'filler joist floors' but during the 120 years or so from about 1800 to 1920, many of the designs for floors with a degree of fire resistance based on cast or wrought iron and, in the latter part of the period, mild steel beams of various configurations and mass concrete of differing formulations were patented. Some included types of the ribbed and hollow terracotta clay sections which later became more commonly familiar as the reinforced concrete hollow pot floors from the 1920s onwards. Names such as Holmans, Frazzi, Monolithic, King and Fawcett became known for their patent floors.

Problems, however, arose form the formulation of the mass concrete. BRE quote a writer commenting on Holman's patent filler joist floors as follows 'the use of fine coke in the aggregate may have seemed at the time a useful way of disposing of a plentiful waste material but it contained sulphur that slowly oxidised and many floors of the period later suffered grave damage through the corrosion of the ironwork with which the coke or breeze made contact'. This problem by no means restricted to Holman's patent floors, was much aggravated in the presence of prolonged dampness from plumbing leaks and in cases where corrosion occurred, the concrete infill would expand and crack along the line of the beams, pushing up the top surface.

As mentioned in the previous paragraph, mild steel beams had begun to be included in patented fire resisting floors in the latter part of the 120 year period 1800 to 1920 from the 1880s onwards. Mild steel which was much more pliable for forming into different shapes, including bars and strips, and has greater and more reliable all round strength, gradually replaced both cast and wrought iron for structural purposes and with the standardisation of sections and the publication of stress tables supplanted both by the 1920s. The ingredients, comprising Portland cement, mild steel and regulations governing the construction were accordingly available from the 1920s for the development by designers of cast *in situ* reinforced concrete fire resisting floors in flat slab, hollow pot or waffle form for use in blocks of flats, either separately in an otherwise masonry constructed block or as an integral part of a reinforced concrete frame. These served well for the purpose and the patented proprietary systems gradually fell out of use until the high rise developments of the 1950s and 1960s introduced various new manufacturer's systems involving, along with other components, floors of prefabricated pre-stressed concrete sections for hoisting into place on site and subsequent bolted connection to wall slabs. The requirement for speed at the time brought into use High Alumina cement and additives, such as calcium chloride, for rapid hardening which, along with the development of lightweight aggregates, produced the problems of carbonation and corrosion of steel reinforcement.

The requirement for an engineer's inspection at regular intervals and what the surveyor should do when instructed to inspect a flat in a block built since the 1950s where concrete floors are encountered and where the construction is otherwise in non-traditional form, has already been discussed in section 4 under the heading of Concrete Walls, Pre-cast Block and Pre-cast Panel. For blocks of flats with reinforced concrete floors built between 1920 and 1950-1960, whether with walls constructed of masonry or with a reinforced concrete frame, advice on the surveyor's actions and what his recommendations should be is given under the heading of Framed Walls, Reinforced concrete.

For blocks with solid floors built before 1920 the advice need not be too dissimilar to that given for blocks built in the period 1920 to 1950-1960 to the extent that if there are no visible defects there would be no need to recommend the engagement of a structural engineer unless substantial alterations were proposed. Such advice, however, can only follow the inspection of those visible parts both above and, whenever possible, below not only to ascertain the mode of construction but also to detect any disturbance or damp staining to finishes, in particular to the floors such as disturbed wood block or parquet, or tiling in kitchens, bathrooms and hallways, but also possibly disturbance to other features such as partitions where defects may have had a 'knock-on' effect. The inspection must, obviously, include the likes of cupboards or storerooms where floors may originally not have been provided with a finish and could, most usefully, be extended to the common parts of landings, staircases and lift shafts. If there is a chance to inspect basement rooms, perhaps old boiler compartments fuel stores, cleaner's stores etc., the opportunity should be taken as the underside of the ground floor may reveal not only the form of construction but more clearly any sign of movement. It is permissible to conclude that the type of solid floor in a block of flats cannot be ascertained but that there are no visible defects. What is inexcusable is to miss those defects if they are visible and fail to advise further investigation by an engineer. With an outline knowledge of the types of construction which could be present according to the age of the block, the surveyor will be in a better position of knowing where to look.

Finishes

For hollow suspended timber flooring, the traditional top layer of the structure has for long been softwood floorboards nailed to the joists, almost universally plain edged or tongued and grooved for dwellings of fine quality and, in the main, between 19 and 25mm thick. It would then be up to the occupant to decide whether to leave the boards bare, have them painted or varnished, or cover them with some form of more attractive, if that was thought to be necessary, wearing surface such as linoleum, plastic tiles, some other type of wood in the form of

parquet or strip, or a woven material in the form of loose or fitted carpet. An underlay may be needed before a wearing surface can be applied, plywood being the most commonly used, fixed to the floorboards to even out any discrepancies in the original boards, before laying tiles or hardwood strip flooring. For carpet, the underlay is usually felt, foam or some form of rubberised ribbed sheet material.

Accordingly, it may not be possible to see the original floorboards. Floorboards, however, depending on the age and the history of the dwelling, can be in good condition but more often than not they will have been taken up repeatedly, cut for the installation of services, frequently re-nailed and heavily worn. This is the reason why most old floors are covered over but if they are left bare, the unevenness, loose sections and in particular any sections which have been attacked by woodworm could not only be extremely unsightly but present a hazard.

What the surveyor can see and discern from his inspection should be noted but whether it

416: The absence of an allowance for expanding movement has led to these floorboards curling at the edges.

falls to be included in his report will depend on the type of report for which the inspection is being carried out, with the exception that where a hazard is noted it will require to be included whatever the type. If such is present in one room but others are covered over then it is necessary to include a warning that similar conditions could become apparent when owner's carpets or the like are removed.

The same comments, in general, apply to the top surface finish applied to floors of solid construction though many of the treatments in the process of application acquire a degree of permanence which renders them non-removable on a change of ownership, wood blocks for example. Most of the defects in the finishes to floors of solid construction are related either to the absence or the lack of effectiveness in the damp proof membrane or the use of unsuitable wood or adhesives. If the finish is of wood then changes in moisture content will cause movement and buckling due to expansion if no allowance has been made. Alternatively rot will eventually set in from the underside. Unsuitable woods can be prone to inconsistent colour change, uneven wear and in some cases a degree of dangerous splintering. Materials impervious to moisture, while keeping damp at bay will not necessarily resist upward pressure or the effects of an accumulation of water below the covering whether in tile or sheet form and will arch, lift or billow upwards. All porous ceramic tiles run the risk of expansion. It is this which causes crazing in the glaze. Movement joints are necessary and many problems are caused by a differential rate of expansion in the tiles, the adhesive material and the base. As with the finishes the surveyor needs to note what he sees and provide warnings if necessary.

Ceilings

The ceiling to the underside of a hollow suspended timber floor in the upper storey, or storeys, of a dwelling can be considered an integral part of the structure in that it reflects the structural condition of the floor. Sometimes that condition may be revealed in a more pronounced fashion in the ceiling below than in the flooring above where, for example, timber floor boarding will provide more resilience to accommodate movement than the homogenous plaster of the ceiling.

Irrespective of whether the ceiling is formed of lath and plaster, 19 to 25mm thick or plasterboard with a skim coat of plaster amounting to about 12 to 15mm in thickness, if the structure is inherently weak or has been weakened by undue cutting, the ceiling may well show signs of bowing or cracking. If the ends of joists built in to external walls are

417: The single storey bay window favoured for many of the smaller houses built in the latter half of the 1800s and the early part of the 1900s needed a beam above to support the remainder of the front wall. Frequently a crack will become apparent in the ceiling of the bay on the line of the front wall as here, even though there is no sign of any really serious movement.

beginning to take up dampness and expand, it could well be that the first sign of this will be cracks in the ceiling plaster and the laths are likely to be the first affected should rot develop.

The undue deflection of a floor in an older dwelling can cause the plaster to lose its key to the laths and become in danger of a collapse. This aspect was dealt with early on in section 3, Roofs, under the heading of Flat Roof Structures, The Inspection, and the advice there on what to do, *viz-a-viz*, the owner, should be followed. With floors, however, it is possible for the surveyor to advise that the floorboards above be lifted to enable an examination of the plaster key to be made before any other steps are taken, such as gentle probing or pressure from the hand which might cause premature collapse. Depending on the outcome of the inspection from above it may be necessary to provide advice to erect an internal scaffold to support loose plaster until repairs are undertaken.

Penetrating damp can be lethal to plaster ceilings. Leakage from showers, generally obvious, can completely destroy the characteristics of plaster and even where staining seems light the plaster can be reduced to the point where it is no longer fit to take a decorative surface. Less obvious however is gradual, but slight, leakage from a pinhole type leak in an old lead pipe or a defective connection in a copper pipe which allows gradual seepage to occur. This can result in the sudden and unexpected collapse of heavy lath and plaster ceilings, added to in layers over the years to compensate for unevenness or, in particular, heavy decorative cornice sections. The inspector should be aware of the possible implications of small, as well as large, stains.

Fungal Attack

Dry Rot

Wood being an organic material is subject to attack by other organisms which can derive sustenance from the cellulose material in its composition. Wood devouring insects have already been dealt with near the end of section 3 on Roofs and the surveyors internal inspection of the dwelling below the level of the roof must pick up any visible evidence of their activity elsewhere such as in floors, the staircase, the understair cupboard being a frequent scene of activity, skirtings and other joinery features.

Wood, however, also provides the source of food for fungi which develop and grow from spores present in the air at all times and which are just waiting for the right conditions

418: When a central heating radiator is placed next to a wall prone to rising damp, a higher level of moisture content than elsewhere will be found in the plaster and adjacent woodwork, as the heat draws damp towards the radiator. It is very necessary in a situation such as this to make full use of the damp meter to check on the moisture content of plaster and skirting. The longitudinal split in the skirting may be an early indication of an attack of 'dry rot'.

to arise. Should those conditions become available in a dwelling and an attack by one particular type of fungus *Serpula lacrymans*, 'dry rot' as it is commonly called, develop, the damage can be considerable and the cost of eradication and repair substantial.

It is important for the surveyor, therefore, to be aware of the places in a dwelling where those conditions are likely to already exist and perhaps where an attack might already be flourishing or if some aspect of the dwelling is not corrected, conditions suitable for an attack could develop.

There are three essential conditions required in the wood for it to be attractive to the spores and susceptible to the deprivations of the 'dry rot' fungus.

1. **Damp.** Wood in a dwelling settles down after construction to a level of approximately 15-18% moisture content of the oven dry weight. This reduces to less than 12% with central heating. If however it rises to 20% the wood becomes susceptible to fungal growth.
2. **Oxygen.** At 40% moisture content there is ample oxygen in the cells of the wood to support growth at its optimum level and maximise the availability of the food supply. Above this level of moisture content the availability of oxygen declines and at saturation level the wood is immune from attack.
3. **Warmth.** The ideal temperature for vigorous fungal growth is 18.5°C (65°F) but the normal range of temperature found in most dwellings will support growth. Exposure to winter temperatures externally will inhibit growth hence the traditional ventilation provided to certain areas of a dwelling likely to contain moist air, under suspended floor construction at ground level for example.

419: As a 'dry rot' attack develops, splitting across the grain begins to appear as the wood shrinks from behind with the loss of moisture. The attack will be extending into the adjacent floor at this stage but a crafty placing of furniture can effectively hide what is happening.

Two relevant points can be made here. One is that wood based products are even more susceptible to attack than the structural soft-wood timber for floors, roofs and partitions and the timber used for joinery. Plywood, chipboard and hardboard are more easily consumed as are related products such as books and papers. Another, is that following a fire or a flood, the drying out process after wood has been saturated brings it down to ideal levels for fungal growth. With the addition of warmth to help the drying out, then very dangerous conditions indeed are created and these are not always appreciated by the occupants after such calamities have occurred.

The most likely places in a dwelling where dry rot is a possibility if warm and damp conditions prevail can be summarised as follows:

a) below parapet or valley gutters in pitched roofs which are poorly ventilated. Boarded roofs with closed eaves are more prone to be affected from slow leaks left unattended and which may not give away their presence on ceilings below

b) where timbers are built in to brickwork and subsequently concealed either as plates, floor joists, lintels, fixing blocks or bonding timbers as part of the construction of walls in the 1600, 1700 and 1800s and in the early part of the 1900s

c) in panelling or matchboarding to walls and in particular around window and door openings where there are shutter boxes, panelled heads and aprons. In construction where the interiors of external walls are battened out and covered with lath and plaster

d) in basements or floor construction at ground level where there is inadequate damp proofing and a lack of ventilation

e) in floors at upper storey levels and below flat roofs where there have been leaks and there is an absence of ventilation

f) in wood floors generally, where there have been plumbing leaks

It is often the case that in the early stages of an outbreak of dry rot there will be either no indications, or very few, of anything amiss. In cases where a dwelling has been neglected, the surveyor needs to pay the closest attention to the likely places set out above, most particularly where there are signs of leaks or overflows, past or present, in rainwater or waste pipes externally and internally around bathroom and kitchen fittings. These latter positions also require close attention in dwellings which have apparently been well looked after. Intermittent or slow leaks from plumbing can remain unnoticed for sufficient time to allow attacks to develop, more rapidly if fittings are boxed in. The surveyor's moisture meter will be calibrated to show moisture content in plaster and wood in excess of safe levels and should be used freely, but judiciously, in the likely places where damp staining will not necessarily be unmistakably visible.

On painted wood panelling, the only visible evidence of something amiss may be slight

420: *An attempt to hide the effects of the damp and this attack of 'dry rot' would be difficult to achieve and it could be that the carpet is effectively spanning a hole in the floor below. It would be inadvisable to tread too heavily in this area!*

421: *The effect of an attack of 'dry rot' on this skirting can clearly be seen where sections have fallen away. The pieces of 'wood' are quite dry and almost weightless and are cubicial in shape due to the shrinkage as moisture is extracted both longitudinally and across the grain.*

opening of the joints and slight cracking to the paintwork as the wood behind initially starts to shrink. On skirtings and other features fixed to walls which are damp, the meter could indicate levels of moisture content in the wood beneath the paint sufficient to support growth. Painted wood surfaces at a later stage might warp or appear wavy before bigger cracks start to appear while unpainted wood may darken in colour.

It is only later that the surface of the wood will provide evidence of cuboid cross cracking as the fungal growth from behind consumes all the goodness in the wood causing it to completely dry out and shrink in all directions. Just occasionally at some stage a pancake shaped soft fleshy fruiting body will grow out from behind an architrave or skirting. This will normally have whitish margins to the top surface and on the underside a ridged and furrowed spore bearing surface, rusty red in colour. If the architrave or other feature is given a thump the reddish spores might drop out from behind. If the feature falls off or a floor can be seen from below, white to grey strands, sometimes up to the thickness of a pencil may be seen. It is these strands or strings which can travel through brickwork and other material for some considerable distance to seek out more timber for attack, which in the case of the 'dry rot' fungus need not be unduly damp for it to surrender, the strings bringing with them a supply of moisture sufficient to encourage growth. In damp and dark places, such as cellars, soft, white cushion like patches may appear, sometimes also lemon yellow coloured and with tinges of lilac. It is at this later stage that a mushroom like smell may become apparent when cellar or cupboard doors are opened after having been left shut for a while.

There is no way of telling how far the 'dry rot' fungus has spread in a dark, damp,

422: The white cushion like fluffy growths which the 'dry rot' fungus can produce in dark, damp and unventilated areas such as cellars and vaults. Apart from the tinges of colour, here lemon yellow and grey although lilac can also be featured, such growths can be likened to the candy floss of childhood memory.

unventilated habitat and the surveyor can only report what he sees and that full opening up will be required to ascertain the extent before an estimate can be obtained for eradication and the necessary renewal of timber and plasterwork coupled with any improvements necessary to prevent it happening again. The estimate should be from a well known experienced firm which is a member of the British Wood Preserving and Damp Proofing Association who will provide a guarantee and operate under a Code of Practice. This should provide for the correct identification of the fungal growth, whether it is active or not and for appropriate treatment. Some outbreaks of 'dry rot' die out of their own accord for reasons which are not always readily apparent and the surveyor needs to be aware that there are many different types of fungi which attack wood. Indeed he will no doubt see, all too apparently, plenty of visible evidence of wood rot in his work. Much of this will not be caused by the true 'dry rot' fungus, *Serpula lacrymans*, but by others commonly known as 'wet rot' fungi. These are not so virulent, quite so damaging or so expensive to eradicate since while the primary measures of eradication are the same, the secondary measures are not necessarily so comprehensive.

Wet Rot

Probably the most prevalent of the 'wet rot' fungi, *Conifera Puteana*, known as 'Cellar fungus', is usually found in very damp situations where there has been a leakage of water and, as its name implies, in vaults and cellars. Darkening of the wood occurs and longitudinal but not usually cross cracking. Deep brown or black thin strings distinguish it from 'dry rot' the strings of which are white or grey and normally much thicker. The underside of the sheet like fruiting bodies are greenish to olive brown as against the russet

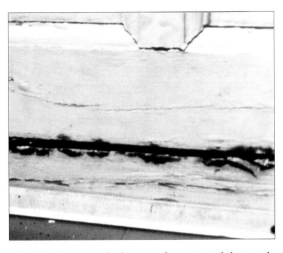

424: *The design at the bottom of an external door and its threshold can either help to keep damp at bay or actively encourage it. The latter case appears to apply here with much 'wet rot' in the bottom rail. At least a new bottom rail will be required with a weather board and a rearrangement of the threshold.*

423: *Softwood windows and doors, and their frames, are prone to attacks of 'wet rot' if the timber contains a lot of sapwood and the paintwork is neglected, allowing movement and damp to penetrate the slightest opening of the joints. Continual overflows from a hopper head have allowed damp to penetrate to the walls and the woodwork in this area so that total renewal needs to be contemplated.*

425: *This photograph is included here not so much as an example of external pipework in a poor state but as a reminder that such are indicative of possible problems internally from attacks of fungal growth. The green staining on the brickwork is a sign of constant leakage, while the white staining in the area of the hopper head is from soapy waste water. It doesn't bear thinking about what could come down from above the missing length of pipe in the corner. Floors, lintels and window framing could all be affected.*

red of 'dry rot'. Other 'wet rots' have different types of fruiting bodies, some have yellow to amber coloured strings, most require a higher degree of moisture than that required for 'dry rot' but can flourish in the open air and in well ventilated areas and one, *Lentinus lepideus*, sometimes found in damp skirting boards, does not produce strings, has fruiting bodies which can be cylindrical in shape and gives off a sweet aromatic smell from both the fungus and the wood affected.

'Wet rots' had a field day in the 1960s and 1970s when economic necessity required the inclusion of much sapwood in the joinery of windows and doors. The building up of sills and other components from smaller sections allowed damp to penetrate on the slightest of differential movement and rot to develop. Much renewal had to be carried out in as little as four to six years and there is evidence to suggest that improvement in the situation since has only been relatively marginal.

BRE publish a useful booklet *Recognising Wood Rot and Insect Damage in Buildings* which should be in the possession of every surveyor inspecting dwellings. The essential requirement for a surveyor, however, is to be aware of where the 'dry rot' fungus is likely to attack timber and actively seek it out if the conditions suggest the possibility, committing to memory the requirements for its development and the visible indications of its activity.

It is pertinent to draw attention here to columns 2 and 3 of Table 1 Identification of Wood Boring Insects on p 150 where damp conditions are said to be preferred for attacks by the Common Furniture and Death Watch beetles and the Wood Boring Weevil. Such conditions also favour outbreaks of fungal growth and it will be found frequently that the two are associated. The softening of the wood by fungal growth encourages the insect attack by making it easier to digest and between them they can lead to its total disintegration.

Staircases

The structural condition of the staircase in a dwelling is more often bound up with the condition, as already indicated, of its support at the lowest level from the floor at that point, and the further support, if necessary, from the strings fixed to partitions. Furthermore the satisfactory trimming to the openings in the floors above at landing level play an important role in maintaining stability. Failures in these components will disrupt the staircase, opening up the joints, loosening carriages and treads from risers and disturbing the support to handrails and balustrades.

The staircase is an area of hazard in any dwelling, BRE estimating that there are roughly a quarter of a million accidents a year on stairs resulting in 600 deaths. Even though the structure of a staircase itself and its supports are sound, the original design may leave a lot to be desired from the aspect of safety and the inspector needs to draw out particular failings in this respect in his report. The Building Regulations for England and Wales 1991 as amended in 1992 at Approved Document K is the standard required for new dwellings as distinct from institutional, assembly or other types of building south of the border while north of the border the Building Standards (Scotland) Regulations 1990 apply. It is to these documents and a comparison with current requirements that the staircases in older building can be related and where many will be found to be well below standard, one of the reasons why there are so many accidents.

Among the current requirements are that if

steps all have the same rise and going the maximum rise and the minimum going should be 220mm. The pitch for the stairs in a dwelling should not exceed 42° and if the staircase has open risers, a feature popular in dwellings of the 1960s and 1970s the treads should overlap each other by 16mm and the open space should not allow a sphere of 100mm diameter to pass through. Small children can fall through open space in excess of 100mm and this amount is also the maximum for the space between balusters on the staircase and the balusters forming part of an open balustrade around landings and stairwells. The staircase in some of the smaller dwellings at the lower end of the price range built before 1900 will often be found to be so steep as to be a hazard to the elderly and the very young.

A headroom of no less than 2m is considered necessary for the stairs between floors, although on the top flight leading to an attic, a minimum height measured at the centre of the stairs of 1.9m will suffice which can be reduced to 1.8m at the sides for an arched bulkhead, a concession useful in the case of loft conversions where space is restricted.

Handrails are a necessary requirement for all staircases, only on one side if the width is less than 1m but on both sides if in excess of this amount. The height should be between 900mm and 1m. Balustrades at the sides to enclose staircases in the absence of a wall are needed not only to provide a handrail of the appropriate height but also to provide sufficient strength for the purpose. This, obviously, also applies to the balustrades around open stairwells so that they do not collapse in the spectacular fashion seen in so many classic Western films. In older dwellings they are often much lower in height than current requirements permit constituting a hazard which should be mentioned in the report.

426: Winders on a staircase, particularly when they are as steep as here in an early 1700s house, present a hazard to children and the elderly, an aspect which should be mentioned in the report, strenuously if they are worn and uneven.

427: A handsome open well staircase in a mid 1800s house, with oak treads and risers and substantial mahogany newel posts, closed outer string, turned barley sugar balusters and moulded handrail. The weight has proved too much for the landing which has settled slightly and, strictly speaking, there should be a handrail both sides if Building Regulation compliance was required.

The layout on plan of staircases is also an important factor in preventing accidents as far as possible. Long flights should be broken up into sections. Landings should be provided at top and bottom of every flight at least to the width and length of the narrowest width of the flight, although this may include part of the floor. There is a need to avoid doors to rooms at the top of a flight opening directly on to the staircase, which can be a particularly dangerous feature. Winders are also a hazard but if necessary, their safety can be improved in use when well lit, if not naturally then by good artificial lighting.

Fireplaces

Until about the 1940s, fireplaces were provided as a matter of course to practically every room in a dwelling, since the burning of wood or coal had been the only and remained the cheapest form of heating until then. This was so even though other forms of heating had been developed from the latter part of the 1800s onwards. Central heating using solid fuel, with the boiler in the basement and heavy cast iron wide bore pipes and radiators was viable for public and commercial buildings but remained expensive and inefficient for the average two storey dwelling. This situation lasted until the 1960s when the development of the electric pump, time controls, small bore copper tubing and steel radiators enabled the, by now, cheaply available gas to be used cleanly and efficiently to heat, and supply constant hot water, to the average flat as well as house.

Fireplaces for bedrooms had already started to be left out of designers' plans but from the 1960s they began to be left out elsewhere as well. However, their absence as the focal point to the main room began to be felt by many and the provision of at least one fire-

place came back into fashion in the 1980s. This coincided with a decline in the provision of new dwellings and it is probably now the case therefore that two thirds of all dwellings still have the fireplaces considered by most occupants as redundant and therefore candidates for blocking up or removal. The first alternative is fine and causes no problem provided the flue is swept first and a vent is left in the panel used to block the opening to allow through ventilation to the opening at the top of the chimney. Without this being maintained there is a danger of staining along the line of the flue from condensation.

The redundancy of the fireplaces tempts some owners to go to the expense of removing chimney breasts to enable use to the made of the space they occupy. To some this seems a rather pointless exercise, costing far more than it is worth. Nevertheless it is done but to do it properly involves cutting away the brickwork and making good all the way to the top, including work to the hearths. If the chimney flue is in a party wall, notices need to be given and in all cases the Building Inspector's approval to the work needs to be obtained. However, very often the work is done 'unofficially' and in an incomplete form requiring adequate support to what has been left behind. It will be recalled that the absence of support to what had been left behind following the removal of a chimney breast formed the subject of the claim in *Sneesby* v *Goldings* (1994) mentioned earlier in section 2, Inspection, under Scope. On a mortgage valuation, the surveyor missed noticing the lack of support which he would have spotted had he opened a cupboard door.

It should be immediately apparent to every surveyor entering a dwelling built before the 1960s whether there is something 'wrong' and that where a chimney breast or fireplace should be, there is nothing. A note therefore in his mind should be triggered to see whether sections of brickwork have been left and if they have, been properly supported. The roof space will no doubt provide further evidence of whether work has, or has not been carried out.

The surveyor will obviously need to note on his internal inspection what he sees in the

way of fireplaces, mantelpieces and hearths. Some will not have been blocked up and the fire parts may still remain. Type and condition need to be noted. Where there is a possibility

430: An early 1900s mass-produced cast iron fireplace for small living rooms with hinged hood, tiled surround, hearth and curb.

428: An elegant marble fireplace surround in classic style with Ionic columns and moulded frieze enclosing a cast iron register grate with Dutch style tiled hearth and oak curb. Possibly earlier than the mid 1800s house in which it is installed and still providing a fine focal point to a living room.

429: A hob grate with high fire basket, cast iron back, side panels with console brackets and hob plates for keeping pots and kettles hot in an early 1700s house. Without closure at the top and a register plate to control heat loss up the chimney, the Reduced Data Standard Assessment Procedure (RDSAP) gives such a solid fuel room heater an efficiency rating of 32%, increased to 42% if fitted with a throat restrictor register plate.

431: One of very many typical late 1800s mass-produced cast iron arched fire grates of a size suitable for bedrooms where the register plate has become a semi-circular flap or damper to be opened when the fire was lit.

of re-use, in principal rooms for example, it would be sensible to check whether the existing hearth meets current requirements, a 500mm projection from the jambs extending for 150mm each side of the opening, that the hearth is adequately supported and that if there is settlement, it is minimal. The establishment of thickness would of course need further investigation. If the surveyor is asked specifically about the possibility of re-use of a fireplace, then he needs to be able to advise on suitable installations, appropriate fuels having regard to Clean Air legislation and systems for maintaining the effectiveness and safety of flues, such as coring or re-lining.

Decorations

The condition of the internal decorations to a dwelling is very much tied up, so far as the walls and ceilings are concerned with the state of the plaster. In older dwellings plaster may have been affected by cracks, subsequently made good, areas may become soft, loose and bulging due to damp and possibly be held together by hefty lining materials. If these are stripped, the plaster may unsurprisingly part company from the backing all adding to the cost of redecoration. The condition of plaster and decorations to walls and ceilings needs therefore to be noted in each room on the internal inspection together with notes on any circumstances which might affect the ease, or otherwise of redecoration.

For example, if there are damp patches, the moisture may have brought with it salts from the ground or from the bricks themselves into the plaster and thence to the surface. These salts are invariably of a hygroscopic nature which although they may not show in dry weather take up moisture from the atmosphere at other times and mysteriously show up as damp patches for no apparent reason. The only way to prevent this happening repeatedly, probably after an initial spoiling of new decorations, is to strip back to the brickwork and render with cement and sand before re-plastering. A buyer will be unlikely to know this unless told by his surveyor. Again a buyer may not realise that it takes so long for a wall to dry out following any form of damp penetration. Depending on weather conditions, it can take anything form nine months to a year for a wall of one brick thickness and proportionally longer for thicker walls.

In a similar way, paintwork if it is of a lightish colour and although a little grubby, if the surface is free of blisters, cracks or flaking areas then repainting is comparatively straightforward. If on the other hand dark brown rough and blistered surfaces meet the eye then the buyer is in for an expensive repaint either financially if a contractor is involved or in terms of time and effort if he does the work himself.

Materials, style and colours for internal decorations are very much a matter of individual, subjective, taste. One man's meat is another man's poison. Comment on these aspects is therefore not thought relevant.

Means of Escape

The average two storey dwelling does not require any special provisions for means of escape. It could be, however, that a report may include special attention being devoted to this aspect should a buyer be elderly and infirm or disabled in a way that would make escape abnormally difficult. This is not to say also that owners and occupiers of two storey dwellings should do much more themselves to reduce the number of deaths from fire by the provision of smoke alarms and arranging for easier escape from first floor windows.

Living accommodation above two floors in new or converted dwellings requires the provision of adequate means of escape by reason of the requirements of the Building Regulations 1991 for England and Wales in Approved Document B, Fire Safety, B1 and in Scotland by the Building Standards (Scotland) Regulations 1990. In general at third storey level and above occupants need to be provided with protected routes of escape. It is these requirements that need to be met for loft conversions to the average two storey dwelling and where there is a division into self-contained flats of former single occupation three storey, or taller, dwellings.

However there are many existing dwellings which do not comply with these requirements, having been converted before they came into force. The surveyor must therefore on considering what he finds now, describe how it falls short, report in outline on what would be needed to be done to bring the building up to standards current at present and advice the client of the potential hazard of occupation should the necessary work not be carried out.

Basement Rooms

Unhealthy basement rooms have long been the subject of legislation at both local and national levels. Many dwellings of the 1700 and 1800s have rooms below ground level with varying degrees of light, ventilation and dampness. Many formerly used for habitation will have had Closing Orders served in respect of their stated inadequacies, their use subsequently confined to storage and utility purposes. The existence of a Closing Order will be revealed by local searches to warn a buyer and the surveyor will no doubt be seeing on his inspection why it was made even though he may not be aware of it at the time. However, as with the circumstances pertaining to means of escape, quite a number of basements will have had no complaints made about them and may still be in use.

The surveyor needs to be familiar with the Building Regulation requirements for England and Wales and those of the Building Standards for Scotland in respect of rooms below ground level for new domestic construction. He also needs to have knowledge of both the national and local legislation on existing basement rooms so that he can include in his report, at least in outline, how the accommodation of the dwelling under consideration falls short of those requirements and what would need to be done to bring it up to standard. It is necessary to include this advice in the report for the avoidance of any misunderstanding that a room currently not used for habitable purposes could be used for habitation in future and one in use at present for that purpose could continue to be so used when ownership changed. Assumptions made by a buyer without such information being made available to him could prove to be costly and he will, with justification, believe that he should have been told by his surveyor about the situation.

Houses in Multiple Occupation

It is not unusual for a surveyor to be instructed to carry out an inspection and prepare a report on a building which contains a number of residential units. It may be that they are self- contained, all of which are let on lease to produce an income. Here, the buyer will be concerned to know that his investment in the property will continue to produce the return which he has been led to believe will be forthcoming. That of course will depend on a number of factors, not least the degree of outstanding repairs and the terms of the leases with regard to the respective liabilities of the tenants and owner.

Should one of the self-contained flats in the building be vacant then the buyer could become the occupier of that flat and acquire the lease as well as overall control of the whole building. In this event the surveyor's task will not be too dissimilar but, obviously, with more attention paid to the vacant flat. After 1 January 2007, however for England and Wales, a Home Condition Report will be required for incorporation into the Home Information Pack. Part 5, section 171, of the Housing Act 2004, Application of Part to Sub-divided Buildings, governs these circumstances and will require consideration as to what will be needed for inclusion.

In other cases, the income may be derived from a building let out part by part, but not necessarily in self-contained units, and with no part vacant. There is a considerable amount of legislation and many local authorities have requirements governing the provision of such accommodation of which the surveyor needs to be familiar if he is to advise his client effectively. Under the Housing Act 2004, there are provisions in Part 2 whereby local authorities in England and Wales will be required to licence both the premises and a 'manager', not necessarily the owner, of Houses in Multiple Occupation, 'HMO's', which are defined extensively in Sections 254 to 259. To be licensed HMO's will have to meet certain prescribed standards, relating to sanitary accommodation, areas for food storage, preparation and cooking and laundry facilities, together with others which might be necessary in particular circumstances. Surveyors will, accordingly, find it all the more essential to be knowledgeable on the standards expected and required in such accommodation.

Section 6 Services

General

The services provided to houses comprising, in the main, water, electricity, gas and drainage, together with the associated equipment which is now more automated and complicated than ever before, differ from the other elements of the dwelling previously described in their ability, singly or severally, to cause instantaneous death or serious injury if wrongly installed or neglectfully maintained. According to the Royal Society for the Prevention of Accidents, an average of 10 people are killed and 750 seriously injured each year in the United Kingdom because of faulty electrical work alone, a far higher figure than in other European countries, while fatalities from defective gas installations, well publicised due to their dramatic suddenness tend to grab the headlines over less publicised health problems due to polluted water and defective drains. Add to this the problems of flats where the communal services of lifts, hot water and air conditioning require skilled management but where the responsibility is in the hands of persons other than the occupier and it can be seen that the services could present a severe test on the surveyor's knowledge and skill in determining the form of advice to be offered.

Surveyors are not routinely trained to test either gas or electrical systems. Some clients might make a forthright point that they ought to be, but it is a fact that they are not and indeed it may not be necessary, on an initial visual inspection, in all cases to provide advice to have further investigations carried out, for example if there are recent test certificates available. For tests, special equipment is involved: they are time consuming and inter-fere with the use of the premises and, most certainly, an owner's permission in writing would need to be obtained in advance. It will be noted specifically, that no tests of services are provided for at present in the Conditions of Engagement, Practice Notes or Requirements for any of the Reports envisaged in this series of books, although all provide for the surveyor to recommend tests if he considers them to be necessary. How long this situation will continue is perhaps a matter for speculation since it is not entirely satisfactory causing much irritation and giving rise to complaints.

The reason for the complaints is that by following the Conditions, surveyors do not always advise tests when they should, giving the misleading impression that because a system works there should be no problems. There is a lack of sufficient stress on safety in all the Conditions. Certainly over the last 20-30 years, the design and specification for the installation of services in new buildings, because they are more expensive and complicated, has tended to pass out of the hands of architects and surveyors and into those of the building services engineer. This has happened not only in connection with non-domestic building but also for the larger individual house and, almost certainly, also in respect of the design input for the volume builders of estates of houses or flats, even though each dwelling will have basically the same services installation albeit with minor variations.

As set out above, there are particular potential hazards in two of the service installations. In addition to electrical and gas installations already mentioned, it could also be said that hot water and heating installations, where present, can also pose hazards from explosion or scalding. For this reason, there is a lot to be said for a seller or buyer employing a services engineer to inspect, test and report upon performance capability as well as safety of at least the electrical, gas and hot water services, those which involve a danger if not kept in good order. The engineer could be a member of the Chartered Institution of Building Service Engineers who can deal with all three systems on a single visit, or, alternatively, a multi-disciplinary practice of engineers could be employed. If the circumstances seemed appropriate, such as a large expensive dwelling for example, when the remit could be extended to include the remainder of the services, particularly if a non-mains water supply and a private system of soil disposal were involved.

Eventually, perhaps, the well rounded Home Inspector may individually be qualified to undertake not only all the requirements of the Home Condition Report but the tests which he might also consider to be necessary. That situation would seem to be quite some way off particularly as 'competence in installation work' is required for persons to be considered qualified to issue and sign Certificates for electrical systems and to be licensed by the Council for Registered Gas Installers (CORGI). However, the opportunity and the day could come for it to be worthwhile for surveyors to organise themselves to provide a 'one stop' service to advise sellers of dwellings in an age of compulsory Home Condition Reports. They could do this by employing or forming a permanent link with individuals or firms qualified to inspect, carry out tests, prepare reports and sign Certificates of Safety for service installations.

At such time as a 'one stop' service of this nature becomes generally available, it would be sensible to split the Home Condition Report into two clearly separate parts. The first part could cover the fabric, roof, walls, floors etc. most of sections A-E and section G, to be reported on by an Inspector appropriately qualified and licensed, the second part to cover all the services, section F and possibly the Energy Report, to be reported on by an individual suitably experienced and licensed,

to report on the safety, capability and performance of the services. The misleading view currently put forward by the RICS that surveyors can provide an 'informed opinion' just by looking, on whether tests are necessary, perpetuated even now by government for the Home Condition Report, can then be buried once and for all.

As to whom should carry out tests in the meantime, there is much use in the current Conditions of phrases such as 'specialists' and 'adequately qualified specialists' without any indication of who such persons might be except that the Terms of Engagement for Home Condition Reports mention, ill advisedly, 'suitably qualified contractors' for the purpose. In this connection by contrast the advice as follows in the RICS Guidance Note for Building Surveys should be noted. 'The surveyor is advised to be wary of making recommendations for further investigations to be carried out by persons who have a financial interest in the implementation of their own advice or recommendations. Wherever possible, all further investigations should be made by those who are qualified and/or experienced for the requisite task and who can illustrate that they can make recommendations that are free of any financial self interest e.g. from the sale of particular products or services.' As stated previously members of professional bodies are best suited to the task but it has to be admitted that not many seem to be geared for the purpose at present. Surveyors could well cultivate an association with local engineers in their area so that a recommendation can be made with confidence and clients are not left to flounder on their own.

Electricity

A further example of where a test is not

considered necessary is where an electrical installation clearly requires renewal. Obviously in this case there would be no Certificate of Safety available. Unlike the components making up the other service installations some of which have a degree of permanence about them, those of an electrical installation have a distinctly limited life. Dwellings of 100 years of age may already have had their installations renewed twice because the covering to the wires carrying the current can degrade with possibly dangerous results. The sight of ancient metal clad light switches, two pin lighting points, a few ancient power points, maybe three pin but the pins round, not square, early on during the surveyor's internal inspection will perhaps confirm suspicions already formed in the roof space. He will then not be surprised when he reaches the incoming main, switchgear and fuses and finds that they are equally ancient with perhaps a few examples of newer wiring tacked on. There is no point in advising a test, since a new installation is required to current standards and the surveyor's report should say so without equivocation, albeit with a preamble giving the reasons, such as wiring of old type, lead or rubber covered, fittings of old round two or three pin design, the system not earthed and a totally inadequate number of power points.

It is where fittings and wiring in PVC covered cable are in reasonable condition that the evidence is not so clear cut. However, in the absence of a signed report on the condition and a Certificate of Safety following an inspection and test carried out within the previous six months, the surveyor must advise in the interests of safety that it is essential for a test and inspection to be carried out and a report obtained. Astoundingly, it has only been from 1 January 2005 that work to install a new, or work to alter an existing, electrical installation has come within the scope of the Building Regulations Part P, for England and

Wales. Until then it was up to owners and contractors to carry out the work in any way they wished. Some owners 'know' about electricity, most do not, but fortunately most contractors were aware of the Wiring Regulations of the Institution of Electrical Engineers now in their 16th edition and incorporated in British Standard BS7671:1992. It has not been made mandatory, however, to follow either until now and therefore there has been plenty of opportunity over the years for divergence from the Regulations, not only by the less responsible contractors but also by the DIY handyman.

All the surveyor can do, therefore, as far as the safety of persons from electrical shock or burn and the safety of the dwelling from fire is concerned, is to advise in the firmest of possible words that a new installation is required or that an inspection and test should

432: The sight of the green and yellow cables used for earthing an electrical installation is no indication that it is safe to use. Only an inspection and testing by an electrical engineer can do this. This installation is over 20 years old but the owner acknowledges that it has never received the periodic five-yearly testing in accordance with the recommendation on the notice above the mains switch and fuses and accordingly an inspection and test is required.

be carried out, coupled with a recommendation that any part of the installation not in accordance with the Regulations should be brought up to standard. If there are electric fittings for the heating of water or space, such as underfloor or ducted warm air, obviously these should be included in the advice for inspection and test. Any other advice is not only dangerously misleading to both sellers and buyers and could rebound with serious consequences on the surveyor. A surveyor is just not qualified to assess the significance of what he can see of the installation as far as safety is concerned.

The Wiring Regulations recommend that domestic installations should be inspected and tested every five years and a report on the condition obtained. There is no requirement, however, for this to be done and the evidence suggests that the vast majority of owners do not follow the recommendation even when reminded to do so by the presence of a notice to that affect on the fuse board. Even in Scotland, where new electrical installations since 1990 have been required to be installed in accordance with the Regulations, there is no requirement for periodic testing.

On the surveyor's room by room internal inspection of the dwelling he is required, however, to identify intake position and the position of mains switch, meter and fuses along with taking note of the position and presence of every lighting and power point and the availability of any supply for cooker, refrigerator, washing machine etc. as part of his description of the facilities available. He should not, however, comment on any matter in connection therewith unless, in his view, there is an immediate aspect of danger. If any such items are found then, in addition to including them in his report, he should report his finding as soon as possible to the owner or his agent. Other less serious matters should not be included in his report, although

433: There is no apparent reason why this double 13 amp socket outlet has come away from the wall, but it represents a distinct danger and the surveyor must warn the owner or his agent as soon as possible of its presence as well as including it in his report. The skirting below looks as though it could warrant attention for signs of rot.

noted, but left entirely for the inspector and tester of the electrical installation to deal with as part of his responsibility. For the surveyor to comment would give the impression of undertaking a specialist's function and could mislead either a seller or a buyer into believing that the surveyor had undertaken that responsibility throughout and that there was no real need for another inspection.

As to inspections and tests, it is preferable for these to be carried out by a member of the Institution of Electrical Engineers, able to report in plain language which can be understood by the average person. Putting such tasks in the hands of electrical contractors is liable to produce a report in technical jargon. It may be that in the background there is the idea of persuading either owner or buyer to part with his money for a complete rewire, when it might be the case that modification on a fairly limited scale is all that is necessary.

Gas

Whereas with an electrical installation, there are two possibilities for the surveyor's advice to take, either renew the installation or test it, with a gas installation there is only one. It has to be for a test to be carried out by an independent engineer unless a gas Completion certificate signed and dated within the last six months in respect of the installation, pipework and all appliances can be produced. The surveyor must describe the fittings he can see room by room on his internal inspection but he is not qualified to comment on the way they were installed, the adequacy of their flue arrangements and whether there is a sufficient supply of air for their combustion, having regard to their size. Nevertheless, he has a duty if he smells a leak and it is not obviously from a gas appliance which has been left on and can therefore be switched off, to report it to the National Gas Emergency Call Centre on 0800 111999. Similarly, and as with electrical installations, if he otherwise sees anything which to him suggests that a hazard could exist, then he must report it as soon as possible to the owner of the dwelling or his agent.

Misplaced suggestions that the surveyor should only recommend a 'specialist investigation' if he sees 'this or that dubious feature' are totally wrong and of no help whatsoever. He should say in respect of every gas installation that an inspection and report is essential, to include all the fittings such as boilers and heaters and associated work on flues and vents, on the grounds of safety and that only an independent engineer can take that responsibility.

One of the main reasons for the necessity of this advice, is that the pipes and connections for carrying gas can have been present in a dwelling for a considerable length of time. Unlike the covering to the wires conveying electricity, which tend to degrade within a comparatively short time, the early tubes for carrying the gas supply were often of cast iron which can last almost indefinitely as long as they are not exposed to undue damp. It must be remembered however, that urban dwellings which were in existence in the 1920s all had a gas installation for lighting and probably cooking purposes and the conversion to electricity for lighting, in general, was only completed by the late 1940s and early 1950s. Even then many dwellings retained a gas supply for cooking purposes and these still remain in place.

Gas for domestic purposes until the 1960s and 1970s was made from coal and was supplied at half the pressure of the natural gas supplied today. It is the old cast iron pipes and their joints which may be affected by more than just a touch of rust and be near paper thin, considering that some could be up to 150 years old. The increased pressure could suddenly be the cause of a leak. Many gas leaks are, of course, attributable to occupants carelessness and forgetfulness in leaving the supply turned on when not required, the explosion causing deaths and the partial collapse of the flats at Ronan Point in 1968 being a case in point. However, the potential for hazardous leaks due to the failure of old pipes to cope with the increased pressure as well as the possibility of joints, even on newer copper tube installations, developing leaks if badly formed, should not be overlooked.

It is for the above reasons that a further investigation of the gas installation should involve not only a visual inspection but also a test on the pipework to a minimum pressure of 0.014 N/mm^2, about twice the current operating pressure. A drop in the level on the gauge will indicate a leak and the use, instead of air, of ammonia and chemically prepared paper will enable the source to be located. Not only can joints be badly formed on newer installations, but pipework can be damaged by building work or carelessness and a statement such as that there is no need for a pressure test on the pipework because it is nearly new should not be accepted. A very small leak can cause a build up of gas to accumulate in a room, for example, over night and it is normally the switching on of a light in the morning that triggers an explosion.

Because of the high cost and lack of efficiency in some respects of electricity and oil as the heat source for the provision of hot water and central heating in domestic premises, more than ever before reliance is placed on gas. The vast majority of installations for these purposes today will be found to be gas fired. As mentioned at the beginning of this Section, because of this, there is every reason for extending the inspection and testing of a gas installation by an engineer to include the safety and performance capability of the

434: Although the incoming service for this gas meter positioned in a cellar is in cast iron tube, the supply to the fittings is by way of copper tubing, much the most commonly found. Although looking fairly new, a pressure test on pipework is still necessary.

components supplying hot water and warmth. If the size of the house and the nature of the installations warrants, the strongest case can also be made for engaging the services of a member of the Chartered Institution of Building Services Engineers to inspect, test and advise on both and, perhaps, even on the other services.

Water

A plentiful supply of clean wholesome drinking water is a prerequisite for satisfactory household management and the water companies in urban locations provide this through their underground mains, to which a dwelling's installation is connected by a stop valve. This is turned on and off by the company dependent on payment of its charges. Within the curtilage of a dwelling, the responsibility for the condition of the mains supply pipe is that of the owner.

In the north, where mains water is stored high up in the hills and therefore there are few problems with the supply or with the pressure, all the fittings in a dwelling using cold water can take their supply 'direct' off the main and there would only be a need for a small storage cistern, minimum capacity 114 litres, to supply the hot water cylinder. This can often be positioned just below ceiling level in an airing cupboard, making it more easily accessible then positioning in the roof space.

In the south, the supply has traditionally been more erratic and with much less pressure in some places, so that an adequate storage cistern, minimum capacity 227 litres, is usually installed in the roof space and the supply of mains drinking water limited to one outlet at the sink in the kitchen. All other fittings for cold water being supplied 'indirectly' by a cold feed from the storage cistern.

The inspection of a storage cistern in the roof space has already been dealt with at the end of Section 3, Roofs, and need not be repeated here. The same degree of attention, however, needs to be paid to the more or less half size cistern which might be found high up near ceiling level in the upper storey when a 'direct' mains supply system is provided.

Surveyors should, of course, know the degree of hardness in the water of the area in which they practice. Most waters contain both 'temporary' hardness, carbonates dissolved in the water which on being heated precipitate out as bicarbonates, causing the furring of kettles and hot water pipes and boilers, and 'permanent' hardness. The latter is due to the presence of sulphates and chlorides in the water which do not precipitate out on heating but can cause corrosion. It is the proportion of 'temporary' to 'permanent' hardness which determines whether water is considered 'hard' or 'soft' i.e. whether it is difficult or not to obtain a lather when using soap. The higher the proportion of 'temporary' hardness present, measured in parts per million, the harder the water. For example, 200 parts 'temporary' to 50 parts 'permanent' hardness would be considered 'hard' while the reverse would be considered 'moderately soft'.

For new, or near new, dwellings BS6700:1987 'Specification for the design, installation, testing and maintenance of services supplying water for domestic use within buildings and their curtilages' is the document available for checking whether what is being proposed or what has been installed is up to modern requirements. Furthermore the water companies are empowered to make, and are required to enforce, bylaws designed to prevent waste of water supplied by them. The Model Water Bylaws introduced in 1986 provided the

framework and the Water Supply Bylaws Guide published by the Water Research Centre, gives much practical assistance. For older dwellings, the Standard and the Bylaws for the appropriate area can provide the detail for a comparison of how far short an installation is of current requirements. This will depend on the date of the dwelling's construction and its subsequent history. Many will fall far short, but unless there is a health hazard there is no legal requirement to bring an installation up to current standards. Most will be giving satisfactory service to their owners and it may be that a surveyor, while noting deficiencies, would not necessarily advise urgent expenditure to alter the situation. Nevertheless, there can be instances where it might be wise to do so.

For example, cold main pipes need protection from freezing within dwellings where this might occur, since a burst pipe when the water is at such high pressure can do considerable damage and terminate, for a while at least, the supply of drinking water until a repair can be affected. Dangerous situations prone to freezing for supply pipes are when positioned against an external wall, in the roof space particularly near the eaves and when run through a cellar. Unlagged pipes in any of these positions clearly need lagging and the surveyor should advise that it be done as a matter of urgency.

Unsurprisingly, in cellars of older dwellings the mains supply pipe will often be found still to be in lead, insecurely supported by a few pipe hooks with sags and bends along its length. Such pipes become crystalline and brittle with age and if disturbed, or in contact with cementatious material, can develop pinhole leaks which, under the pressure, soon enlarge. As lead pipes are no longer permitted in new work because of the health risk, no repairs are allowed to old pipes and only total renewal provides an acceptable remedy. Even if there are no leaks, most owners would consider it advisable to renew and this has to be the surveyor's advice, most particularly in soft water areas. In hard water areas there is much less danger since, over time, lead pipes acquire a lining of calcium carbonate.

In other old dwellings, the mains supply pipe in the cellar might be found to be in cast iron of advanced age and exhibiting signs of rust, leading to the conclusion that renewal is the best option. From the 1930s onwards, dwellings in most cases would be provided with thin wall copper tubing, for the plumbing system, with its capillary and compression fittings for jointing. The dangers of bi-metallic action should always be borne in mind where there is evidence of a mixed use of copper and other metals in plumbing systems. The pre-eminence of copper over steel or any other metal for plumbing systems over a considerable time is widely evident, 90% coverage as estimated fairly recently. However, polyvinyl chloride, PVC, tubes for cold water supplies are now in considerable use with the endorsement of BS6700:1987, which devotes many pages to the various ways they can be connected and jointed to tubes of other material. They have the advantage of resistance to corrosion, lightness in weight and a degree of flexibility. They are not, however, suitable for hot water services, softening too readily, although tubes of a high grade thermoplastic, polybutylene, are satis-factory for this purpose. There are British Standards for copper, steel and the two types of plastic tube which will enable the surveyor to check the quality of materials used on new, near new or improvement work.

Heating

A dwelling to be considered fit for human habitation is required to have systems for the heating of water, cooking and for the heating of space. This can be achieved in a number of ways using different types of fuel, electricity, gas, oil and solar energy. The latter source for heating water, all that it is suitable for at the present time has already been dealt with at the end of section 3, Roofs. The fossil fuels can be used separately or in combination to supply both hot water and space heating from individual appliances or central units.

As already discussed, it is necessary to advise clients to have both electrical and gas installations, along with any fixed appliances, inspected and tested for safety and condition if a report signed by a qualified engineer and dated within the previous six months cannot be produced. Installations relying on oil as their source of energy which have not been mentioned previously, are infrequently encountered in urban locations, requiring as they do a storage tank in an enclosure with wall tanking up to flood level plus 10% in case of leakage, various safety provisions and access for fuel delivery. Such installations are more likely to be found serving the larger detached dwelling, perhaps in the outer suburbs but more likely so in country areas. In view of the aspects of safety inevitably involved, advice to the client has to be the same as for gas installations that an inspection, testing and a report by a specialist is necessary.

To an extent therefore it could be argued that the surveyor's inspection of these services borders on the unnecessary. Nevertheless the Conditions of Engagement as currently phrased for all the types of report envisaged by this series of books, even the Home Condition and Single Survey Reports, requires the surveyor to inspect and describe the installations and the fixed appliances providing the means to deliver hot water and heat space. Furthermore a European Union Directive of 2002 on the Energy Performance of Buildings requires the generation of an Energy Performance Certificate whenever a building is built, sold or let. The requirement is that the Certificate must indicate the individual energy performance assessment, benchmark that performance against an ideal and make recommendations for improvement. The Directive applies to all buildings, whether brand new or existing, of all types and obliges all member states to develop and apply a methodology for the calculation of energy performance. The United Kingdom Government for its part, initially, intended that all mortgage valuations for dwellings after a certain date, would include an Energy Performance Report but this proposal was talked out during a debate in Parliament a few years ago. It now sees the compulsory introduction of Home Condition Reports in England and Wales and Single Survey Reports in Scotland whenever a dwelling is marketed, both including an Energy Performance Report and Certificate, as a means of partly fulfilling its obligations under the Directive.

The reason for the Directive is the view, not universally shared and over which arguments rage, that the increase in the emission of 'greenhouse' gases, primarily carbon dioxide, is causing the world's climate to warm up. It is said that a 50% reduction in carbon dioxide emissions is needed by 2050 if 'catastrophic consequences' for the planet are to be avoided. While no one would deny that the world's climate changes, witness the Ice Age, and that the weather can be the cause of major disasters, it is maintained by some that there is not a shred of evidence to suggest that the actions of mere humans over such a short time span makes a hoot of difference one way or the other. The proportion and type of

gases in the atmosphere has only been recorded in the last hundred years or so and the proportions seemingly fluctuate in different areas and in a manner quite unrelated to the numbers of humans, cars, animals or anything else on the surface of the planet.

Nevertheless, whatever the reasons for the Directive, saving energy by using it more efficiently and thereby reducing emissions fulfils at least two significant purposes. First, it lowers the level of pollution in the cities, better for health and also buildings for that matter and, second, perhaps the most cogent reason of all, it can save money. There can be no objection, therefore, to providing home buyers, and eventually those renting dwellings, with the typical running costs and suggested improvements to reduce thosecosts while improving comfort and the performance rating at the same time. It is, after all, government policy and the law but there is no compulsion to follow up the suggested improvements. The government estimates that over a quarter of all the United Kingdom carbon dioxide emissions originate from its dwellings and has set an ambitious target of a 20% reduction by 2010, based on 1990 levels.

The government's methodology for producing Energy Performance Reports for buildings includes the Standard Assessment procedure (SAP) which has been available for some years and is used for new dwellings where the availability of the specification and the drawings enables all the necessary data to be assembled and, from 2005, the Reduced Standard Assessment Procedure (RDSAP) which is used for existing dwellings. This recognises that all the information ideally required may not be obtainable and that, in those circumstances, certain inferences may need to be drawn.

RDSAP is appropriate to use for 'Standard'

dwellings which are defined as those of mainstream size, construction and style with standard heating systems and there are permutations to allow for extensions, integral garages, attic conversions and heated conservatories. Dwellings outside the limits of what are considered normal are termed 'Non-Standard' and additional data needs to be collected for these to enable the Standard Assessment Procedure (SAP) to be used. Other dwellings, because of size, construction or unusual heating systems are termed 'Excluded' and need to be assessed differently along the lines of non-domestic buildings. The surveyor is required to identify dwellings which are 'Non-Standard' and excluded from detailed published specifications and, accordingly to know when to collect data additional to that required for 'Standard' dwellings.

While the data required to be assembled for the RDSAP to be used for 'Standard' dwellings is extensive, it is no more than can be ascertained by the inspection of all visible and accessible parts of the dwelling, as required for all the five reports envisaged by this series of books. Constructional details are required along with insulation levels and information on appliances. It does, however, include the measurement of room heights, floor areas and the length of perimeter walls for heat loss at each floor level, whether 'fascia cladding' is present and banding by age, both for the main structure and any extension.

It is not apparent from the RDSAP why the surveyor has to allot the dwelling into either a pre-1900, or one of eight bands covering the 1900s, particularly where there are only four years from the beginning to the end of one, 1991-1995 and only six years for another two, 1975-1981 and 1996-2002. It would be interesting to know the significance of the dates used for the age bands, but it is to be hoped that they are not intended to influence

the outcome too greatly. While they probably relate to changes in the Building Regulations affecting insulation levels which were first set in 1966 and increased in 1975, 1982 and 1990, and perhaps other matters, the provision of such could be made, or the existing features improved, at any time. A computer programme to assess the energy efficiency of a particular dwelling is more likely to be on a sounder basis if fed and strictly related to the construction and installations on a set day of inspection and not to the dwelling's age, whether known or assessed. Perhaps age is to be used in the software in the event of an absence of other information, but there is no specific mention of this.

For 'Standard' dwellings ie those with 'standard' heating systems, the RDSAP also requires details of all forms of energy consuming appliances in place at the time of the inspection, from the old to the new and from the most basic to the more complicated, differentiating between 'main' and 'secondary' heating systems. Since the most basic is an open fire for heating space, this is included. As already stated in Section 5 under the heading of Fireplaces, the open fire in a grate is given an efficiency rating of 32% but with a throat restrictor it rises to 42%. If it has a back boiler for a supply of hot water, but no radiators for space heating, efficiency rises further to 55%. A closed stove for solid fuel, on the other hand, has a 60% efficiency rating, rising to 65% if fitted with a back boiler, but again not supplying radiators. These are just five examples and efficiency ratings are also included for gas (8 types) and electric room heaters, electric storage heaters, solid fuel, gas, oil and electric boilers and many other appliances and systems since 'standard' in regard to heating systems is interpreted widely. At the other end of the efficiency scale, for example, ground-to-water wet system heat pumps are rated at 320%. It is to be hoped that the inspector will recognise the difference between this type of heat pump and one of the air-to-water type which is only rated at 250% efficiency!

It could, perhaps, be that there is a steep learning curve involved here for surveyors in recognising the difference between the 88 different forms of appliance listed for heating space and water, let alone the 38 different types of controlling systems. It is not surprising that surveyors on a trial run said that the two days training period provided by BRE was inadequate.

All the required data collected, the information is then passed on to a government approved organisation for processing, to produce the four page Energy Performance Report. This gives a summary of the dwellings energy performance related features under the headings of Construction, Heating and Hot Water, the Energy Rating, Typical Running Costs, carbon dioxide emissions in tonnes per year and the Annual Energy consumption in kWh per square metre. To comply with the EU Directive an Energy Certificate is provided which gives the Energy Efficiency Rating of the dwelling on a scale of one to 100 contrasted with the Average New Build Rating.

Automatically created by the software in the processing is a suggested list of improvements to reduce running costs, improve comfort under 'Lower Cost' and 'Higher Cost' headings, providing also the typical cash savings for each year and the Performance Rating which would apply after each improvement had been implemented. Finally, there are suggestions for further improvements to help the environment, an example given being the installing of panels to the roof for solar water heating.

The software uses 'default logic' in compiling its list of suggested improvements. For example, if the main roof is accessible and loft insulation at joist level is a mere 100mm thick, the suggestion will be to increase it to the recommended amount of 250mm of mineral wool or equivalent. As already discussed in section 3, Roofs, the consequences for future inspections by surveyors, plumbers and others if this is carried out without consideration being given to access provisions is horrendous. A phrase such as stepping on to a cloud comes to mind and proper consideration of this suggested improvement in many instances would take it out of the 'lower cost' measures and place it among those of 'higher cost'.

The surveyor is required to examine the list of suggested improvements and is empowered to remove those which are genuinely inappropriate to the dwelling, although he must give his reasons for doing so. For example, while the software is programmed not to recommend the provision of cavity fill insulation for a solid wall, it could do so for an unfilled cavity wall even where the dwelling was in an area of 'Severe' or 'Very Severe' exposure as defined under BS5628:Part 3: 2001 Categories of Exposure to Wind Driven Rain, and where such a recommendation would be entirely inappropriate. Another case might be where double glazing is suggested as an improvement, but where the dwelling is listed or in a conservation area. There are quite a number of possible suggested improvements where it is considered that the surveyor would have no cause whatever to remove them from the list. One could be the suggestion to insulate a hot water cylinder and another to upgrade the controls for a heating system or linking a community heating charge system to usage.

Considering the range of information which can be derived from an energy report for comparatively little further work by the surveyor, a recommendation to the client to incur the extra cost would be a sensible inclusion for reports other than the Home

Condition and Single Survey, particularly the Homebuyer and Building Survey reports. However, this is not mentioned in either the Practice Notes for the former or the Guidance Note for the latter as published by the RICS.

As to the requirements of all five reports for the installations to be inspected and commented upon, a brief mention should be made first of the unit appliances for the provision of space heating and hot water which were the predecessors of what might be expected to be found today. For the average sized dwelling, these were the open fire in a grate burning solid fuel for living rooms and bedrooms, the kitchen range for cooking, with boiler and tap or small storage tank for supplying hot water to the sink, the copper for boiling clothes and laundry purposes and for filling the zinc bath hung by the back door. These were all that were generally available until the early 1900s when gas cookers began to be more common for cooking and gas fires for heating and the 1920s when electricity was added for both purposes as a further alternative to solid fuel, although that remained the cheapest until the 1950s. The kitchen range was thus superseded and often replaced by an open

435: The modern equivalent of the old solid fuel kitchen range but which now can be run on electricity, gas (both North Sea and bottled) and oil as well. Much valued by some for cooking, the larger versions, as here, can supply domestic hot water and heating.

fire or closed stove with back boiler. The remnants of such systems might still be found in some dwellings, hence their inclusion in the RDSAP, and may need to be classed, in some cases, as the main system but are more likely to be considered as 'secondary heating' or just as 'room heaters'.

While the installation of a cold water supply system with reserve storage, where necessary, can be treated separately because it is a basic requirement not in itself dependent on the provision of any other service, this is not the case with the provision of a constant hot water supply and central heating by means of hot water. These are not only obviously dependent on the provision in the first place of cold water, but ought to be considered together, if the overall aim of the satisfactory provision of both is to be achieved in the majority of cases. The only circumstances where the provision of hot water can be considered separately is where overall heating is provided by some means other than by hot water, for example by electric storage heaters, convector heaters or a system providing warm-air circulation. It is therefore appropriate to consider briefly the ways in which a supply of hot water can be provided independently of any heating system.

The design of any hot water supply system needs to provide, so far as is practicable, hot water at the locations, in the quantities and at the temperatures required by the user at the least overall cost taking into account installation, maintenance and fuel costs. User requirements can be specified if known but BS6700:1987 'Design, Installation, Testing and Maintenance of Services Supplying Water for Domestic Use Within Buildings and Their Curtilages' provides data on which an assessment, based on size and type of building, can be made by way of the necessary design flow rates for domestic fittings at the appropriate temperatures. It is this Standard by which new or near new installations in dwellings need to be judged and which should form the basis of any consideration of an older system. The Standard recommends that initially a choice should be made between water heating by one or more instantaneous heaters, a water-jacketed tube type heater or a hot water storage system.

The rate of flow from hot water instantaneous heaters is limited and depends upon the power rating of the appliance which needs to be relatively high requiring an adequate gas or electricity supply. For example, the electrical loading for a satisfactory shower heater needs to be 6kW to deliver the recommended flow rate at the appropriate temperature while for a heater suitable for hand washing, it is 3kW. There are no instantaneous electric water heaters available capable of delivering the requirements for kitchen use or baths and consideration should have been confined, therefore, to gas-fired heaters. What most people think of as instantaneous electric heaters are, in fact, small heavily insulated electric, thermostatically controlled, water heaters.

Single or multi-outlet gas-fired instantaneous heaters are often installed, although the latter are only considered satisfactory if they can be restricted to one outlet being used at a time, which can be the case when there is single person occupation. For economy in the use of fuel and water, a multi-outlet, instantaneous gas heater, when installed, should be located nearest the most frequently used outlet, usually that of the kitchen sink.

Even with gas-fired heaters many of the required design flow rates quoted in the Standard are not achievable and, although they deliver hot water continuously and do not require time to reheat, the slow flow

tends to be irritating. Many will have come across old heaters of this type delivering water at such a rate that a bath is almost cold before it can be filled to the required level. This factor combined with their bulky appearance and the need for satisfactory flue arrangements has counted against their use as a means to satisfy current requirements, unless the restricted rates of delivery are acceptable by the client, which is thought to be unlikely. Another factor counting against their installation is that some have a limit to their recommended periods of continuous use in hard water areas because of scale formation.

Dependent on type and heat output of the associated heat source, water-jacketed tube heaters deliver hot water at rates comparable with the design flow rates referred to in BS6700: 1987 but at a slowly reducing temperature. The cold water feed for such heaters may be from the main or from a storage cistern while the water drawn off for use passes through a heat exchanger in a reservoir of primary water heated by an integral or separate boiler. The size of this reservoir, the rate of heat input to it and the characteristics of the heat exchanger determine the amount and rate of flow of hot water that can be provided without unacceptable temperature drop.

Clearly if this type of heater has been installed, considerable care needs to be taken to ascertain the performance characteristic from the manufacturer and the installation instructions need to have been followed to the letter for the selected type. BS6700:1987 states that such an installation supplied directly from a mains supply pipe shall accommodate expansion of water so that there is no discharge from the system except in emergency situations. This form of connection is unlikely to have been made since the presence of a cold water storage cistern is envisaged in all the circumstances covered. If, however, such a connection had been made for an existing installation then running an expansion pipe to discharge over the cistern would satisfy this requirement or, if there was indeed no cistern present, the pipe would need to be taken to discharge into a gully below the grating.

Storage type hot water heaters can deliver hot water at rates which comply with all the design flow rates recommended in BS6700: 1987 until all the stored hot water has been used when, ideally, the system will switch itself on and replace the amount used. The amount of hot water stored in domestic premises is related mainly to the number of bathrooms on the principle that if there is more than one, all will be filled in succession, but that hot water will still be required for kitchen use.

Storage type hot water systems are available in a variety of different forms and, after having obtained Building Regulation approval, it is

417: An unusual design of hot water radiator from the late 1800s in a country mansion with large diameter flow and return pipes indicating a gravity system.

now possible to install systems which are unvented. BS6700:1987 summarises the main differences between the traditional vented storage systems and the now permitted unvented systems as follows.

(a) Vented domestic hot water service systems are fed with cold water from a storage cistern which is situated above the highest outlet to provide the necessary pressure in the system and which accommodates expansion of the water when it is heated. An open vent pipe runs from the top of the hot water storage vessel to a point above the cold water storage cistern into which it is arranged to discharge. The main characteristics of vented systems are:
(1) explosion protection is provided by the open vent pipe and the cistern involving no mechanical devices
(2) the storage cistern provides a constant low pressure and a reserve of water in case of supply failure, but needs to be protected against the ingress of contaminants and frost.
(b) Unvented systems can be fed from a storage cistern, either directly or through a booster pump, but usually are fed from the mains supply pipe either directly or, more commonly, via a pressure reducing valve. The main characteristics of unvented systems are:
(1) explosion protection is provided by safety devices that need periodic inspection and maintenance
(2) mains-fed systems have no reserve of water but higher pressures are available if required
(3) the elimination of the storage cistern reduces the risk of frost damage and removes the source of refill or float-operated valve noise
(4) the safety aspects of unvented storage-type, hot water systems are subject to the requirements of the Building Regulations.

While there are owners who will no doubt

have been attracted to the saving in costs by the omission of a cold water storage cistern and associated pipework, the reduction of noise in the system and even perhaps the availability of the higher pressure from the main, it is considered that all owners would be well advised to insist on the reserve of water provided by a storage cistern, not only for the supply of drinking water when boiled in the event of supply failure, but also because the reserve enable the continued use of the hot water supply system, at least on a limited basis, and avoids the necessity for shutting down the system, provided always that the interruption to the supply is not too lengthy.

It is thought that most owners would also find further good reasons for cistern retention or provision when consideration is given to some of the factors brought out in the commentary and recommendations set out in BS6700:1987 under the heading 'Prevention of Bursting'

The production of steam in a closed vessel, or the heating of water under pressure to a temperature in excess of 100°C can be extremely dangerous. Water heated in this way flashes into steam when it escapes to atmospheric pressure, with a corresponding large increase in volume. If such water escapes in an uncontrolled way, as would result from the rupture of the containing vessel, an explosion will occur.

For this Standard a key requirement is that the highest water temperature does not exceed 100°C at any time at any point in the system.

Successful and continuing safe operation is in practice dependent on having the right equipment correctly installed in a well designed system that is properly maintained and not exposed to misguided interference. The reliability and durability of the equipment

on which the safety of the installation depends should be considered bearing in mind the conditions under which it will operate. Systems dependent on good maintenance for their continued safety should not be installed without reasonable expectation that this will exist. Maintenance and periodic easing of temperature relief valves is particularly important. The selection of all equipment, its location and even the choice of type of system will be influenced by these factors.

What is of concern is that unvented systems rely upon good maintenance and the periodic easing of temperature relief valves for their continued safety and should not be installed without reasonable expectation that these operations will be carried out. Even with notices displayed and the availability of operating manuals, experience suggests that it is all too easy in the average household to overlook periodic tasks and even the regular maintenance which should be carried out to the likes of gas-fired boilers is often forgone under the pressures of time or financial constraint.

The consequences of a lack of maintenance with traditional vented systems is usually no more than a breakdown and an inconvenience to the occupiers. Indeed many such systems often function with no attention whatever for quite lengthy periods of time. In the case of sealed unvented systems the consequences could be an explosion. Why, one might ask, incur the risk of an explosion and the likelihood of injury, or worse, when the installation of a storage cistern renders the need for an unventilated hot water storage system unnecessary?

The extra cost of a cistern is a small price to pay for the avoidance of that sort of worry: insulation can help with any refilling noise, which most occupiers get used to anyway and which is far less than the noise from water-hammer which often occurs when there is only a mains supply to fittings. As to the absence of the higher pressure from the main supply, miniature pumps have overcome the problem of the requirement for something like a 30m head of water to provide a satisfactory performance from a shower fitment.

It is recommended, accordingly, that in view of all the risks involved the surveyor should give serious consideration as to whether an unvented hot water storage system should be permitted in domestic premises. The surveyor needs to stress very strongly the need both for periodic checks and adjustment of the system and for its regular maintenance. If a new owner considers that there should be no problem in providing such regular attendance then the surveyor could consider that at least he has been given full information on the possible dangers.

Perhaps the simplest type of electrically powered hot water storage heater is the non-pressure inlet controlled type, the one already mentioned and often thought of as an instantaneous heater. When, in one of its smaller incarnations, a heater of this type is fitted above the kitchen sink it bears a superficial resemblance to a gas-fired instantaneous heater of similar size. In fact such heaters come in sizes from 6 to 136 litres, can be fitted over a sink, bath or wash-basin and are supplied with cold water from either the main or a cold water storage cistern. As the name implies when the inlet valve on the cold supply is opened the cold water expels the stored hot water through the open outlet. There is no need for an expansion pipe as space is left at the top of the storage vessel for the water to expand and there is an anti-drip device to prevent the unwanted escape of hot water from the outlet.

Such heaters are also known as open-outlet-type water heaters and under no circumstances can the outlet be controlled by a valve or tap as it must remain unobstructed at all times. Because of this, such heaters have to be installed above the fitting to be supplied and accordingly the larger sizes are difficult to support on walls and being bulky are not attractive in appearance. Two heaters would be required for a house or flat with a single bathroom and it is considered that either of these types of property would be better served by the following type of pressure or outlet-controlled electric heater. For a small house or flat, compactly planned as regards the relationship of kitchen and bathroom the 'UDB' or 'under the draining board' variety of this type of heater is very suitable. Installed near the sink for greater efficiency and with a capacity of between 100 and 150 litres one of these well-insulated and thermostatically controlled water heaters can provide an unobtrusive but comprehensive hot water service to the taps of sink, basin and bath. Heaters of this type are fitted with two electric immersion heaters, the smaller for general use and usually left on all the time, while the larger is switched on for laundry or bathing purposes. Heaters with a capacity of 200 litres are available for use with cheaper off-peak electricity but, obviously, greater care needs to be exercised so as not to exhaust the store of hot water as it cannot be replenished during peak periods which many would consider a disadvantage.

Connection can be made to either a cold water storage cistern or the main if permitted by the supply company and provided the heater is selected to withstand the appropriate mains pressure. Some heaters although called pressure-type heaters are only designed for storage cistern feed and care must be taken to verify the capability of the heater to withstand mains pressure if that is the form of connection. In all cases, it is necessary to run an expansion pipe to discharge over the storage cistern or in the case of mains supply to a safe spot such as below grating level at a gully.

The advantage of this type of heater is that it can be accommodated on the floor, out of sight, and being a manufactured fitting arrives on delivery encased and fully insulated ready for connection to water and electricity supplies. No servicing is required and such heaters will operate for many years satisfactorily without any attention.

As with the electrically heated variety, self-contained hot water storage heaters with thermostatically controlled integral gas firing are available in a range of sizes. Smaller versions of a capacity from 9 to 23 litres can be fitted above a sink and have a swivel outlet. Larger versions can be floor mounted and can have a capacity, typically of around 120 litres, sufficient to supply the needs of a compact small house or flat. Cold water supply is from a storage cistern to which must be run a vent or expansion pipe to discharge over the top in the usual way.

Because such heaters require regular servicing, they are less convenient than their electrically powered equivalents, even allowing for more favourable unit cost in use, and more obtrusive because they require air for combustion, adequate flue arrangements, except for the older smallest sink heaters, and can some-times smell if ill adjusted. As such they are by no means as useful as electric hot water storage heaters in similar situations.

Where the accommodation is less compact or larger and where the facilities served are more extensive than a kitchen sink and a single bathroom with bath and basin, the storage capacity required will be greater and it becomes necessary to provide a hot water

storage cylinder of appropriate size and a separate heating unit. Where gas has been the fuel selected this can either be a circulator or boiler. However, in either case it will have been necessary to decide before installation whether the water in the cylinder is to be heated directly by means of flow and return pipes from the circulator or boiler or indirectly by means of a coil of pipework within the cylinder, this coil only being connected by what are known as primary flow and return pipes to either circulator or boiler. Where the installation, as envisaged here, is limited to the supply of hot water for domestic purposes only the decision is governed by the degree of temporary hardness in the local water supply.

The continual heating and drawing off of water, and its replacement for heating by further cold water with a high degree of temporary hardness, causes the build-up of scale which can, in due course, reduce sub-stantially the bore of pipes and accumulates in boilers and cylinders. This build-up can be minimised in the boiler, the primary flow and

418: A six column type radiator providing a large surface area for increased heat transfer by convection but also a veritable dust trap.

return pipes and in the coil within the cylinder if the same water only is allowed to circulate. It will be abundantly apparent that these pipes form a circuit. Indeed they are those which when extended supply the heat emitters of a hot water central heating system. Hence an indirect cylinder is a basic requirement where both hot water and central heating by hot water are installed and supplied from one boiler.

Provision must be made in the circuit for topping up with cold water and for the expansion of the hot water. Because of the degradation of the water in such circuits the water supply companies insist that a separate small balancing and expansion cistern is installed for all such situations and that there is no connection with a cold water storage cistern used for other domestic purposes. It needs to be mentioned again here that it is permissible under the Building Regulations to provide an unvented, i.e. sealed, primary circuit but, for the reasons already stated, such a system can be considered undesirable for domestic premises.

For the less compact house or flat the separate heating unit can be a gas-fired circulator. This type of unit is defined in BS5546:1990 'Specification for Installation of Gas Hot Water Supplies for Domestic Purposes (1st , 2nd and 3rd Family Gas)' as a boiler, with a rated input not exceeding 8kW, designed primarily for the supply of domestic hot water in conjunction with a separate storage vessel. Circulators are fitted with their own flow and return pipes for connecting to the storage cylinder, which is supplied with cold water from the domestic cold water storage cistern and to which the expansion pipe is taken to discharge over the top in the usual way. The operation of the circulator is controlled by a thermostat in the cylinder and the appliance should be fitted as near to the sink as possible, although it can be fitted in an airing cupboard or elsewhere provided there is an adequate air supply and flue arrangement and suitable access for servicing and adjustment.

A three-way economy valve is usually fitted so that, on one setting, only sufficient water for sink and basin use is heated. Where these units are used it is important that the circulation head and the size of the flow and return pipes is sufficient to give an adequate stored water temperature and to avoid noisy operation. The manufacturers should provide information on the minimum circulation head necessary and also on installation requirements where an excessive head is unavoidable.

For a large house in a soft-water area a direct cylinder system for the supply of domestic hot water alone would be satisfactory together with a boiler of appropriate size to provide the better response than would be available from a circulator. The system could be designed for gravity circulation with flow and return pipes of not less than 25mm bore between boiler and storage cylinder, run as directly as possible.

The cylinder needs to be located at a sufficient height above the boiler to give adequate circulation, and certainly not less than one metre, which should be sufficient to prevent the circulation of hot water from the cylinder to the boiler when an electric immersion heater is fitted for summer use. In hard-water areas, as mentioned previously, an indirect cylinder system is required as it would be in all cases where a hot water system is installed to provide central heating as well. It is, of course, when this is required that the services of a heating engineer or specialist contractor should have been engaged to design the system and where an inspection and test by an independent heating specialist needs to be recommended by the surveyor on the grounds of safety and for assurance on the adequacy of performance.

BS5449:1990 'Specification for Forced Circulation Hot Water Central Heating Systems for Domestic Premises' recommends that the consulting engineer, heating contractor or other person responsible for the design of a hot water and central heating installation should have considered the following matters:
a) thermal characteristics of the building for the calculation of heat requirements and possible improvements for energy conservation
b) fuel to be used
c) position of the boiler, bearing in mind access for maintenance, means of flueing and provision of combustion air
d) type, location, dimensions, construction and suitability of chimney and flue terminal, where required
e) location and size of fuel storage and access thereto, where required. For solid fuel, ash removal and disposal will also have required consideration
f) position of feed and expansion cistern
g) facilities for filling and draining the system
h) requirements for domestic hot water supply
i) position of any domestic hot water supply equipment, e.g. hot water storage cylinder
j) temperatures required to be maintained and the manner in which the dwelling and system are to be used, bearing in mind ventilation and condensation
k) type and position of heat emitters
l) system control of heating and hot water including frost protection
m) route and method of installing pipework
n) the need for compliance with relevant Building Regulations, Gas Safety (Installation and Use) Regulations, Regulations for Electrical Installations, Water Bylaws and BS Codes of Practice.

It is the response to these matters as embodied in the system installed which will require the attention of the specialist engaged to report on the system.

A surveyor should have been present when these matters were being considered so that the owner had the benefit of his experience. The owner was unlikely to have a knowledge of the merits of the alternatives available and while the consultant engineer or specialist contractor should, of course, explore and explain all the possibilities the owner may well have been a trifle confused by the wealth of information available.

For example it is perhaps all too easy for owners to have been seduced by the financial attractions of early gas-fired combination boilers which provided hot water for both central heating and domestic use without the need for either a cold water storage cistern or a hot water cylinder. Without any storage, however, delivery of domestic hot water could have been delayed and even then may have been so slow as to take time to fill a bath, if there were competing demands.

The designer of the installation ought to have obtained sufficient details of the premises to enable heat losses to be calculated so that he can then draw up his written specification stating the type and output of the boiler and heat emitters and the room temperatures that will be attained at stated design conditions.

The specialist should also have given consideration to the following matters if they had not already been dealt with:
a) minimising air leaks through the structure from old boarded floors, badly fitting external doors and unused open fireplaces
b) the possible insulation of ceilings and roof
c) filling of cavity walls with insulating materials
d) internal insulating linings and external cladding
e) double glazing

f) reflective or insulating surfaces behind radiators on external walls

The surveyor or specialist will no doubt have checked the retention of a ventilation opening to any of the remaining chimney breasts internally, at least 300mm above the fireplace base and the external provision of a chimney terminal with ventilation below. These should prevent dampness from penetrating or condensation from forming within the flue itself.

It should be remembered however, that it is totally undesirable to exclude ventilation within the dwelling and BS5449:1990 says that care and judgment should be exercised to prevent carrying out excessive measures to minimise air leaks since an air-change rate of about one room volume per hour is necessary for fresh air supply and odour removal when rooms are occupied, even more in the case of kitchens for the removal of steam and odours from cooking and laundry.

Allowance should also be made for the supply of combustion air necessary for any other appliance such as an independent fire, cooker or water heater in the room in which a boiler is situated.

Boilers should be positioned as close to the hot water storage cylinder as possible, so that heat losses from the primary flow and return pipes are reduced to the minimum. The boiler and cylinder should be placed in a central position to reduce the length of the secondary circuits to the various hot water draw-off points. These may circulate hot water by gravity or by pump and the pipes must be insulated to conserve heat. The boiler should, obviously, be installed in accordance with the manufacturer's instructions and to the appropriate British Standard, but the specialist should ensure that the structure upon which a boiler is supported, and its immediate surroundings, are adequate for the weight imposed and, where necessary, be suitably protected against the risk of fire or damage from heat. Sufficient ventilation should also be allowed for, although the use of balanced flues has reduced the need for some requirements in this regard. For the highly efficient (to the order of 85%) condensing type boiler the surveyor should check that provision has been made for the condensate drainpipe usually of 20mm plastic to run into an internal soil stack or waste pipe via a 'U' trap.

The specialist will have particular concern in relation to the flueing and air supply of boilers. Chimneys and flue pipes should be constructed of materials appropriate to their location to avoid overheating of combustible structures and the effects of condensation and should be terminated in a freely exposed position to minimise the risk of downdraught or the creation of a nuisance. Any existing chimney flue liner should be checked for its soundness and freedom from obstructions.

It is probable that gas fired boilers will be the most frequently encountered. The RDSAP lists 15 types grouped into sets of five. The first group is of post 1998 types with efficiency rates ranging from 65% for gas room heaters with back boilers to 83% for condensing boilers including combination boilers with automatic ignition. The second group covers pre 1998 types with fan assisted flues and efficiency ratings of 68% to 85%, the latter for the plain condensing type, the highest rate quoted for boilers using gas as the energy source. The combination version is only one percentage point behind. The third group also covers pre 1998 types but with balanced or open flues. All except one type have efficiency ratings of 65%, the odd one out being pre 1979 floor mounted being types, having a lower rating of 55%.

Whether types of boiler with high efficiency ratings but relying on fan assisted flues and automatic ignition prove to be more economical in use is debatable. The energy savings may well balance unfavourably with the higher CORGI engineer service and maintenance costs when things go wrong and the more frequent need for replacement due to corrosion.

In old buildings, particularly having regard to the varying and various floor levels, the specialist needs to check on the positioning and supporting of pipes. Due allowance for free movement for expansion and contraction should have been made and the pipes should not be allowed to sag or form local high points in which air could accumulate, particularly where the circulation is by gravity. Notches and holes in structural joists should only have been cut in accordance with the rules set out in the Building Regulations while pipes passing through brickwork and masonry should be sleeved to prevent corrosion and allow movement.

BS5449:1990 recommends that pipework shall be routed so as to prevent contact with electric cables. The Standard also recommends that pipework should not be buried in concrete. Pipes are sometimes buried in small shallow wavering grooves in concrete flooring, with subsequent expansion and contraction allowing the pipes to break through the cement screed. If the floor is of concrete, ducts of adequate shape and depth at least 25mm from the finished surface are desirable, so as to allow for the pipes being adequately protected from damage and corrosion.

Pipes if possible should have been fitted clear of timber joists and floorboards but where this is not possible, suitable pads of material resistant to damage by insects and vermin should have been fitted between the pipe and structure to minimise noise and notches should be lined with felt or similar material.

Floor surfaces should ideally be marked to show pipe runs and protective saddles employed to prevent damage by nails when laying floorboards or carpets.

The pipework should allow for the free circulation of air behind emitters and where dust from convection is likely to discolour decorations, the client should be advised as to the fixing and sealing of deflecting shelves. Due provision should be made for access to valves and the air release cock, however, when these are fitted.

BS5449:1990 contains the recommendation that heat emitters should be located on outside walls preferably beneath and for the full width of windows to offset the cooling effect. The latter part of this advice overlooks the fact that most householders, certainly in their principal rooms, prefer to have full-length curtains which if present would, of course, negate the value of the emitter.

Finally the specialist should check that all the accessories specified are in place, correctly labelled with diagrams for the householder. The range of fittings is now very wide and the British Standard specifications warn against the use of the wrong fittings as the consequences can be disastrous. These range from the many types of valves such as safety valves, key-operated or automatic by-pass valves and balancing valves. Pressure gauges and indicators together with automatic or manual air vents and the arrangements for temperature controls should be checked including thermostats and the efficiency of feed, overflow and expansion pipes together with the arrangements for draining down the installation when necessary.

Drainage

Sanitary Fitments

Water whether cold, hot or both, having been delivered where required, consideration needs to be given to the fitments where it is used. On the room by room inspection of the dwelling, every sanitary fitment needs to be examined, described as to type, name of manufacturer if marked, the material used and the condition noted.

A surprising range of sanitary fitments will be encountered, some of them almost conceivably of the age of the dwelling itself. However, the aim to be 'up to date', an attitude typical of the 1930s and the improvement process applied to older dwellings carried out since, has meant that many have been replaced by efficient fitments made by reputable manufacturers in accordance with British Standards. Some are not so keen to have 'standard' fitments so that now, the surveyor may meet up with reproductions of fitments previously considered obsolete, along with those of novel design. In any household the water closet, particularly if there is only one, is probably the most critical, as those that are insanitary and do not operate efficiently can be irksome and a source of irritation to the occupants and an extreme embarrassment to visitors. The surveyor needs to recognise the different types.

In some of the terrace houses of the 1700 and early 1800s examples of the valve closet may still be found. Invented by Joseph Bramah in 1778, this had a large water area in an earthenware pan with a central outlet which opened on the pull of a handle in the seat which also operated the almost noiseless flushing arrangement. With its double water seal, it was looked upon quite favourably for a long time and retention of perhaps one

example, for historical reasons, in a household where other more modern alternatives are also available would seem a reasonable possibility. There are drawbacks, however, with many mechanical parts which can go wrong and a casing, usually of wood now considered insanitary, making access to those parts difficult.

The advent of the combined pedestal and water seal trap, one piece wash down water closet pan pioneered by Thomas Crapper in the mid 1850s was a great improvement, with its nine litre high level cistern fixed about two metres above the pan. The release of the water in the cistern forced, by gravity, the contents of the pan via the trap into the drain if the fitting was at ground level, or into the soil pipe at upper floor levels. The shape of pans and traps went through a number of evolutionary stages to varying degrees of success with 'long' and 'short' hoppers, 'D' traps and the like, culminating in BS5503:Part 1:1977 for vitreous china wash down WC pans for use with nine and 13 litre cisterns and Part 2, covering use with the smaller seven litre cistern. When the cistern was fitted at low level, sometimes necessary if the WC was positioned in front of a wide window, ideally the large size cistern would be used to improve the force of the flush, along with a greater diameter flush pipe than would be the case with a cistern fixed at high level. The current fashion for totally enclosing low level flushing cisterns leaving only a handle or push button visible for operation can produce the problem of much dismantling when the ball valve needs adjustment or renewal. Easy access should be available at all times.

The wash down type closet remains simple, cheap and seldom liable to blockage but it is, however, noisy in use, particularly when used in combination with a cast iron bell type flushing cistern. A better and much quieter performance could be obtained from the siphonic type of

438: A wash down type WC pan with low level flushing cistern. The overflow from the flushing cistern is connected to the flush pipe which is one way of overcoming the difficulty of when the route to the exterior is obstructed. The capacity of the cistern and the size of the flush pipe determines how well such installations operate.

439: A Shanks's 'Combination Levern' close-coupled siphonic WC suite with matching mahogany cistern enclosure and seat, about 100 years old but still giving good service. Pulling the handle on top of the cistern forward and releasing it, operates the mechanism satisfactorily without any need to 'pump'.

closet which became available from around the 1900s for use with a high level cistern. Single and two trap versions were available, both, by creating a vacuum, drawing the contents of the pan out to the drain or soil

pipe. These later developed into the neater, close coupled, suites covered by BS7358:1990 which were designed to operate with a maximum flush of seven litres. They are, however, more prone to blockage than wash down closets.

It is to British Standards that new, near new or upgraded installations should be related and if fitments are not supplied to those Specifications or their European equivalents, then a firm actionable guarantee should have been obtained from the supplier or through the installer which can be transferred to a new owner, having first been examined by the surveyor. This is particularly important in regard to WCs as efficient performance depends very much on the shape of the pan and trap, the size of the flushing cistern and flush pipe, if there is one. This is why it is so important that the components of close coupled suites are installed exactly in accordance with the maker's instructions. Slight departures and, most certainly, the employment of a cistern of a different make, can make all the difference between good and irksome, irritating performance.

In practice, the British Standard Institution indicated that WCs should successfully cope with six pieces of newspaper each 150 by 150mm. Certainly many of the cast iron high level cisterns operated by means of a cast iron bell on a pull handle will be so rusted as to be incapable of delivering a fast effectual flush. Often even to obtain a flush, delicate skills honed over the years are required, skills not available to the uninitiated visitor. As a replacement, a cistern with a piston actuated siphon which functions with less noise, requires less force for its operation and is more reliable should be used.

The materials from which sanitary fitments are made should be hard, durable, smooth, easily cleaned and be non-absorbent. Some materials satisfy these requirements better than others.

Of the ceramic materials used, which for their strength and degree of permeability depend upon the composition of the clay and the temperature of firing, vitreous china is undoubtedly the best for WC pans, bidets and wash hand basins. It is similar to good quality tableware enabling sharper detail to be produced and even without its glaze, which is applied purely for appearance and easy cleaning, is completely impermeable. It is, however, more expensive and not quite as strong as glazed fireclay and glazed earthenware which accordingly are often favoured for the manufacture of sinks. These have to cope with harder useage, although both materials are also used for WCs and lavatory basins in dwellings at the lower end of the quality scale, where more rounded detail will suffice. The disadvantage of both is that for impermeability they rely solely on their glaze which if chipped or damaged by a hard knock, allows liquid to penetrate, eventually producing a stain and an insanitary condition.

For ceramic fitments that have to take the toughest of treatment, sinks for example in dwellings, glazed stoneware provides another alternative, like vitreous china not relying on its glaze for impermeability, witness its use nowadays unglazed, for drainpipes, but not capable of much subtlety in design. All the ceramic materials used for sanitary fitments provide no resilience so that glassware, jars or bottles usually break on contact. However, if heavy, jars and bottles have been known to remove the bottom out of a vitreous china wash hand basin, often themselves surviving intact.

Hardwood, and early versions of vitreous enamelled steel draining boards, used in conjunction with the rounded edges of ceramic sinks tend to produce unpleasant and unhygienic conditions where they meet and this led to the development from the 1930s of one piece sink and drainer units, first in porcelain enamelled mild steel and then in stainless steel. The enamelled surface of the former tends to chip too readily, allowing corrosion to develop in the steel, with eventual perforation, and limiting the life, a problem not solved despite their continuing availability. Accordingly, stainless steel tended to secure a dominance for such fitments which began to be fitted on top of kitchen units as distinct from being supported on brackets and later set in to the top of a continuous work surface.

440: Better by far in normal use than the installation at 438, is this close-coupled siphonic suite, cistern and pan manufactured to operate together. However being more complicated in shape than the wash down type the pan is more subject to blockage by abnormal use.

441: The plumber, or owner, seems to have mistaken the function of this bidet by providing it with pillar taps, perhaps for use primarily as a foot bath, instead of valves for the supply of hot and cold water to the rim and the ascending central spray.

Separate round sinks and drainers in stainless steel also became available, the drainer with a separate outlet and trap. Not to be outdone, the ceramic ware manufacturers extended their ranges from the traditional London and Belfast pattern sinks, the latter type having a British Standard, BS1206, and in one rather expensive form was provided with an integral grooved ceramic drainer, to other types of sink and combined drainers which, similarly, could be dropped into the top of a work surface. The edges of such ceramic fitments are inevitably bulky and difficulties with sealing the joint between sink and worktop, which was usually of plastic laminate at this time, led to damp penetration resulting in the plastic lifting from the block or particle board to which it was glued. To overcome this problem, one piece combined sink, draining area and worktop units were developed to provide totally seamless surfaces made from a combination of materials including quartz and acrylic based synthetic stone in various finishes, including marbled effects, and in a variety of sizes and colours. Some are very heavy necessitating sturdy base units, or the rein-forcing of those not so strong, to ensure level and firm support otherwise there is a danger of induced cracking. They are said to resist scratching, heat and household chemicals. In practice some have been found to stain easily, suffer from inherent cracks which were not visible when new and be difficult to clean.

In new or near new dwellings, or where kitchens have been upgraded and unorthodox fitments have been installed, the surveyor will need to obtain and examine as much detail as he can, together with any guarantees which are available before providing advice on the shape, characteristics and likely performance of the fitments. Ideal arrangements to secure the satisfactory performance of sanitary fit-ments, established over the years through British Standards, are jettisoned for the sake

of appearance leaving the surveyor in a difficult position. He needs to be wary of expressing approval even when there is a guarantee available, pointing hopefully to the fact that the fitments are from an old established, well known and reliable manufacturer, installed by a company which knows what it is doing not by sub-contractors of unknown reputation, and, in the event of problems, redress is likely to be available for the sake of the maker's and installer's reputations. The opposite would apply, obviously, where information and guarantees are lacking or sketchy, the fitments are of dubious provenance and installers bordering on the fly-by-night type.

The remarks in the previous paragraph apply equally to less than orthodox, non-standard, fitments in bathrooms. Materials have already been mentioned for use in respect of WCs and wash basins, of which there is a British Standard, BS1188, and the same remarks apply to bidets and shower trays, although in the case of the latter glazed fireclay and earthenware are also acceptable. Although baths will on occasions be found to be in glazed fireclay, they are expensive and extremely heavy. Even the traditional porcelain enamelled cast iron baths, introduced around 1880, are moderately heavy compared with their modern equivalents, although with their curved ends, roll tops and ball and claw feet, still find favour. More frequently found will be the mild steel version, available in the same shapes, including the roll top, and the same porcelain enamelled finish. The roll top and curved shapes always allow much water from splashing to run down and wet the floor from wherever the bath is positioned with possible consequences from rot. This was sometimes counteracted in larger dwellings by the provi-sion of a lead or zinc tray below the bath with its own outlet.

442: This cast iron roll top bath is unusual in that it is square at one end with pillar, instead of the usual globe, taps, soap trays, a pop-up waste but no provision for overflow. At one time the bath stood on a lead tray but most of the lead has been removed leaving the floor and, particularly, the ceiling below, very vulnerable to careless use.

The 1930s saw the development of squarish baths with flat topped edges for fitting tightly against a wall which, with tiling brought down on top, could produce a reasonably water tight joint when well formed and the bath was positioned on a firm floor. The bath could then be provided with side and end panels if required in various alternative materials, wood, plastic, asbestos or vitrolite are examples, an arrangement which well suited the small bathrooms of the 'three-up-two-down' semi-detached dwelling of the period. A British Standard, BS1189, covers square top baths in both cast iron and mild steel.

Although better in both respects than the ceramic version, the metal bath apart from its still considerable weight, has another disadvantage in that it is cold to the touch and takes a lot of heat out of the bath water. On the other hand, baths made from acrylic perspex or glass fibre reinforced polyester, fibreglass, are much lighter in weight and have a lower thermal conductivity than metal baths, being warmer to the touch and allowing the water to stay hotter longer and are, there-

fore, better in these respects. They can be obtained in many colours and supplied moulded into the traditional shapes, including roll top models and, a considerable advantage, can be obtained as corner units. When first introduced acrylic baths were often provided with inadequate support and had a disturbing tendency to soften under hot water requiring additional cradle supports but, nonetheless, resulting in distortion and pulling away from adjacent wall surfaces. Modern versions are made of thicker material and adequate support has been developed by the manufacturers. Glass fibre reinforced polyester did not share these problems, being much stronger but, accordingly, more expensive. Unlike metal baths which can usually be restored by re-enamelling if worn, scratched or damaged, plastic baths have to be discarded and replaced by a new bath.

An important item to check on the visual inspection of the sanitary fitments is that each has an overflow. Furthermore the overflow must have the capacity to cope with the amount of water entering the fitment when the supply tap or taps are discharging at full bore, Surprisingly, not all fitments have overflows and those that do not, need to be drawn to the client's attention with an appropriate warning. Whether the other overflows which do exist are of sufficient capacity is not always easy to gauge. Many of the overflows integral with sinks and wash hand basins are clearly inadequate and must be thought of as tokens rather than as serviceable appendages to be relied on if waste outlets are plugged and supply taps left on full bore. At the risk of seeming over cautious, warnings need to be given in regard to these. After all, the seller can always demonstrate how wrong the surveyor is if he believes everything to be in order.

As to overflows from flushing cisterns, it is

usually a comparatively easy matter to check that an overflow is present, of the appropriate size in relation to the incoming supply, runs straightforwardly through an outside wall to discharge as a warning in a visible place. It should be well clear of the wall below and fitted with a hinged flap, often omitted but desirable, to prevent any trickle overflow in the pipe from freezing and blocking the pipe.

Access should always be available, without creating undue disturbance, to check what happens at the outlet to all sanitary fitments. There is not only the joint of fitment to pipe, direct in the case of WCs, to be checked but also in the case of waste fitments, access to check that a water seal trap is provided, its type and whether, if needed it is vented properly in some instances, and then a further joint to the pipe beyond. In the case of all fitments, there is the need to see that joints are not leaking and that they have not been leaking in the recent past to cause, perhaps, an out-break of rot in wood flooring below. The

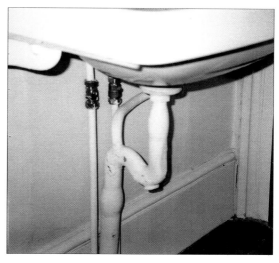

444: The surveyor should examine below all sanitary fitments for slow or intermittent leaks and to check that a suitable trap is provided. Here, this wash hand basin has a lead waste pipe and trap with cleaning eye and a lead anti-siphon to maintain the water seal at all times.

443: An early type of shower rose fitted over a bath, the temperature of the water controlled by a mixer valve which should preferably be replaced by thermostatic control to avoid scalding.

presence of bath panels can present a problem. Even if the panels are secured by visible screws to framing, not always the case by any means, the surveyor is not permitted to remove them without the owner's permission. If traps and joints cannot be examined, this needs to be stated in the report. This is also particularly relevant in the case of shower trays at upper floor levels where outlets and traps have often to be arranged in awkward places. An examination of the ceiling below might show signs of a slow leak, always a serious danger and providing the necessity for the surveyor to advise the need for opening up for closer examination.

It might seem logical when checking on waste outlets and traps, say in a first floor bathroom, to dash outside and continue to trace what happens to the pipes from the various sanitary fitments until they connect with the under-ground drains. While to do this is always possible, it could be that it would be more time consuming, particularly if the dwelling is of some size, than following a process starting at the other end, so to speak, i.e. where the drains of the dwelling join on either to a public sewer or, in rarer cases, to the dwelling's own soil and waste water storage cesspool or its own disposal system, and then work backwards.

For example, an inspection chamber giving access to the underground drains of a dwelling can have any number of branches, depending on need. Viewing those branches once the cover has been lifted, a requirement for all five reports where accessible and safe, and gauging what each one does having full knowledge of all fitments internally is much simpler than dealing individually with the destination of the discharge from each fitment. For this reason on the inspection, it is preferable to deal with the dwelling's drainage system before giving consideration to the vertical pipework on the dwelling' elevations.

Drains

The underground drains of a dwelling, 'out of sight', ought, too, to be 'out of mind'. In other words they should fulfil their function without the owner's attention being drawn to them for many years at a time. Although a British Standard Code of Practice recommends that they should be inspected 'at regular intervals' and BRE advise that a water test should be carried out 'every two to three years' this is very much a counsel of perfection. There is no reason why drains, provided they have been designed as they should have been to be self-cleansing, should require any examination or maintenance for years. Indeed when a dwelling changes hands, say on average every 10 years, is probably as suitable a time as any between inspections.

Whether the drains of a dwelling perform in

the desirable manner mentioned above can only be established through evidence given by owners or tenants who have been in occupation over a period of years and, of course, on the assumption that they are being truthful. Many dwellings built during the 50 year period from 1880 to 1930 do in fact achieve such a performance. This was a period of good design for drains, particularly in relation to laying at adequate self-cleansing gradients, good workmanship and primarily being laid at a depth substantially greater than during later periods. A water test properly carried out, while establishing whether a drain leaks or not, is a severe one on both pipes and joints even when involving a pressure, as recommended, at no more than around 0.007 N/mm^2, equivalent to about a 600mm head of water at the top end but which can in practice amount to much more at the lower end of a long drain. In real life this is only likely ever to occur should the drain be totally blocked and very much greater than when the drains are in normal use. It would be unreasonable to apply such a test, even at this pressure, which is less than half that usually adopted by local authorities for testing new drain runs, to drains 70 to 120 years old. If carried out it would have to be expected that many lengths would leak, at least slightly. Some would argue that a water test might actually induce a degree of leakage in elderly drains and that before advice to carry out such a test is given cogent reasons should be established in justification and, most certainly, owners should have the risk of damage explained before their permission is sought.

Possibly in recognition of the above argument none of the Conditions of Engagement for the five types of report envisaged as being covered by this series of four books requires a test of any sort to be carried out on a dwelling's drainage system. All five, however, require those parts of the drains which can be seen and are accessible to be inspected. This is extended to lifting, where it is possible to do so by one person and without causing damage, the covers of drainage inspection chambers, or manholes as they are referred to in parts of the relevant British Standard Code of Practice, and examining the insides. This provides only limited information but in conjunction with taking a few measurements and walking along the surface above the probable run of drains, can provide evidence to suggest whether or not the drains will give relatively trouble free performance over the years. It will also indicate whether further investigations ought to be carried out, which might contain advice to commission a water test on the whole or perhaps only part of the underground drains and whether some work ought to be done in the short term towards the achievement of that goal.

Knowing or assessing the age of a dwelling will not, in itself, tell the surveyor much about its drainage system. Whereas the surveyor could be inspecting and reporting on a dwelling built 250 years or so ago, exhibiting most of the characteristics of that time of building and still basically little altered, this will hardly be so with the drains. The drainage system is unlikely to be even half that age because the remains of earlier systems, probably based on the Dry or Conservancy principle, of soil and waste disposal from before the latter part of the 1800s will have long since been cleared away and replaced. What enabled the subsequent water carriage system of drainage to be developed was the spread of a piped water supply to individual dwellings from about the middle of the 1800s. Dwellings built after the Public Health Act of 1875 generally still have the drainage systems with which they were built. Those from before that time were improved by the construction of up to the minute drainage systems, embodying the same newish principles.

Thus most dwellings built before 1950 have drainage systems embodying the principles developed in the mid 1800s, both in design and character, and constructed in the same materials which were developed around that time. For example, until around 1950 the only choice of material for drain pipes lay between, ceramic, in either fireclay, earthenware or stoneware form, and cast iron. As pipes and fittings in cast iron cost twice as much as those in the ceramic materials, it is not surprising that practically every dwelling built up to 1950 is equipped with ceramic drain pipes. It is only after 1950 that dwellings will be found with alternatives such as concrete, asbestos cement or uPVC pipes. Pitch fibre were also available but these were infrequently used, being thought of as too insubstantial. The fireclay, earthenware and stoneware pipes came to be salt glazed, more of an advantage to the two former varieties to make them impermeable but not really needed for stoneware which is impermeable even without a glaze. The current view is that both earthenware and fireclay as well as stoneware can be used unglazed for underground drains in certain circumstances.

British Standards were developed for all the pipes and fittings used for drainage installations from the 1930s onwards and for inspections and reports on new or near new dwellings, where the surveyor has the opportunity to examine drawings and a specification, it will be to the assurance that materials and fittings to British Standards have been used that he will look. As to design, the British Standard BS8301:1985 'Code of Practice for Building Drainage' is the document which should have been followed for all new dwellings built after 1985 and to which installations constructed earlier should be related and compared.

The publication of the 1985 edition of the

Code of Practice replaced the earlier 1971 edition, which was withdrawn, but which in itself had been produced to reflect the considerable amount of research work done on drainage systems since an even earlier edition of 1950. It included practice covering both rigid and flexible pipes, rigid and flexible joints, the different materials which had come to the fore and the prospects of deterioration, depths of pipes in the ground and loads thereon, support from both concrete and granular material, depth of cover, gradients etc. Site characteristics were to loom large in the considerations governing design, particularly on establishing earlier usage of the site. Supporters of different types of systems fell into two camps and surveyors should be aware of the opposing views of each. The 1985 edition of the Code sensibly tried to form a bridge between the old and the new methods developed in the 1960s and 1970s. It considered that an approach was needed that was both practical and knowledgeable and extended to the belief that where features in older systems might be found to be obsolete, even incorrect, usage in domestic situations was far less intense than in other circumstances, such as in schools and hospitals, and that as long as systems worked and there was no major leakage drains should be thought of as satisfactory. However, frequent blockages, leaks causing insanitary conditions or damage to structures would necessitate action. A visual inspection of the system can help to determine whether or not that further action is necessary.

For an inspection of a dwelling's drainage system it is useful to record the information obtained on a sketch plan showing the position of the dwelling and its relation to the boundaries of the site on which it stands. This is best prepared on lightly printed graph paper, roughly in the correct proportion if at all possible. Reference to the sketch plan will make it easier to compile the paragraphs for inclusion in the report.

For the typical suburban dwelling of the 1930s, and for quite a few from other periods as well, it may be that the surveyor will have observed near the gate to the front garden the cover to the last inspection chamber of the dwelling's system before the main drain crosses the boundary to connect with the local authority sewer. It is useful to start at this point and with this type of dwelling, which usually has a fairly long front garden, there is unlikely to be more than two items to record within the chamber, apart from its condition and the condition of the cover. It is possible that there could be another branch drain entering the chamber from a surface water gully in the driveway in front of a garage. Its presence would indicate that surface water from roofs and yards is combined in the same system of drainage as foul and waste water. Its absence, but with a gully still present, would be the surveyor's first intimation that he would need in addition to give consideration separately to surface water disposal, with possibly a catch pit on the site and eventual disposal to soakaways, though visible signs of these may be minimal.

With other types of dwelling, for example the inner city multi-storey terrace with basement, the last inspection chamber of the system could be fairly close to the front wall and the main drain could receive quite a number of branch connections at this point.

The last chamber in a dwelling's system is often referred to as the 'intercepting chamber' because it has long been thought that foul air from the sewer should be prevented from entering the dwelling's system. This can effectively be achieved by the provision of an intercepting trap positioned at the outfall from the last chamber. The trap has a dual function in that it also deters rats from entering the household system. The interceptor trap needs to be provided with a rodding eye, closed off by a secure stopper, to enable the clearance of a blockage between the chamber and the sewer. The provision of an interceptor trap is a requirement of most local authorities, particularly in urban areas where it is considered that there is an unduly high concentration of offensive gases in the sewers. However, interceptor traps have their disadvantages. They interrupt the smooth flow, often retain foul matter in the water seal and they are liable to be the cause of blockages. It is for this reason that the best form of stopper for the rodding eye is one provided with a lever locking device and galvanised chain which, when pulled, releases the stopper and allows the contents of the chamber, usually full to the brim by the time a total blockage in the interceptor is noticed, to be released towards the sewer. Anything less, such as the ceramic Stanford stopper or any other type which needs unscrewing and therefore entry by an owner or plumber into a full chamber is totally inadequate and warrants replacement. Possible blockages in the interceptor are also the reason why the British Standard Code of Practice recommends their omission in new drainage schemes unless they are a specific requirement by the local authority. Surveyors should therefore not be surprised if there is no interceptor in the last chamber.

Where an interceptor trap is provided in the last chamber of the dwelling's system, it is essential to provide for the entry of fresh air to ventilate the whole of the system. This is usually achieved for the average suburban house by the provision of a low fresh air inlet situated behind the front garden fence and connected by a pipe to a high level point in the chamber. The fresh air inlet draws air in through a light mica flap valve, secured behind a protective metal grille, and when air passes across the top of a ventilating pipe fixed high at the head of the system produces a through current of air. Low level fresh air inlets are prone to being damaged, and always need renewal if they are damaged; otherwise there is a risk of foul air escaping. The ventilating system can always be reversed if necessary, with a high level vent at the front and a low level at the rear, although in close urban situations high level inlets at both front and rear may be preferable. Where there is no requirement for an interceptor trap in new systems, the high level ventilating pipe at the head of the system is usually sufficient to introduce a through current of air.

A routine should be established by the surveyor when examining inspection chambers to ensure that essential information is not

445: An unusual treatment for the fresh air inlet to an interceptor chamber, of which there is no sign externally. Despite the probable original intention of avoiding damage, which would have been better achieved by taking the vent up to roof level, the brass grille and mica flap valve are missing and need to be replaced. Drain air could, at times, be circulating around the entrance door welcoming visitors. Even with grille and valve renewed the pipe buried in the brickwork will eventually corrode and leak foul air.

missed. For example, in the typical intercepting chamber, there is the inlet where the size and material of the main drain can be ascertained and the level of the invert measured. This can be the position where any differential movement between the drainpipe and the side of the chamber may have affected the true alignment of the drain, introducing a visible ridge or sinking in the channel and possibly cracking or breaking the collar of the pipe. Channel sections should be smooth and neatly jointed and there should be no deposits suggesting either blockage from slow running sewage or muddy water suggesting infiltration with grit or gravel from the earth outside. The presence of any root growth should be noted. Well formed smooth sloping benching should be provided on either side of the central channel to enable a person to stand and do what is necessary to clear a blockage. This needs to be in good condition and not likely to crumble away to cause a possible blockage as does any rendering on the walls of a chamber or if not rendered, the situation favoured by the Code of Practice, then the pointing to the brickwork needs to be kept in good condition. Step irons, soundly built in, need to be provided for any chambers where rodding cannot be done from ground level. Cover to inspection chambers need to be readily removable. For this they should be provided with lifting handles set flush with the top. If they are missing, advice to renew the cover should be given, otherwise the need to lever up with a long strong screwdriver or bolster and hammer to an extent negates their purpose. On the other hand, difficulty experienced by the surveyor in lifting covers because they are partially covered by earth or for other reasons does provide evidence that there may have been no need to disturb them for a while because no troubles have been experienced.

Access to the last chamber of the dwelling's system will provide the surveyor with an indication of where the penultimate chamber might be by following the line of the inlet. Having found the next chamber and lifted the cover it is a good idea, provided there are no children or the elderly in occupation and it is safe to do so, to continue the procedure and track down all chambers on the site, lifting the covers of each and leaving them off temporarily. This will enable the surveyor to check that the main drain traverses a straight line between chambers and will enable him, progressively, to allocate each branch drain connection to a particular sanitary fitment or group of fitments, if necessary via a trapped gulley. Hopefully all can be accounted for but if not then it is possible that connections are made to the main drain without access being obtainable at a chamber. Along with the possibility of the main drain taking a convoluted route between chambers, these represent a risk of substantial problems should blockages occur.

At each chamber where there are branch connections, the direction of the inlet should indicate the purpose when related to the sanitary fitments in the dwelling or in relation to the gullies externally. If the indication is not very clear and the surveyor has an assistant or he is able to enlist the co-operation of the owner, the running of taps, the pulling of chains or the operation of levers may need to be invoked. If not he may need to do what is necessary himself to establish what function each branch fulfils. The use of dyes can be a help. It will be found that some branches practically double back on themselves to join the main drain in the direction of flow. For these cases it is important that the correct angle bends are used to enable a smooth and free flow to be obtained. If this is not being achieved, there may be signs of heavy fouling to the benching and even to the sides of the chamber which can lead to insanitary conditions, particularly if the branch is to the base of a soil pipe where the force of the flush from an upper floor WC fitment, compounded by the drop, leads to a rapid discharge into the chamber.

Waste water fitments, baths, shower trays, wash hand basins, bidets and sinks in contrast to soil fitments are taken to discharge over trapped gullies and branch drains to these tend to flow at a gentler rate. Signs of grit or gravel deposits, fouling by grease or misuse from the deposit of building materials can provide evidence that all may not be well and that partial blockages may be causing a back up of liquid to the next chamber above. The surveyor will know the size of the main and branch drains and he will have the invert level for each. Given a fairly level site, pacing the length of main drain between chambers and the length of branch drains and with the aid of McGuires rule for the inclination of drains to produce a self-cleansing velocity, ie 1 in 40 for a drain of 100mm, 1 in 60 for a 150mm drain and 1 in 90 for a drain of 225 mm diameter, it should be possible for the surveyor to check that the system is laid to reasonably correct self-cleansing gradients. With a sloping site, of course, levels would have to be related to a datum, which work would be beyond the scope of the type of inspection envisaged here. Pacing the length of underground drains can usefully be combined with an examination of the ground above. Suspicions about the condition of the drains might well be reinforced by signs of subsidence or by the presence of trees or shrubs nearby. This could be so particularly if the sub-soil is of shrinkable clay and the tree or trees are of a variety indicated by BRE as likely to cause damage, as discussed in section 4, or again what appears to be the excessive growth of vegetation nearby. Of course, if the drains are running near a part of the dwelling where movement has taken place, then straightforward leakage of the drains washing away the support to the foundations could well be the cause.

One of the objectives in lifting all the covers to inspection chambers is to identify the purpose of each branch drain with a visible connection to the main drain, while also identifying those branches which are connected without the benefit of access as they may prove to be troublesome in the future. In comparatively rare cases there may also be a connection to the main drain which has no identifiable source. If the branch looks dusty and disused it could, of course, be from a fitment now removed, an outside WC for example, but on occasions it may appear to emanate from adjoining premises. The attention of buyers and their legal advisers needs to be drawn to these together with any information which the seller may volunteer on the purpose of the branch and any information on liability for maintenance and repair. Of course, there is yet another possibility. It could be that the dwelling's main drain enters the last chamber within the boundary and instead of continuing on and connecting with the local authority sewer transpires, in itself, to be a branch drain connecting to another underground drain within the dwelling's boundaries. This is particularly likely to be the case if the dwelling is one of an estate. The surveyor needs to establish, if he can, the status of this 'drain' but more likely a study of the title deeds will be necessary by the clients legal advisers as the position can be complicated by various factors, including the location of the dwelling.

An underground pipe is either a drain or a sewer but which of the two it is depends on a number of factors. The rules are different in the London area as against other districts. Outside London any pipe taking sewage or surface water from within the curtilage of one property is a drain and as such is in private ownership. All other pipes are sewers. The responsibility for the repair and maintenance of a drain does not end however at the boundary line of the property but extends up to the point where the drain connects to a sewer. Sewers however fall into two categories, public and private. Private sewers can only be found outside London with groups of properties constructed after 1 October 1937. In every matter of maintaining, renewing and cleansing, private sewers are the responsibility of the owners concerned, the proportion of liability being fixed at the time of construction by the local authority. All other sewers are now public sewers be they below land owned by the local authority or below private land. However, where such sewers before 1 October 1937 were classed as 'combined drains' the local authority, although responsible for carrying out the work, can recover from the owners of those properties which utilise the sewer any costs of repair, renewal or improvement, but not the cost of cleansing.

In London, however, the position differs in that a drain not only includes a pipe used for the drainage of the premises within the same curtilage, but also a pipe used for draining a group of buildings, provided the drain was constructed after 1848 with the sanction of the appropriate authority. Thus it will be seen that any combined drain serving properties constructed before 1848 has become a sewer and is the responsibility of the local authority both as to any works required to it, and to the payment of costs. After 1848, any pipe constructed by private owners to drain a group of properties remains a combined drain repairable by those owners unless taken over by the local authority. Since 1937 the local authorities in London have been empowered to make an order requiring new groups of properties to be drained by a combined operation in which case the combined drain so constructed remains the responsibility of the owners concerned. However, the order does not specify the proportion of liability as between owners for the cost of any necessary work, leaving a wide margin for possible

dispute. There are no 'private sewers' by definition in London. Pipes in this category within the London area are known as 'combined drains'.

The surveyor is accordingly advised to ascertain and verify the status wherever possible of all pipes and sewers within the boundaries of the dwelling and to inquire from the local authority as to the position and depth of their sewers, together with any other useful information in connection with both the sewers, local conditions and any difficulties which the local authority may have been experiencing in this area. All this information should, of course, be brought to the attention of the client's legal advisers.

Nothing has been said so far on the length of drain between the last inspection chamber of the house system and the connection with the local authority sewer. Although the owner of a property is liable for the repair or renewal of this length of drain, it can perhaps be considered in a rather separate manner from the remainder of the house drains. The connecting drain is normally situated at a reasonable depth below ground to render it relatively free, in comparison with the house drains, from outside pressures. On the other hand there are occasions when it is affected by road subsidence or vibration and, in particular, roots of trees planted in the footpath have been known to cause substantial movements.

More often than not a defect is brought to the attention of an owner by an unclearable blockage in the connecting length, immovable either by rodding from the interceptor chamber or by the council's sewer men from the sewer. If the defect and blockage is caused by tree roots sometimes the authority's insurance policy can be invoked so that the expense of the necessary work of relaying will be defrayed by the insurance company on a firm line being taken by the owner that the council is responsible. If the blockage is due to any other cause then the house owner is likely to have to bear the cost of the work. Although local authorities can insist that the connecting length of drain should be watertight, as long as the original connection was passed they would seem to do very little to check on this point subsequently.

Connecting Pipework

With consideration given to the sanitary fitments together with any fittings associated with their outlets, such as traps and anti-siphon pipes, and the underground drains there remains, for examination, the vertical plumbing which connects the two.

A surveyor needs to be aware of the principles of the two-pipe system, where soil and waste are piped separately, and the one-pipe system where they are conveyed together. Both these systems entail the use of ventilating pipes, thus confusingly doubling the number of pipes used at times. He will also be aware that the even simpler single-stack system is in use for housing purposes, in which separate ventilating pipes are not used.

The seal of water in a trap can only be broken if the pressure changes in the branch pipe leading from the trap are of a size and duration to overcome the head of water in the trap itself. Such pressure changes can be brought about by liquid flow in the main stack, that is to say by induced siphonage, or by liquid flow in the branch pipe, that is to say self-siphonage. Induced siphonage depends on the flow load in conjunction with the diameter and height of the stack while self-siphonage depends on the design of the appliance and the length, fall and diameter of the branch pipe.

Both the two-pipe system and the fully ventilated one-pipe system virtually eliminated the risk of seal breakage, the first by interrupting the continuity of the system at hopper head and gully, the second by ensuring, with the help of vent pipes, that the pressure in branch pipes never deviated appreciably from atmospheric. With the single-stack system the combined soil and waste stack alone is relied on for ventilation and the system needs to be of good design and workmanship to ensure that it functions correctly. This system is now much in favour for new small dwellings of one or two storeys particularly on larger estate schemes due to the saving of pipe material.

In houses of some age the surveyor will invariably find the two-pipe system. This of course involves the soil stack taking discharges from the WC, or WCs, and the waste pipe or pipes, taking the waste from the baths and basins. This system ensures that the sewer gases cannot enter the house. The ventilated soil stack and the double break at the hopper head and at the gully to the waste pipe, with its trap, provide all that is necessary for an effective barrier. Even so, the surveyor will observe that the hopper head often becomes foul and offensive and the same criticism has also been levelled at the gully

For new or near new dwellings British Standard BS5572:1978 'Code of Practice for Sanitary Pipework' provides the basis on which systems should have been designed for dwellings built since its publication. It is also the basis on which systems installed in dwellings built before that date can be compared but only if they were provided on the one-pipe or single-stack systems. Its value is therefore somewhat restricted for surveyors carrying out inspections of dwellings of all ages since a considerable proportion, probably well over 60%, will have installations based on the two-pipe system.

The single pipe system only came into use in the 1950s and while it is now commonly used for two storey housing on estates, there needs to be a close proximity of fittings coupled with care in the design and running of pipes and the formation of connections. Furthermore, the earlier one-pipe system from which it was developed was mainly used, from its introduction in the 1920s up to the 1950s, for the drainage of multi-storey blocks of flats. Its use for that purpose has continued, particularly in the conversion of multi-storey dwellings into separate units, but it did not feature prominently in the drainage of the two storey semi-detached dwellings typical of the period, which continued to employ the two-pipe system.

Accordingly, it will be found that most dwellings built up to around the 1950s will be drained on the two-pipe system above ground which for two, or even three, storey dwellings is reasonably satisfactory for the purpose unless major improvements are proposed. If that is the case, it will be necessary for them to be planned on the basis of the one pipe system in view of the banning of the use of hopper heads for new or the renewal of existing installations.

What needs to be looked at critically on the elevations of two or three storey dwellings built up to around the 1950s is the condition of all visible pipes carrying either soil, waste or rain water and note taken of whether there are any cracked, damaged or loosely fixed sections, whether there are any signs of leakage at the joints together with the condition of the hopper heads, which is likely to be dependent on the adequacy of size and positioning to cope with the number of pipes discharging into them, hopefully without splashing and staining the adjacent wall. In

446: Typical two-pipe system plumbing on a 1930s two storey semi-detached dwelling but not usually on the front elevation. Here the small first floor bedroom has been converted into a bathroom. The cast iron soil pipe has been correctly extended to roof level as a vent but is of insufficient height to comply with the Building Regulations and the wire balloon is missing. Waste water from bath and wash basin, along with rainwater are taken to the hopper head.

particular, the surveyor should examine and, wherever possible, feel behind cast iron hopper heads and downpipes for rust and holes since repainting over the years is often neglected due to the problem of access with the result of damp penetration. At the base of the waste pipes, which can, of course also be carrying rainwater if the systems are not separate, connection needs to be made to a trapped gully, preferably by means of a back or side inlet below the grating but above the level of the water in the seal. Alternatively, generally more often but less satisfactorily, the discharge into the gully is by means of a shoe above the level of the grating which in the case of waste pipes from sinks can lead to smelly and greasy conditions compounded by leaves and other debris when the shoe is carelessly fitted and even when the curb to the gully is in reasonably good condition.

447: Gullies on the two-pipe system for breaking the continuity of direct connection between the traps of waste water fittings and underground drains should be arranged with adequate curbs to prevent undue splashing and if taking a sink waste ease of cleaning. This one at the foot of a rainwater pipe needs repair and re-rendering although the shoe to the pipe is set at a suitably low level.

Finally, high level ventilation to the drains needs consideration and in this respect BS5572:1978 'Code of Practice for Sanitary Pipework' recommends and the Building Regulations 1991 require, that high vents should in all cases be taken to a point above the eaves or flat roof and not less than 900mm above the head of any window or other opening into the building within a horizontal distance of 3.0 m from the vent pipe and be finished with a cage or other perforated cover which does not restrict the flow of air. Anything less, and very often top lengths of vent pipes and wire balloons are missing, can leave the possibility of drain air being blown into the dwelling more likely.

Where the two-pipe system was used in such dwellings there was seldom any need to ventilate fitment traps to combat possible siphonage. Provided there was only one WC serving the upper floor, or floors, and branches were of reasonable length for waste fittings not more than about 3m and of a size, not less than 32mm for wash hand basins and

40mm for baths and sinks, siphonage was unlikely to occur. Even if siphonage did occur it may not amount to more than the smell from a hopper head. It is when additional fitments are installed and little or no consideration given to the effect on the existing that problems can arise. The surveyor should keep this aspect in mind when there are signs of additions to an older plumbing installation. Odours and gurgling when appliances are brought into action can be a clue that there are siphonage implications.

In larger dwellings and blocks of flats built up to the 1920s, the surveyor will often see the tangle of pipes usually associated with a two-pipe installation. The trap to nearly every fitment will have an anti-siphon pipe because of the reduction in the number of hopper heads where they would have been inaccessible for easy cleaning and maintenance. Both soil and waste stacks will be taken up high to roof level as vents along with their associated anti-siphon pipes although, of course, these could be turned in to connect with the respective soil and waste stacks at high level, above the topmost fitments. It is in these instances that an earlier British Standard Code of Practice CP304 'Soil and Waste Pipes Above Ground', dated 1953 can provide the benchmark by which existing systems can be compared and judged.

As with smaller dwellings, it can be the addition of extra fitments, far in excess of those envisaged when the block was built, and their installation, without consideration to their effect on others, that can cause problems. The addition of one particular type of fitting in the 1960s to flats, the washing machine, produced the problem for those occupying flats at the lowest level of foam emanating from gullies and flooding what little open space was attached to the flat or a communal area. Arrangements to overcome

449: A 100mm uPVC combined soil and waste stack on a one-pipe system taking WC and waste water branches and fixed to the brickwork with clips. Although there is a sharp bend at a change of direction to the near horizontal with a liability to blockage, generous cleaning eyes are provided. There are no external signs of ventilation to traps which is essential to avoid siphonage. Resealing traps may have been used but their presence needs to be checked if they have not already been picked up internally.

448: A range of pipes added over the years to a multi-storey mid 1800s dwelling, divided into flats. Looking from the right, the first pipe is a 100mm light iron rainwater pipe fixed by clips to the brickwork. Next to it is a 50mm cast iron waste pipe fixed through ears on the sockets to the brickwork, as are all the other cast iron pipes on the elevation, and with four 32mm branch waste pipes in lead and an anti-siphon pipe of the same size, also in lead. Both waste pipe and anti-siphon pipe are taken up to just below eaves level to act as vents, less effectively than they might have been if taken further to above eaves level. To the left, between the pair of windows at ground floor level there is a 50mm lead anti-siphon pipe with three branches which is connected at top floor level above the highest fitting to a 100mm soil and vent pipe. This is in lead for the top two floors and is satisfactorily carried up as a vent to above eaves level, but lower down is in cast iron, with branches in a mixture of lead and cast iron and with visible cleaning eyes. All the lead pipes on the elevation are fixed to the brickwork with cast iron tacks soldered on to the lead.

this problem will not be found in the erlier Code of Practice since it did not exist at the time it was written. However, as the problem of foaming also affected flats where the one pipe system had been used for the drainage installation in an even worse way, by coming out of the soil and waste fitments within the flat, it was covered in the subsequent British Standard, BS5572:1978 'Code of Practice for

Sanitary Pipework'. If the surveyor is instructed to carry out an inspection and to report upon a flat in a multi-storey block situated at such a level that foaming and possibly flooding of this nature could happen, then he should seek evidence of any previous occurrences. In the case of a two-pipe system installation, the consequences would be smelly and unpleasant but in the case of a one-pipe system, they would be positively insanitary as well. If there is evidence, he should seek to establish whether rectification work in accordance with BS5572:1978 has been carried out but if he can find no such indications because of casings and the like obstructing the view, then he should recommend that further investigations be put in hand to ensure that the risk has been removed.

In the one-pipe system, the omission of the gully and the hopper head throws the whole burden of ensuring an effective barrier to gases from the sewer on to the seals in the appliances and it is necessary to ensure that the traps are kept adequately filled at all times in order to achieve a satisfactory performance. Trap water seal of 75mm, as against the 40mm which is effective in two-pipe plumbing, are considered necessary, together with the venting of all traps except the highest to prevent any siphonage or blowing of seals.

Although often more expensive than two-pipe work for two storey houses, the one-pipe system has increasingly tended to replace the two-pipe on improvement schemes, particularly in houses of more than two storeys. Nowadays with the Building Regulations requiring pipes to be run internally on dwellings above three storeys and total abandonment of waste hopper heads, it must be considered standard for such work. It is particularly suitable where there are bathrooms one above the other in three, four or five-storey converted houses and where kitchens can be arranged adjacent.

Single-stack plumbing (i.e. a true one-pipe system, with the appliances ventilated only by the stack), becomes comparable in cost for two-pipe work on the old style with hoppers and no trap ventilation, but it is vital that the seals remain intact under normal conditions of use. The BRE now says that the risk of siphoning and back pressure are not so great as had been once thought but single-stack plumbing is mainly suitable for single units of not more than two storeys and will only achieve its full value and economy when used in a large estate scheme of single or two storey houses.

It is possible however, that, as for the one-pipe system, a surveyor will be faced with a problem of an incorrect installation. It is here that the Code of Practice, BS5572:1978 will be of value for checking an installation against the good practice recommended today. Among those practices are those set out below relating to the layout and running of the combined soil and waste stacks, now

increasingly known as 'discharge' pipes and stacks as they are virtually the only means of connecting fitments to underground drains in use at the present time in both the one-pipe and single-stack systems. These have superseded the two-pipe system, although on the many examples of this system which still remain the answer to problems can often be found through the application of the principles governing those practices, developed as they have been over many years through experiment based on experience. Ignoring them can produce pungent odours and blockages difficult to clear. The surveyor will do well to spot the places where these are likely to occur because at the time of his inspection, the seller could well have arranged for sweet aromas to be wafted around and all traces of persistent recurring blockages scrupulously cleaned.

The basic principle is that water flowing in discharge stacks will cause air pressure fluctuations. Suction can occur below discharging branch connections and offsets, causing water seal loss by induced siphonage from appliances connected to the stack. Back pressures or positive pressures can occur above offsets and bends in stacks causing foul air to be blown through the trap water seal and, sometimes, seal loss. These seal losses will be affected by the flow load, the height and diameter of the stack, the design of pipe fittings, changes of direction in the wet portion of the discharge stack, provision or otherwise of a ventilating pipe, surcharging of the drain, and the provision or otherwise, of an intercepting trap in the drain.

Suction produced in the discharge stack below branch inlets when in use is affected by the radius or slope of the branch inlet. A large radius or a 45° entry will tend to minimise the amount of the suction but a near horizontal entry with a small radius will tend to have the opposite effect. Branch inlets which are significantly smaller in diameter than the stack are not so critical in this respect. Sharp bends at the base of a stack can cause large back pressures due to restriction of the air flow and similarly, offsets in the wet part of a stack can produce large pressure fluctuations. Changes in stack direction can also cause foaming of detergents and consequently pressure fluctuations. If the drainage run to which the discharge stack is connected is surcharged, the normal flow of air down the stack during discharge is interrupted and high back pressures can occur necessitating additional stack ventilation.

In a situation where a single discharge stack is connected to a drain, fitted with an intercepting trap in close proximity, large pressure fluctuations can occur again necessitating additional stack ventilation.

Wind blowing across roofs can produce pressure fluctuations in the vicinity of parapets and corners of the building. If discharge or ventilation stacks are terminated in these areas, unacceptable pressure fluctuations can be developed in the discharge system.

The internal diameter of a discharge stack should not be less than that of the largest trap or branch discharge pipe connected to it. The discharge stack above the topmost appliance connection should be constituted without any reduction of diameter to the point of termination except for one and two storey housing where, in certain cases, a 75mm vent pipe can be used. Bends at the base of a discharge stack should be of large radius, at least a 200mm radius to the centre line, but preferably two 45° large radius bends should be used. The distance between the lowest branch connections and the invert of the drain should be at least 750mm but for low rise single dwellings, 450mm is considered

adequate. Offsets in the wet portion of a discharge stack should be avoided and if the drain, to which the discharge stack is connected, is likely to be surcharged, a ventilating pipe or stack should be connected to the base of the stack above the likely flood level. Ventilated systems may require larger ventilating stacks.

Branch pipes should not be reduced in diameter in the direction of the flow. WC branches of 75mm or 100mm size do not normally require venting whatever the length or the number of bends included in the run. Bends, however, should have as large a radius as possible to prevent blockage.

Wash basins are normally fitted with 32mm discharge pipes and the length and slope of the pipes and the number and design of the bends should be strictly controlled if venting is to be avoided. In certain conditions a suitable resealing trap may be fitted. If a vertical 32mm discharge pipe is used with a 'P' or 'S' trap, venting or a resealing trap will probably be necessary. Sinks and baths are normally fitted with 40mm discharge pipes. Self-siphonage is not a problem because of the trap seal replenishment which occurs at the end of the discharge due to the flat bottom of the sink or bath. Therefore length and slope of the discharge pipe are not so critical and venting is not normally required although the maximum length should be restricted to 3m to reduce the likelihood of blockage from deposits. Venting to wash basins is only likely to be needed where ranges of basins are concerned.

A trap which is not an integral part of an appliance should be attached to and immediately beneath its outlet and the bore of the trap should be smooth and uniform throughout. All traps should be accessible and provided with adequate means of cleaning and there is considerable advantage in providing traps for sinks and basins which are capable of being readily removed or dismantled.

The internal diameters of traps should be 32mm for basins and bidets and 40mm for sinks, baths and shower trays. Traps of WCs should have a minimum water seal of 50mm. The size of ventilating pipes to branches from individual appliances can be 25mm but, if they are longer than 15m or contain more than five bends, a 32mm pipe should be used. Connections to the appliance discharge pipe should normally be as close to the crown of the trap as practicable and within 300mm.

As for the materials from which pipes, stacks and traps are made the surveyor will encounter traditional cast iron spigot and socket soil, waste and ventilating pipes, both sand cast and spun, lead and alloy pipes and capillary and compression tube fittings of copper and copper alloy. Steel tubing is also encountered, but rarely.

450: While lead pipes with their jointless flowing form have the advantage in appearance over heavy cast iron pipes with their obtrusive joints, bending around obstructions can be taken to extremes and there is always the risk of damage from ladders. If the stone band course had not been cut back, as here, there would be a serious risk of blockage.

A wide variety and range of thermoplastic pipes and fittings are now to be found. The most common are probably those of unplasticised PVC (polyvinyl chloride) but these should not be employed where large volumes of water are discharged at temperatures exceeding 60°C. Some washing machines discharge water in excess of this level. Jointing of unplasticised PVC pipes is usually by means of synthetic rubber ring joints or by solvent cementing, in which case expansion joints must be incorporated in the system. Thermoplastic pipes are also available in polythene, polypropylene and acrylonitrile butadiene styrene (ABS). These pipes are more resistant to higher temperatures and are more flexible than unplasticised PVC pipes. In particular, polythene pipes are less liable to damage from impact and are more resistant to damage from freezing. All polythene pipework needs to be adequately supported, however. Thermoplastic pipes are light in weight and easy to handle. They are also highly resistant to corrosion. The chief danger, however, is that their coefficients of expansion are much higher than those of metals and they are not invariably suitable for use without great care in selection and fixing.

The joints between pipes of the same material such as caulked joints for cast iron pipes, wiped joints for lead pipes and the compression and capillary joints for copper pipes are well known but BS5572:1978 which replaced CP304 gives in Tables 11 and 12 no less than 25 recommended joints between pipes of different materials and sizes.

External pipes requiring painting or other protective coating need a free space for access all round the pipe. The clearance with the wall should be 30mm and it is not desirable for pipes to be set in angles or in chases. The walls of houses are often so formed that pipes are tucked in angles where they are exceptionally difficult to inspect, let alone maintain.

Where pipes pass through walls or floors or other parts of the structure they are best provided with an aperture formed with a sleeve of inert material to protect them. Many existing pipes are not provided with adequate support. The most vulnerable are of course horizontal pipes of polythene or lead.

Cast iron pipes are normally fixed by means of ears on the sockets with or without distance pieces. Lead pipes require double or single cast lead tacks, soldered or lead burned to the pipe, fixed to the structure with galvanised nails or gun-metal screws while sheet lead milled tacks are also used. Asbestos cement pipes are fixed with galvanised mild steel holder-bats for building into or screwing to the structure. Copper pipes may be fixed with copper alloy holder-bats or strap clips.

Disposal

Up to now in this section it has been assumed that disposal of sewage will be to a drainage authority sewer. It is now necessary to consider those circumstances where this is not the case and where sewage has either to be stored for collection in a cesspool or is taken to a small domestic treatment plant, with waste and surface water taken to a sump or soakaways.

Cesspools are very often used, particularly in country districts, for one or two isolated buildings of insufficient size to justify a treatment plant. Their use however should be discouraged for more than say eight persons. Cesspools are, of course, employed due to the fact that drainage to sewers is not possible within easy reach.

The chief difficulty with cesspools, however,

is that they have to be emptied at intervals and in considering whether it is wise to advise retention of a cesspool the surveyor should check up on the nature of the services provided by the local authority for emptying, together with details of their charges.

Cesspools are defined in BS6297:1983 as 'a covered watertight tank used for receiving and storing sewage from premises which cannot be connected to a public sewer and where ground conditions prevent the use of a small sewage treatment works including a septic tank'. It is essential that cesspools are and will remain impervious to penetration from ground or surface water and to leakage.

The available local facilities for continual emptying whether by public authority or private contractor should be investigated by the surveyor since the cost of emptying by tanker vehicles may be high. The Code says that an average household of three persons will produce $7m^3$ (the capacity of a typical tanker) in about three weeks and this necessitates some 17 journeys each year.

The cesspool should not be sited so near any inhabited building as to be liable to become a source of nuisance or danger to health (a minimum of 15m is recommended) and it is essential that no well, stream, spring etc., becomes polluted. The cesspool should be sited on ground sloping away from any effect of the prevailing wind but within 30m of vehicular access. Constructional considerations will probably have limited the economic capacity of a single-tank cesspool to a maximum of about $50m^3$. The Building Regulations prescribe a minimum of $18m^3$. The British Standard Code of Practice suggests, as a general rule, that a capacity of not less than 45 days storage should be allowed.

The surveyor should bear in mind that the

local authority are empowered under section 48(1) of the Public Health Act 1936 to examine and test cesspools in their area. The provisions include such matters as repairing leaking cesspools by default or requiring this to be done. It used often not to be considered a particularly serious crime in the remoter country districts for an overflow pipe from a cesspool to be taken out on to neighbouring field but this is not now permitted.

While the surveyor on his inspection can check on aspects of siting, as discussed above, and on the covers, frames and ventilation to the cesspool along with the adjacent interceptor chamber, as an intercepting trap is a requirement in all cases where the connection is not to a public sewer and can also make enquiries from the local authority on emptying facilities and whether there have been any problems, it is inconceivable that he would not, furthermore, advise a buyer or seller to obtain a full report on the installation. This would have to follow emptying, cleaning examination and testing for leakage and the frequent practice of one owner moving out and another moving in on the same day would seem bizarre for a dwelling involving a cesspool unless the seller could produce a satisfactory report in advance. Accepting assurances at face value would be risky in the extreme.

All surveyors will have no doubt learnt at some time in their studies the principles behind sewage treatment and many would no doubt be able to draw a reasonably approximate section through a septic tank, filter bed and humus tank, since it is a frequent enough examination question. Notwithstanding this, however, it is doubtful whether many surveyors really have sufficient contact with buildings where treatment to sewage is carried out to be really familiar with the requirements for successful operation.

It is a field which borders on the specialist and a study of BS6297:1983 'Design and Installation of Small Sewage Treatment Works and Cesspools' shows why. There may well be surveyors who have proceeded beyond the normal ambit of the general practice or building surveyor to study the subject of sewage treatment from the municipal engineer's point of view and they would certainly be qualified to deal with testing and supervising repairs or renewal to a small treatment works.

However, for the general surveyor it is outside his scope and such matters should be entrusted to a specialist. This would apply also to surveyors in the public service whose local authority purchased, say, a large country house for institutional purposes where such plant might exist, though he, at least, should have the facilities of the borough engineer's department to fall back upon.

The Code itself indicates that there is no substitute for the taking of skilled engineering advice based on a knowledge of sewage works practice and of the local conditions.

In regard to an existing system functioning and in use it is of course possible that it may have been installed by specialist contractors and be maintained by them. Such a firm would no doubt be able to advise a surveyor or owner on the present condition. On the other hand, it may have been installed to the specification of a consultant by a general contractor. If the consultant is recalled, he or his firm could be asked to report on the performance and present condition. If there are no indications of this nature available then the local authority may be able to advise on the selection of a reliable specialist contractor in the area or of a consultant familiar with local problems. Although in these circumstances a little knowledge can be a dangerous thing and specialist contractors or a consultant should be responsible to the client a few salient points could well be set out in regard to maintenance tests and one or two special points so that the surveyor will be familiar with the problems and if necessary instruct contractors on behalf of the client.

As to maintenance, regular weekly inspections should be made, giving particular attention to the filter bed which may become clogged with organic matter so that the liquid does not percolate, causing ponding. If loosening of the surface of the bed does not overcome this, other lines of investigation and treatment are needed together with further consideration as to whether the septic tank has been de-sludged at sufficiently frequent intervals (normally about every six months), whether there has been an excessive use of detergents, which have now become a problem in sewage treatment, or whether the filter capacity is adequate.

Distributor pipes require cleaning with sufficient frequency to prevent obstruction to the flow of effluent. The same type of test can be used for any tank used in a sewage treatment plant to ensure that it is watertight as for a cesspool. In addition it is desirable for some means to be available whereby a sample of the effluent can be taken for testing.

The Building Regulations require a septic tank to have a minimum capacity of $2.7m^3$ and, as for cesspools, to be impervious from within and without and to be sited having regard to various factors (see the guidance to satisfy Requirement H2 from Part H of Schedule 1 to the Building Regulations 1991).

It is probably in regard to suitable siting in relation to local conditions and in the specification of a suitable plant taking these factors into account that the skill of specialists or consultants are fully exercised.

The disposal of waste water from baths, basins and sinks together with surface water when domestic sewage is taken to a cesspool or septic tank is usually by means of the surface filtration method.

This involves taking these categories of waste water in ordinary jointed pipes to a sluice box and from the sluice box through a series of open-jointed agricultural drains to a sump. The method is also sometimes used for the final disposal of effluent after treatment from a small domestic sewage treatment plant.

Consideration of site conditions again is vital as it is essential to avoid pollution and the system must be in open land away from trees, otherwise the roots will quickly clog up the system. Care is necessary too in laying to prevent the system from becoming silted up, the sluice box being used to divert water along the different lines of drain.

It is difficult however to ascertain the amount of land required for this method and to observe the results. The level of the under-ground water table has a distinct bearing on the performance and if it is nearer the surface than 2m in winter it is inadvisable to adopt this method. Since access is seldom provided the indications of trouble would be waterlogging of the ground and a too frequent necessity to bale out the sump.

There is little that can be recommended in the way of repair, apart from cleaning if access is provided, otherwise a reconsideration of basic principles in accordance with the current British Standard Code of Practice would probably be required with a view to re-laying.

Although it is far more preferable for rainwater and surface water to be taken to either a combined or separate local authority sewer, in many cases of estate development where a combined drain is the method adopted at the rear to pick up the discharge from each house, rainwater from roof slopes at the front and surface water are taken to a soakaway below the front garden. Failures in the arrangement and performance of these soakaways are usually indicated by flooding or a waterlogged condition and sometimes also by subsidence if the structure of the soakaway fails. BS8301:1985 'Code of Practice for Building Drainage' defines a soakaway as 'a pit from which water may percolate into the surrounding ground'. Small pits may be unlined and filled with hardcore for stability or the soakaway may take the form of seepage trenches following convenient contours.

Larger pits may be unfilled but lined, for example with brickwork laid dry, jointed honeycomb brickwork, perforated pre-cast concrete rings or segments laid dry and the lining surrounded with suitable granular material. An unfilled pit should be safely roofed and provided with a manhole cover to give access for maintenance. A soakaway can be used most effectively within impervious subsoils such as gravel, sand, chalk or fissured rock, and where it is completely above the water table. In ground with low permeability, a number of soakaways can be provided, linked at overflow levels, by piped seepage trenches. The size of a soakaway is governed by the amount of surface water it will have to deal with and the rate of percolation, but a common method of assessing size is to allow a water storage capacity equal to at least 13mm of rainfall over the impermeable area. If the suitability of the ground is in doubt, absorption figures can be obtained by digging trial pits and measuring the rate at which the water soaks away. Alternatively, disposal to a watercourse or by means of storage vessels might be considered.

The drainage of the subsoil is essential on

certain sites and in certain soil conditions to divert the natural flow of water away from foundations, particularly on the uphill side of a house, or to reduce the level of the groundwater.

There are five reasons given in BS8301:1985 for the drainage of subsoil water which is water occurring naturally below the surface of the ground (the upper surface of which being the 'water table') in contrast to surface water which consists of the run-off of natural water from the ground, whether paved or not. The five reasons are:
a) to increase the stability of the ground
b) to avoid surface flooding
c) to alleviate subsoil water pressure likely to cause dampness to below ground accommodation
d) to assist in preventing damage to the foundations of buildings
e) to prevent frost heave of subsoil which could cause fractures to structures such as roads or concrete slabs.

In the construction of a dwelling the question of subsoil drainage may or may not have been taken into account. If it was and adequate drainage at the time of construction was provided, confirmation of this might be obtained by the presence on the site of a soakaway, a catchpit with pipe discharging into a ditch or watercourse or a connection through a reverse action interceptor to the local authority surface water drainage system. If the building is not too old the local authority may have records of what was done originally.

What is unlikely however is that there will be any record of the run of the drains or any access to them. Field drain pipes or French drains consisting of a shallow trench filled with rubble or clinker are notorious for becoming clogged with silt or obstructed by tree roots and the absence of access for ensuring that they are free of blockage often means that over a period of years the system either in whole or in part ceases to function effectively. The result may well be a return to those conditions existing on the site before the work was carried out with the consequences of possible structural damage to the building, surface flooding and unpleasant conditions for the occupants.

Although the evidence of flooding would make the diagnosis of this problem relatively simple there can be cases where there is no surface flooding and if subsoil water movement is suspected as being a reason for structural damage, the surveyor would need to advise the obtaining of additional evidence before being certain. This will involve ascertaining the depth of the water table and the direction of flow of the subsoil water. This information can best be ascertained by the digging of trial holes. The depth of the water table varies with the season, the amount of rainfall and the proximity and level of natural drainage channels. It is desirable to ascertain the level of the standing water in the trial holes over a considerable period so as to enable the seasonal variations to be recorded, but in particular the highest water level, and it may conceivably be possible to relate structural movement to periods when the water level is at its maximum.

The movement of subsoil water can usually be inferred from the general inclination of the land surface and confirmation obtained from the trial holes. If the problem is verified as being one of inadequate subsoil drainage the solution will probably lie in the provision of a new system, of a necessity as the old system assuming one exists, will almost certainly be inaccessible for repair. New subsoil drainage systems are described in BS8301:1985. The design considerations are covered together with five systems of laying pipes, one of these

being particularly apposite to the repair field called the moat or cut-off system designed to intercept the flow of subsoil water and thereby protect the foundations of buildings.

The choice from the remaining systems will depend on the character of the site but in particular on whether the whole of the site need be drained or whether it is only a problem in relation to the building. If the whole site is being redrained then it is important to divert carefully into the new system any of the drains of the old system that may be cut in the process because it is possible that parts of the old system still function to some extent however slight.

Section 7 Surroundings

Boundaries

When considering whether to pursue the purchase of a dwelling it is usually the internal layout and the condition that are paramount in a buyer's mind. The boundaries do not feature prominently, indeed may have received no more than a glance, yet when in occupation they can take on a degree of importance not only as a matter of convenience and comfort but of expense as well.

The case of *Bolton v Puley* (1983) will serve to illustrate the aspect of expense in relation to boundaries. It concerned the survey of a house in Somerset purchased in 1977 for £27,000. This had a 200 year old random stone boundary wall 55m long. Over 17m of this length it served as a retaining wall for a patio some 3m above the adjoining roadway. There was a partial collapse of the wall in 1980 and engineers were called in. They established the rate of movement in the wall, which by then was leaning 155mm out of plumb, and the judge considered half this amount would have been visible in 1977. He decided that the surveyor should have mentioned this along with the bad condition at two points and should have advised a partial repair which would have cost £5,900 at that time. This would have provided the purchaser with a better wall and accordingly he reduced the damages to £4,425 plus £500 for distress, vexation and worry. The trial took 12 days and costs in addition must have been considerable, the whole of which would have had to be met by the losing defendant.

In hilly areas the boundary will often comprise a retaining wall holding back the soil of the garden. Indeed in such areas retaining walls may form a feature of terracing in the garden itself. Collapse of such walls can be a very serious matter particularly if they adjoin a public thoroughfare, but equally so if they are

the cause of death or injury to the occupant or his family.

The surveyor does well to err on the side of caution when reporting on retaining walls. Old walls rely on the mass weight of stone or brick to prevent the soil exercising an over-turning motion. The foundations, if any, and the thickness of the wall at its base are just those items which cannot be seen.
The surveyor must therefore use his eyes to make a careful inspection along the full length for cracks and fissures and use his plumbline to check whether the wall is upright or not. A fractured and leaning wall must be a candidate for rebuilding and even a severely leaning wall without a fracture must be considered ready for rebuilding because the appearance of a fracture would probably coincide with collapse. The surveyor

451: The surveyor needs to view what he can see of retaining walls very carefully, looking along the entire length for leans and bulges. If these are combined with cracks there is a danger of collapse and the surveyor needs to advise shoring, screening and rebuilding as soon as possible. If there are no fractures and the leans and bulges are of a moderate degree, monitoring at regular intervals is required. This retaining wall has been extended upwards, is out of plumb and free of cracks but has been tied back. It may be that the surveyor can conclude that there is no danger but a seller or buyer may need to be advised to take a second opinion if doubt exists.

must clearly advise the taking of further advice from an engineer in the case of suspect retaining walls.

If the dwelling is of any size the surveyor, having completed his inspection of both the interior and the exterior along with the services, might well consider moving outwards to inspect the boundaries before dealing with immediate concerns around the dwelling. This will make for a bit of variety and may even allow a little time for reflective thought on what has been seen so far and, of course, a longer distance view of the dwelling itself, which can be useful for considering the relationship of the various parts.

It will be found that it is not only retaining walls which are candidates for possible collapse. Prior to 1900 brick walls were most commonly provided to form boundaries some five or six feet high and the local brick of the district was invariably used even if not always wisely. It is a tribute to many of the local bricks produced that considerable numbers of these boundary walls, if carefully maintained, are in as good condition today as when they were built. When neglected, however, they can prove troublesome. Insufficient foundations, the lack of vertical movement joints and the absence of damp-proof courses, together with open joints and long unstiffened lengths of brickwork can cause leans, bulges, fractures and settlements to a spectacular degree. The surveyor must carefully determine whether a boundary wall in such a condition can be saved or whether it is essential to rebuild on the grounds of safety, perhaps reusing the old bricks.

The Office of the Deputy Prime Minister points out that garden and boundary walls are among the most common forms of masonry to suffer collapse and one of the commonest causes of deaths by falling

452: *A boundary wall on level ground fractured by lateral pressure from a tree, now removed, but remaining upright. Stitching the fracture should ensure stability.*

masonry. Should such walls be neglected, it is always possible that an insurance claim could be rejected. Besides general deterioration and ageing, it lists four possible ways in which a wall might be affected later in life. The first is by an increase in wind loading or driving rain following the removal of a building nearby, the second is the felling of mature trees close by or the planting of new trees adjacent to the wall. The third is by changes in traffic patterns leading to greater vibration and the fourth by alterations to the wall itself, particularly the removal of part or an addition to the height.

For inspecting garden or boundary walls, the Office of the Deputy Prime Minister lists aspects to be considered which have already been dealt with in relation to the dwelling's external walls but which assume a far greater importance when a wall is entirely free standing and exposed to the elements on both sides. These include crumbling of the surface and cracks in the brickwork, defective pointing, adjacent tree action both by the roots and from lateral pressure, penetration of vegetation such as ivy, loose copings and impact damage from vehicles. It also suggests further investigation if a half brick, one brick or one and a half brick wall leans out of plumb by more than 30, 70 or 100mm respectively and provides a table prepared by BRE for safe maximum heights for brick and blockwork walls, on level ground, when situated in four United Kingdom Zones. As adapted these heights are set out in Table 7. All these aspects need to be checked by the surveyor, very much in the interests of safety.

Although there is no law which requires an owner to enclose his land, most choose to do so to keep casual interlopers at bay and avoid arguments that they or their goods might

453: A boundary wall of lumps of chalk, clay and straw, known as 'witchert', with tile coping and about 2m high but which has lost its rendering coat. The render is essential to avoid erosion from driving rain, since the wall is exposed on both sides, which is already taking place and will eventually lead to collapse.

Table 7: **Maximum Heights for Free Standing Walls on Level Ground**

UK Zones	Maximum Height for Wall Type and Thickness in mm					
	Brick			**Block**		
	$\frac{1}{2}$ (100mm)	1 (215mm)	$1\frac{1}{2}$ (325mm)	100mm	200mm	300mm
One	522	1450	2400	450	1050	2000
Two	450	1300	2175	400	925	1825
Three	400	1175	2000	350	850	1650
Four	375	1075	1825	325	775	1525

Zone 1:   SE England up to a line from Birmingham to South coast

Zone 2:   E Anglia, Eastern, Central and Western Counties of England into Mid Wales, excluding Cornwall

Zone 3:   Cumbria, Eastern and South Western Counties of Scotland up to Aberdeen, West Wales, Cornwall

Zone 4:   Scotland North of Aberdeen and the Western Counties

Note: In very sheltered conditions and where piers are used, taller walls may be satisfactory

inadvertently be encroaching on a neighbour's land. Some will be happy to bear the cost of erecting walls and fences for this purpose on all sides, but within the boundary, proudly proclaim their ownership of the constructed features and be prepared to accept the entire cost of maintenance, invoking the Access to Neighbouring Land Act 1992 if opposition is encountered to entering a neighbour's land to carry out the necessary work. Soundly built walls or fences of sturdy character, constructed in the correct positions according to measurements on deeds represent the ideal for any owner keen to avoid what can amount to distinctly unpleasant disputes with neighbours involving both the position of boundaries and the ownership and repair of walls and fences. A surveyor will be surprised if he can advise a buyer that he will be entering into such a pleasant situation because all too often it will be that he has to indicate that boundaries are ill defined and the ownership of walls and fences unclear.

The surveyor must inspect and report on what he can see of the type and condition of boundary walls and fences, on what he is told and on whether there is any evidence on site to suggest confirmation of what has been said to him on ownership, such as the presence of piers to a brick wall, horizontal arris rails on a close boarded fence or concrete posts for a chain link fence, all on the dwelling's side of the boundary. However, walls and fences get put up wrongly and occupiers are not always in possession of the correct information. In the absence of the surveyor being handed a copy of the deeds before his inspection to enable a check to be made on measurement and the incidence of the traditional 'T' mark on plans to denote ownership, the surveyor may have to advise reference to the legal adviser for investigation. Of course, between gardens, brick walls may be built on the line of junction and ownership

be deemed to be shared by the owners on either side but, in the absence of firm information on the deeds, they may have to be treated as such under the Party Wall etc. Act of 1996 if alterations or repairs are required.

One or two points need to be borne in mind about brick boundary walls which can sometimes be overlooked. In view of their cost at the present time, most of those encountered will tend to have been built before the 1920s but without a damp proof course. Where abutting the dwelling they can be the source of damp penetration and may need the insertion of a vertical damp proof membrane at the point of contact. Another aspect concerns the detailing at the top of the wall. While brick on edge and tile creasing may be an ineffective barrier to downward penetrating damp, it is usually more secure than other forms of capping such as clayware, stone or concrete sections which are often loose and can be easily dislodged. The presence of a large heavy gate without proper strengthening of the wall by piers can loosen sections by frequent use, and even if piers are provided, impact from vehicles can dislodge and render them unsafe. Failure to provide a brick relieving

454: Three types of boundary enclosure. To the right, and by far the best, a brick boundary wall in good condition, in the centre a sturdy close feather edged boarded timber fence with, on the left, a less than sturdy waney edge lap panel fence.

arch above the spreading out of roots at the base of a growing tree can lead to severe fractures and a danger of collapse.

For timber fences, a note on the type, condition, approximate age and probable future life should be given. For example on close boarded fences, where arris rails are bowed, split or discoloured green from moss or dampness the strength is bound to be affected and the need for renewal may not be too far off. The most common defect with timber fences of this type, however, is that the wooden posts have become rotted at their bases and the surveyor should slightly rock the fence to see if it is stable, as it will have to withstand gale force winds and driving rain. The surveyor may consider that the provision of concrete spurs is sufficient to extend the life of the fence for a number of years ahead or he may feel that more radical repair work is required. Very often an apparently safe-looking close-boarded fence may be on the point of collapse.

Fence quality can vary considerably, from the good quality contractor erected close feather edge boarded with arris rails and 100mm posts set in concrete, by no means cheap to replace in long lengths, to the flimsy woven panel or waney edged lapped panel type nailed to 50mm uprights set in short 'metposts' and put up by the DIY handyman.

Brick walls and timber fencing are generally considered suitable enclosures for gardens providing privacy, but, their height being limited by the Planning Acts over the last 50 or so years to 2m, not unduly interfering with the

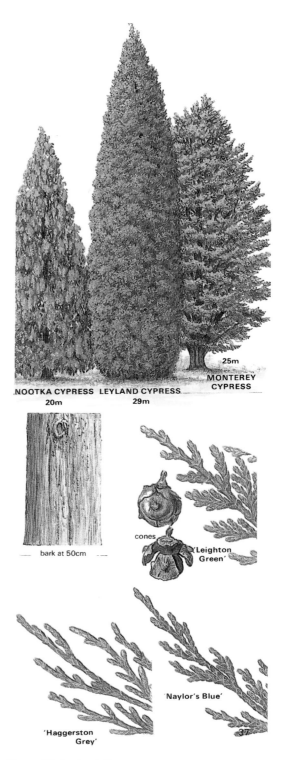

Figure 63: There are many varieties of the Cypress species of tree, all of which are evergreen and of which three are illustrated here with a Leyland Cypress in the centre. The leylandiis are a hybrid distinguished for their rapid growth. The typical foliage of flattened sprays are all of leylandii types.

Figure 63: Leyland Cypress

light to gardens and neighbouring dwellings. This is not so, however, with natural vegetation which can grow to inordinate heights and become a nuisance to neighbours. Without co-operation on cutting back or reducing height, sadly lacking in many cases, nothing could be done about this until the coming into operation in England at the beginning of June 2005, of Part 8 of the Anti-Social Behaviour Act of 2003. This enables those complaining of high hedges and neighbours refusing to reduce their height, to refer the dispute to the local authority who, if necessary, will issue a formal notice to have the hedge reduced in height by its owner or, on failure to comply, will enter, do the work and recover the cost. The spur to the passing of this Part of the Act was the planting by some owners of a line of Cypress Leylandii evergreen trees, to screen their garden from neighbours to the north, or thereabouts. While an isolated mature specimen of this evergreen in a corner or on a large lawn is

not unattractive, see Figure 63, its main characteristic is its unusually rapid growth and planted as a hedge it is a different story, see 455. It soon becomes too high for cutting back from a pair of steps and can deprive a neighbour's garden eventually of all sunlight.

Apart from brick or stone walls, timber fencing or hedging, another form of enclosure, railings, will be found, particularly along the frontage of dwellings built in the 1700 and 1800s to protect persons, owners and visitors, from falling down into the basement areas, so much a feature of the multi-storey terrace dwellings of the time. For their purpose, they need to be strong, securely fixed and set close enough together to prevent small children from slipping through. The surveyor needs to check on these aspects for his report, applying a little pressure to see that they are secure. If not they will need either repair or replacement to ensure that accidents do not happen, rust at the base and elsewhere being the principal cause of disrepair.

As to the materials used for railings through the ages, these are wrought and cast iron and

455: The end of a long hedge of Cypress Leylandii evergreen trees, planted only a few years previously and allowed to grow well beyond the 2m considered reasonable for height. The trees will continue to grow to the annoyance of neighbours unless cut down.

456: The surveyor needs to note loose sections of balustrading around basement areas and to the sides of entrance steps and landings, common features of terrace housing in the 1700 and 1800s. Loose and missing sections of nosings, as here, to the entrance steps are a particular hazard to this early 1800s dwelling.

mild steel. Both wrought and cast iron, the former from 1735 and the latter from 1714 as recorded, were used for railings up to around 1830, cast iron for the stouter heavier support sections and wrought iron for the subsidiary and decorative members. From 1830, with the Industrial Revolution in full swing, cast iron reigned supreme and on many sections of castings in the style of the times, the makers stamp will still be found. Mild steel became available in the late 1800s and took over from cast iron almost entirely from the 1920s onwards. The sections are very much thinner than those in cast iron and tend to look distinctly inferior. Cast and wrought iron are affected to a lesser extent than mild steel by corrosion but all need coating with a paint system and for this to be renewed at regular intervals.

Gardens

No description under the heading of surroundings will be complete without some mention of the garden itself. Usually some general note is sufficient to describe the condition of the garden, but considerably more attention is desirable where trees are present and it may well be necessary for each of these to be described separately, their condition noted if unstable or with loose elements likely to cause injury, even apart from cases where spreading roots exert a pressure on the house or outbuildings under review. The chief danger with trees is that they absorb the moisture from the sub-soil and, in the case of a survey in an area of shrinkable clay soil, the surveyor should take great pains to gauge the probable effects of any trees he may find on the site and of any that he can catch sight of nearby.

Enough has been said about trees in the subsection of that title in section 4, External

Walls, between pp 317 to 330 not to need repetition here. For trees likely to cause damage the mature height is given on each illustration within which distance a tree of that species, if present, could well be a danger but with the proviso that for Poplars and Willows the distances should be increased to 35m and 40m respectively.

Sufficient has also been said in the subsection on Ground Movement between pp 330 and 337, following the subsection on Trees, about other aspects which may be seen in the garden such as sloping ground, areas which have sunk perhaps because of leaking drains, cavities below ground, mining subsidence or settlement of sites filled with unsuitable material. The boundary between long since consolidated unfilled land and comparatively recently filled land can sometimes provide spectacular effects but all indications of movement need recording and including in the report, conclusions drawn and, if necessary, a recommendation given to obtain further advice. Of particular concern are the effects of such movement on steps and pathways which may have rendered them dangerous.

Visible evidence of any wildlife in the garden should be included in the report on the basis that it might be of concern to a buyer. The persistence of moles in digging up a lawn can be considered by some as a nuisance and badgers while not an endangered species are protected by law and have active and valuable support in the form of the National Federation of Badger Groups. The antipathy between dogs and badgers is well known, and disturbance of setts and blocking of trails is more than just frowned upon, so that buyers with dogs as pets will wish to know of any indications of their presence.

Foxes are to be expected in country areas where they are considered a pest and even if now safe from hunting with hounds, can be shot. The urban fox prospering from the scraps and rubbish left out by restaurants and by dwellings is equally a pest and difficult to get rid of. A family of them can ruin a garden by fouling and digging and the noise made at night has to be heard to be believed. Evidence of their presence should be included in the report because some buyers will be dissatisfied if not told and it is best not to risk that sort of complaint if there is evidence at the site, even though it might be arguable whether a buyer would succeed in an action at law against a surveyor.

Some clients will press a surveyor to include comments on environmental matters such as noise from road vehicles, railways or air traffic. One surveyor asked to comment on possible noise from aircraft said that the dwelling was 'unlikely to suffer greatly from noise, although some planes would inevitably cross the area', probably a fair enough comment if that was his view after a single day time inspection. The buyer after moving in considered the noise excessive and sued, *Farley* v *Skinner* 2001 when he was able to show that if the surveyor had asked the aviation authorities he would have been told that 'aircraft stack over the area for two periods a day' and secured an award of £10,000 in damages. The Court of Appeal reversed this decision saying there was no evidence of physical discomfort or inconvenience and the nature of any harm was not compensationable. The determined complainant took the case to the House of Lords where it was held that the surveyor's accepted obligation to report on noise was an important and exceptional one and became a contractual guarantee overriding the obligation of reasonable care, the purchaser requiring an expectation of peace and pleasure, and the original award of damages was reinstated.

The lesson to be learned form the above case is that it is unwise for a surveyor to accept an invitation to comment on an aspect of a dwelling which can only be determined by living in it for a period of time. Second best for the buyer would be for him hiring a room in an hotel or pub nearby and living there for a while, or again by visiting the dwelling's immediate location, frequently and at night time and weekends. Alternatively, in some areas as mentioned in section 2, there are firms that can be employed to report on neighbours and the neighbourhood, visiting the area when the surveyor is enjoying himself at the weekend or safely tucked up in bed. To know about them and to be able to pass their names on to clients, taking care not to recommend, is useful and could be helpful.

Outbuildings

Returning towards the house and having obtained an overall view of the site from the boundaries the surveyor needs to examine and record the state of repair of the various outbuildings adjoining the house and within the periphery of the grounds. The first out-building of consequence that will require inspection may well be a garage. Garages will vary from small asbestos structures at the rear of a terraced or semi-detached house built in the 1920s and 1930 to a large detached imposing structure of brick and tiles, perhaps with a flat or playroom above. Inspection of the garage will, of course, follow in the same way the rules for the inspection of its parent structure and does not require any exhaustive description at this point.

It is worth mentioning, however, that any defects in materials, for example facing bricks or roofing tiles, which the surveyor has found in the main structure may well be repeated on the garage and outbuildings. The surveyor will often find many examples of defects that are likely to be caused from neglect. Outbuildings are never maintained with the degree of care given to the dwelling house itself and it is commonplace to find advanced attacks of beetle in the timbers that have never been treated, or troubles from dampness caused by blocked gullies and fractured gutters that have never received attention.

The consequences of skipping lightly over outbuildings when carrying out an inspection on a comparatively modest three bedroom detached house were brought home to the surveyor in a case, *Allen v Ellis & Co* 1990, concerning the separate garage used as a utility room. The surveyor described it as 'brick built in 9" brickwork and in a satisfactory condition'. The judge found that the garage should have been described as of 'breeze block construction with an asbestos sheet roof which was brittle and fragile, likely to split and crack, scantily supported, much repaired and at the end of its useful life'.

The judge concluded that 'if the roof had been accurately described ... the plaintiff would never have been in peril of suffering the injuries which he did in fact suffer' when he fell through the roof on investigating a leak. In this case damages were awarded not only on the basis of diminution in value on account of the wrongly described garage but also in respect of the injuries suffered by the plaintiff. Fortunately for the surveyor, the plaintiff's injuries were not as severe as they might have been. It matters not that every surveyor knows, or should know, that it is unsafe to walk on any asbestos roof, new or old. The plaintiff in this case was not told.

The surveyor should also look for faults of design that can cause trouble, due to insufficient attention to detail when the

garage was built. Even though every care may have been devoted to designing the dwelling house, the garage might not have received the same degree of attention. Alternatively it might have been erected after the main house was built and such matters as awkward roof gutters that are not adequately drained and no provision for wash down gullies might have been overlooked. For smaller garages lack of drainage is no particular disadvantage.

Other exterior structures, varying from conservatories, greenhouses, summer houses and tool sheds to detached barns, will rarely have any common characteristic apart from the fact that they are very often constructed partly of timber and that this will be generally neglected. A thorough soaking in preservative may extend the life of these structures for a surprising period, but very often they are beyond repair and can be remarkably costly to replace.

Environment

Mention has already been made of the availability of two websites to provide information about the dwelling for which the surveyor has instructions to inspect. These are www.imagesofengland.org.uk, from which it can be ascertained whether the dwelling is listed as one of architectural or historic interest and www.nrpb.org.uk for finding out whether the dwelling is in a 'radon affected area'. The outcome of a search of the former is a straightforward matter of fact. The dwelling is either listed or not, but it also enables the surveyor to advise, if necessary, on the implications of listing, both favourable and unfavourable.

The information derived from the second of the websites is more controversial but could lead the surveyor to advise the client, if a seller, to obtain the dwelling's average radon concentration level over a three month period since this could possibly be a question asked by a buyer. If the client is a buyer, he should ask the seller for the figure. Depending on the outcome both, with the help of the surveyor, need to decide whether to ask the surveyor whether to do anything further having regard to the availability of the conflicting views on the subject, as set out at the end of section 4, which the surveyor should be prepared to put forward and discuss without taking a stand either way.

There is much more 'information' available on websites which the surveyor needs to access to confirm or supplement his background knowledge concerning the area in which he practices. Such information is, of course, available to the general public and when considering the purchase of a dwelling, a buyer is quite likely to have already checked by entering the postcode and, in some cases, the specific address. The surveyor will be well advised to do this himself as a matter of routine before he carries out his inspection, even if there is no mention by the client of having done so at the time the instructions are given. The websites are no doubt designed to be helpful but can produce information which can be alarming to a prospective buyer. It could be said that the dwelling being in a 'radon affected area' is one such, but, there are others relating to flooding and pollution for example which could be equally so.

If the client has already carried out a check it could well be that he will ask for clarification or wish to know the implications of the information he has obtained. It is for this reason that the surveyor needs to be aware of what is available so that he can provide re-assurance or, alternatively, advise on what course of action needs to be taken. This is of particular concern where his own information is at variance with that given on the website, or where he considers that what is presented is speculative in the extreme, too wide ranging or omits a contrary view which is available elsewhere.

Websites can be contrasted as to those produced by government departments and agencies which are usually accessible free of charge, one exception, however, being the Land Registry's www.landregisteronline.gov.uk /ad where a charge of £2.00 is made for obtaining the most recent sale price of a specific dwelling. Others on which the information is freely available include the Environment Agency's www.environment-agency.gov.uk for a check by post code on flood risk, past and present contamination, landfill, mining, dangerous gases and subsidence. Another is the Office of National Statistics at www.ons.gov.uk/neighbourhood/ which through post code and more specifically street or estate, access enables statistics to be obtained for the nearest 125 households, for example how many have more than one car, how old the children living nearby are and, perhaps surprisingly, how many of the neighbours have paid off their mortgage. It is limited to some extent and passes no view on whether the neighbours are noisy or unpleasant. This would require either the employment of a private investigator at probably a high fee, currently indicated at something to the order of £250-£500, or a DIY operation by a prospective buyer himself visiting the area at odd hours and sitting in the car and watching what goes on. Another website produced by the National Society for Clean Air and Environmental Protection www.nsca.org.uk provides advice on how to deal with neighbours and is linked to www.mediation.org.uk, to obtain a clickable map to identify the location of a mediator.

Non-government organisations recognising that there is a market for dwelling specific environmental information produce it for a fee. The RICS has suggested that home buyers should be prepared to incur the standard £39.00 disbursement to obtain a 'Landmark Home Envirosearch Report' from the Landmark Information Group at www.home-envirosearch.com or www.landmark-information.co.uk. Many legal advisers obtain these reports but as they pass on the contents to buyers, in most cases without comment, it would be better if they were obtained, by agreement with all parties, by the surveyor, or at least passed on through him for comment if already obtained by the legal advisers on behalf of a buyer. For a seller it should be a routine matter for the surveyor to obtain a report of this type, always provided the seller is prepared to meet the fee, or compile it himself from the appropriate websites. Another website www.freeagents.co.uk offers a full background environment report for £49.00 and modestly priced reports on limited aspects, for example an air pollution report for £3.50. Other websites involved in this type of information gathering are www.homecheck.co.uk and www.fish4homes.co.uk some claiming to have information not available to others. For clients interested in the history of the London area into which they are moving, or for that matter have moved to already, Charles Booth on line Archive at www.booth.ise.ac.uk provides a detailed snapshot by street of life in 1886 with maps colour coded according to wealth.

The commercial websites which charge for their information are providing a service by collecting together, for the convenience of users, what is produced by diverse government departments and agencies, either freely or for a nominal charge. The success of the websites and their continued availability will depend, as in all commercial activity, on how their services are perceived in regard to usefulness, accuracy and value for money. It is inevitable that some will fall by the wayside and others will take their place, so that the scene will be continually changing. To an extent, of course, so will the information over time and the surveyor must do as best he can to keep abreast of changing developments if he is to keep his customers happy.

Appendix: Building Stones

Appendix: A Selection of Building Stones of the United Kingdom by Old (pre1980's) County from North to South.

Notes: Locations of use for a few stones have been taken from various books and pamphlets, a feature which could be supplementd by surveyors from their own local city and town guides. In particular, a 1999 publication *'Building Stones of Edinburgh'* by the city's Geological Society provided much information of the type useful for this purpose and could well be emulated elsewhere. As to availability, quarries in many cases open and close intermittently according to demand but the information shown has been checked from the 2001 British Geological Survey map of Building Stone Resources of the United Kingdom.

County (Old)	Location of Quarry/Mine	Name and Colour	Type, characteristics and locations of use	Available Yes/No
Moray	Clasbach Quarry, Hopeman, Moray.	Clasbach. Yellow to buff.	Sandstone. Even grained. Museum of Scotland (extension) Chambers Street, (1998), Scottish Widows, Morison Street, (1997), Refacing of Paton Building, 1-3 York Place, (1998), 202-254 Canongate, (1958-1966), Chessels Court Canongate, (1969) all in Edinburgh.	Yes
	Hopeman, Moray.	Greenbrae. Fawn.	Sandstone. Even grained.	No
	Spynie Quarry, NE of Elgin Moray.	Spynie. Yellowish-grey, cream or pinkish.	Sandstone. Fine grained. Mouldings at 83-89, Great King Street, (1982), and 39, Howe Street, Edinburgh.	Yes
	Cutties Hillock (Quarry Wood). Elgin, Moray	Cutties Hillock.	Sandstone.	Yes
	Rosebrae Quarry, Elgin, Moray.	Rosebrae. Cream or pinkish	Sandstone. Fine grained.	No
Aberdeen	Boddam, South of Peterhead Aberdeenshire.	Peterhead. Grey through to brilliant red.	Granite. Coarse grained. Old Waverley Hotel, Princes Street, columns around windows, (1883), Palmerston Place Church, pink arches and columns internally, (1873), Edinburgh. Prudential Assurance, Ramsden and New Streets below terracotta. Royal Bank of Scotland, columns, Market Square 'Nawaab' restaurant, Westgate, pink, all in Huddersfield.	No
	Kemnay NW of Aberdeen, Aberdeenshire.	Kemnay. Light speckled silvery grey.	Granite. Medium grained. Marischal College, Aberdeen.	Yes

County (Old)	Location of Quarry/Mine	Name and Colour	Type, characteristics and locations of use	Available Yes/No
Aberdeen (continued)	Quarries in Aberdeen. Aberdeenshire.	Rubislaw and Lower Persley. Light to dark bluish grey.	Granite. Fine to medium grained,Fine to medium grained, coarsely crystalline. Extremely durable. Royal Bank of Scotland, 42 St. Andrew Square grey granite base (1936). Most of the notable buildings of Aberdeen and much exported.	No
	Alford, E of Aberdeen, Aberdeenshire.	Corennie. Light to dark salmon pink. Also grey.	Granite, Coarse and medium grained. Blocks can sometimes be half and half the two colours. Lloyds TSB, The Cross, Worcester, columns.	Yes
	Bucksburn, suburb of Aberdeen.	Sclattie. Light bluish grey.	Granite, medium grained.	No
	Quarry near Aberdeen.	Dancing Cairns. Light grey.	Granite. Medium grained.	No
	Dyce, NW of Aberdeen, Aberdeenshire.	Dyce. Dark grey.	Granite. Fine grained.	No

Note: the above three stones are not recorded on the BGS map, but all are said to have been used for general building work in the 1930s.

County (Old)	Location of Quarry/Mine	Name and Colour	Type, characteristics and locations of use	Available Yes/No
Argyll	Isle of Mull, Argyll and Bute.	Ross of Mull. Pale to deep red	Granite. Coarse, fairly even grained. Trustee Savings Bank, 120-124 George Street, pink paving in the entrance hall and atrium, (1986) Edinburgh.	Yes
Fife	Between Kirkcaldy and Burntisland,	Newbigging. White, cream Fife.	Sandstone. Fine grained freestone. National Library of Scotland, 33 Salisbury place (1984) to buff Causewayside Building, Edinburgh. Also used in Glasgow and Dundee.	Yes
Midlothian	West Edinburgh, Edinburgh.	Craigleith. Whitish grey to brown.	Sandstone. Fine grained, good for ashlar. Hard and durable. Good examples, City Chambers, High Street, (1761), Register House, Princes Street, (1778), 8, Queen Street (1771), Leith Town Hall, Constitution Street (1827) North side of Charlotte Square (1794), all Edinburgh. Exported to England, USA and Europe. Fully worked by about 1900.	No
	Quarries at Binny, Uphall, West Lothian.	Binny. Orange-brown	Sandstone. Fine grained freestone. Good for ashlar, but traces of oil. Good examples, Royal Scottish Academy (1831), Scott Monument (1840) National gallery (1850), Bank of Scotland (1802), City Observatory, Calton Hill, (1818), all Edinburgh.	No

County (Old)	Location of Quarry/Mine	Name and Colour	Type, characteristics and locations of use	Available Yes/No
Midlothian (continued)	Slateford, SW of Edinburgh, Edinburgh.	Hailes. Available in three tints: white, fine grained, hard, blue, medium grained but marked lamination and pink,with intermediate properties.	Sandstone. Fine to medium grained. Very hard. Used largely in rubble form throughout Edinburgh and surrounding area since about 1700. Good weathering properties. Examples, Blue, Royal Infirmary, Lauriston Place, (1872), Scotch Whisky Heritage Centre, Castlehill, Royal Mile (1896) Dalry and Sciennes Primary Schools, Dalry and Sciennes Roads (1876 and 1889). Pink, Coates Crescent, front, (1820), 1 Cluny Gardens, (1880), Roseburn Primary School, Roseburn Street, (1843) ashlar, all Einburgh. Some exported to London.	No
Lanark	High Blantyre, NW of Hamilton, South Lanarkshire.	Earnock. White and grey.	Sandstone. Good working, very hard and durable.	No
	Cleland, between Airdie and Wishaw, North Lanarkshire.	Auchinlea. White, yellowish to cream. Often flecked with brown.	Sandstone. Medium grained freestone. Roseburn Terrace, (1882), tenements at South Buchanan Street (1878), and Villas at South Buchanan Street (1878), and Villas at Trinity (1883), all Edinburgh.	No
Dumphries	Gatelawbridge. Thornhill, Dumphries and Galloway.	Gatelawbridge. Bright red.	Sandstone. Fine to medium grained. Jenners Store, Rose Street (1890 and 1902), Saltire Court, Castle Terrace (1991), with buff sandstone from Stainton, Barnard Castle, County Durham, Edinburgh Solicitors Property Centre, 85 George Street, (1980s), all Edinburgh. Much used also in Glasgow.	Yes
	Thornhill, Dumphries and Galloway.	Closeburn. Light to Dark Red	Sandstone. Fine grained, Good for working. King's Theatre, Tollcross, (1904) (repaired with Corsehill 1980s), Candlish Church Tower, Merchiston (1915) Edinburgh.	No
	Cornockle Quarry. North of Lochmaden. Dumphries and Galloway.	Cornockle, Red.	Sandstone. Fine to medium grained. Scotch Whisky Heritage Centre, Royal Mile, (1888), Leith Academy, Lochend Road Annex (1885), Royal Bank of Scotland, Nicholson Street, (1902), Milton House School, Canongate, (1886).	Yes
	Lochabriggs, NE of Dumphries. Dumphries and Galloway.	Lochabriggs. Pink to red.	Sandstone. Fine to medium grained. Free working. Fairly durable. Caledonian Hotel, (1899), Lauriston Place Fire Station, (1897), College of Art (1906) and Extension (1972).	Yes
	Annan, SE of Dumphries. Dumphries and	Corsehill. Pastel pink to warm red. Galloway.	Sandstone, close grained, good working. Durable. 70 Princes Street (upper part) (1886), Roseburn Primary School, Roseburn Street,(1893) dressings with 'pink' and grey Hailes. St. James Episcopal Church, Inverleith Row (1885). Royal Hospital for Sick Children, Sciennes Road, (1892), all Edinburgh. St. Helen's Church, Fish Street, Worcester, porch.	Yes

County (Old)	Location of Quarry/Mine	Name and Colour	Type, characteristics and locations of use	Available Yes/No
Kirkudbright	Creetown. Dumphries and Galloway.	Creetown. White when hammered. Bluish white when polished.	Granite. Fine grained. National Library of Scotland, George IV Bridge, grey base (1937-1955), Edinburgh. St. Andrew's House, Calton Hill, base course, (1936), Edinburgh.	Yes
Northumberland	Elsdon, NW of Newcastle Northumberland.	Blaxter. Buff	Sandstone. Medium grained. Burton, 30-31 Princes Street (1906), Arden Street, Marchmont, houses (1905), National Library of Scotland, George IV Bridge (1937-1955), Sun Alliance, 68 George Street (1955), Royal Museum of Scotland, Lecture Theatre, Lothian Street (1958) Edinburgh.	Yes
	Wooler, S of Berwick on Tweed, Northumberland.	Doddington. Pink to purple grey.	Sandstone. Fine grained. Methodist Central Hall, Tollcross (1899), General Post Office Extension (1908), George Watson's College, Colinton Road, (1930), Edinburgh.	Yes
	Heddon on the Wall, W of Newcastle, Northumberland.	Heddon. Light Brown	Sandstone. Medium Grained. Good for general building purposes.	No
	Fourstones, Hexham, Northumberland.	Prudham. Brown.	Sandstone. Coarse grained. Newcastle Central Station (1846). McEwan Hall, Teviot Place (1888), Crown Office, Chambers Street (1886), Balmoral Hotel, Princes Street (1902), Tenements, Marchmont (1876), St. Andrew Square, Bus Station (1970) Edinburgh.	No
Cumberland	Moat, NE of Longtown, Cumbria.	Moat. Bright Red.	Sandstone. Fine grained but weathers badly. Couper Street School (1889), National Portrait Gallery (1885) Edinbrugh.	No
	Penrith, Cumbria.	Penrith Red. Deep rust, bright red, pink to buff.	Sandstone. Moderately coarse grained. Plain details fare better. All the older part of Penrith and throughout the neighbourhood.	Yes
	St. Bees, S of Whitehaven, Cumbria.	St. Bees. Pink to bright red.	Sandstone. Fine grained. Good working properties. Very durable. Furness Abbey.	Yes
	Three quarries working NE of Penrith Cumbria.	Lazonby. 'Red', light terracotta, 'White' light yellowish pink.	Sandstone. Medium to coarse grained. Difficult to work.	Yes
Cumberland/ Westmorland	Various quarries listed as working eg Kirkstone Green, Elterwater Green Broughton Moor, plus others. Cumbria.	'Westmorland'. Green.	Slate. For both roofing and cladding. Natural History Museum, London roofing slates, alternating with Burlington Blue Grey slates. Nat. West Bank, New Street, Huddersfield, panels below windows.	Yes

County (Old)	Location of Quarry/Mine	Name and Colour	Type, characteristics and locations of use	Available Yes/No
Westmorland	Shap, Cumbria.	Light and Dark Shap. Greyish pink to dark brown.	Granite. Medium grained with characteristilarge crystals of pink feldspar with black mica and quartz. Entrance to Post Office, St. Aldates, Oxford (1879). Eddisons Estate Agents Huddersfield with local sandstone and Rubislaw grey granite from Aberdeen. St. Mary's cathedral Palmerston Place, Edinburgh, portico columns (1917).	Yes
Ulverston	Kirkby-in-Furness, Cumbria.	Burlington Blue Grey. Blue through to black.	Slate. For both roofing and cladding. BHS, Eastgate, Gloucester, panels, Natural History Museum, London, roofing slates, alternating with Westmorland green slates.	Yes
County Durham	Springwell,	Springwell. Yellow. Gateshead.	Sandstone. Medium grained. Fairly easy to work. Durable. Apex House, Leith Walk, (1975). 8-11 Royal Crescent, restoration (1979) both Edinburgh.	Yes
	Winston between Darlington and Barnard Castle. Durham.	Dunhouse.	Sandstone. Fine grained. Durable. Egglestone Abbey, Bowes Museum. M&S, 104 Princes Street (1980), Exchange Plaza, Lothian Road (1997), 1-8 Atholl Crescent, repairs (1985), Edinburgh.	Yes
	Stainton near	Stainton. Buff. Barnard castle,	Sandstone. Fine to medium grained. Saltire Court, Castle Terrace (1991), Standard Life, Durham. Lothian Road, (1997), Sainsburys, Queensferry Road, (1993), John Lewis, Leith Walk, extension cladding, Edinburgh.	Yes
	Hewworthburn, South Tyneside.	Hewwworthburn. Bluish grey.	Sandstone. Fine grained but not easy to work. Very hard and durable.	No
Lancashire	Near Blackburn. Blackburn and Darwen.	Butler Delph. Buff	Sandstone. Coarse grained, very hard and durable.	No
	Farnworth, near Bolton, Bolton.	Edgefold. White to biscuit.	Sandstone. Fine grained, easily worked. Very durable.	No
	Rainhill, East of Liverpool, Knowsley.	Rainhill. Red to orange	Sandstone. Fine grained.	No
	Appley Bridge, NW of Wigan, Wigan.	Appley Bridge. Blue	Sandstone. Fine texture, fairly easy to work, very hard and durable. Much of Oldham, Rochdale, Colne and Wigan.	No
	Woolton, SE of Liverpool, Liverpool.	Woolton. Red to orange.	Sandstone. Fine grained. Anglican Cathedral, Liverpool.	No
Yorkshire	Aislaby. South West of Whitby, North Yorkshire.	Aislaby. Light brown.	Sandstone. Fine texure. Whitby Abbey, Guisborough Priory, Houghton Hall, Norfolk.	No
	Wass, East of Thirsk. North Yorkshire.	Wass.	Limestone. Fine texture. Byland Abbey.	No

County (Old)	Location of Quarry/Mine	Name and Colour	Type, characteristics and locations of use	Available Yes/No
Yorkshire (continued)	Anston, E of Sheffield, Rotherham.	Anston. Light brown to cream.	Magnesian limestone. Fione grained. Used on Houses of Parliament but unsuitable for decorative features.	No
	Shipley. Bradford	Appleton. Blue to brown.	Sandstone. Coarse grained and hard to work but durable.	No
	Bradford	Bolton Woods. Light greenish brown.	Sandstone. Fine grained.	Yes
	Ackworth, Near Wakefield, Wakefield.	Brackenhill. Light grey to brown.	Sandstone. Fine and coarse textured.	Yes
	NW of Leeds, Leeds.	Bramley Fall. Light brown.	Sandstone. Coarse grained. Very strong and durable. Suitable for engineering uses. Weston Hall (Banquet House) ruins of Kirksall Abbey. Leeds Town Hall.	Yes
	Cadeby, Near Doncaster, Doncaster.	Cadeby. Ranges from white to yellow.	Magnesian Limestone. Fine grained. General Accident Building, York.	Yes
	Near Huddersfield, Kirklees.	Crossland Hill. Light brown.	Sandstone.	Yes
	Elland, NW of Huddersfield, Calderdale.	Elland Edge. Brown to grey.	Sandstone. Fine grained and fissile, a 'York' stone. Much used in Huddersfield, Halifax and Bradford regions.	No
	Guisley, near Leeds, Leeds.	Guisley. White but variegated.	Sandstone. Even texture, free working. Strong and durable.	No
	Morley, near Leeds, Leeds.	Howley Park. Light brown.	Sandstone. Fine grained. Easily worked. Durable. A 'York' stone much used.	No
	Sherburn-in-Elmet, near Leeds, Leeds.	Huddlestone. Cream.	Magnesian limestone.	No
	Outibridge, NW of Sheffield, Sheffield.	Middlewood. White to brown.	Sandstone. Medium to coarse texture. Durable.	No
	South Milford, near Leeds.	Park Nook. Cream.	Magnesian limestone. Moderately good working. Much used locally.	No
	Farnley, South West of Leeds, Leeds.	Park Spring. Light brown.	Sandstone. Medium grained. Much used in Leeds. A 'York' stone.	No
	South east of Leeds, near Wakefield, Wakefield.	Robin Hood. Bluey and greenish grey.	Sandstone. Fine grained. Durable if carefully selected. A' York' stone.	No
	East of Rotherham, Rotherham.	Roche Abbey. Cream.	Magnesian limestone. Fine grained but rather soft and subject to discolouration.	No
	Near Halifax, Calderdale.	Scout Stone. Light brown to greyish white.	Sandstone. Hard and durable.	No

County (Old)	Location of Quarry/Mine	Name and Colour	Type, characteristics and locations of use	Available Yes/No
	West of Tadcaster. North Yorkshire.	Tadcaster/ Thevesdale. Cream.	Magnesian limestone. Fine grained. York and Beverley Minsters, Selby Abbey.	Yes
	Barton. Near Darlington to SW. North Yorkshire	Swaledale. Light brown with many fossils.	Limestone. Barton marble also produced.	No
	Thornton, near Bradford, Bradford.	Thornton Blue. Bluish grey.	Sandstone. Fine grained but costly to work. Very hard and durable.	No
	Huddersfield, Kirklees.	Wellfield. Light brown, occasional quartz pebbles.	Sandstone. Fine grained, free working. Very durable.	No
	Keighley, Bradford.	West End. Brown.	Sandstone. Fine texture. Very hard and durable.	No
	South West of Sheffield, Sheffield.	Stoke Hall. Cream to brown.	Sandstone. Fine grained. Sheffield Town Hall.	Yes
	Morley, near Leeds. Leeds.	Woodkirk Blue. Blue.	Sandstone. Fine grained. Good working. Durable.	No
	Morley, near Leeds. Leeds.	Woodkirk Brown. Grey to brown.	Sandstone. Fine grained in massive beds.	Yes

Note: The generic name of 'York' stone is applied to the products of a number of quarries to the south of Leeds and Bradford and from around Halifax, much used in the area for general building purposes and, because some of the stone is highly laminated also used widely for paving, steps, sills and copings. However, due to its porous nature it is not so suitable as some other sandstones for roofing purposes A good example in ashlar is the railway station, St. George's Square, Huddersfield (1850).

County (Old)	Location of Quarry/Mine	Name and Colour	Type, characteristics and locations of use	Available Yes/No
Cheshire	Pott Shrigley. NE of Macclesfield, Cheshire.	Berristall. Cream.	Sandstone. Fine grained. Free working. Durable.	No
	Kerridge. NE of Macclesfield, Cheshire.	Kerridge. Buff.	Sandstone. Fine to medium grained. Good working properties. Very hard and durable but also fissile and used for roofing.	Yes
	Weston, near Runcorn. Halton.	Runcorn, Red. Mottled.	Sandstone. Coarse to fine grained. Easily worked. Durable.	No
	NE of Macclesfield, Cheshire.	Windyway. Blusih grey.	Sandstone. Fine to medium grained. Good working qualities. Durable. Available in large blocks.	No
Derbyshire	NW of Matlock. Derbyshire.	Birchover Moor. Pink to yellow buff.	Sandstone. Medium grained. One of the 'Dale' stones. Highly suitable for ashlar.	Yes
	East of Chesterfield. Derbyshire.	Bolsover Moor. Warm, yellowish brown.	Magnesian Limestone. Fine grained. Good durable building stone.	No
	Dene Quarry. Near Matlock, Derbyshire.	Derby Dene. Creamy grey with fossils.	Limestone. Fine grained.	No

County (Old)	Location of Quarry/Mine	Name and Colour	Type, characteristics and locations of use	Available Yes/No
Derbyshire (continued)	Bakewell. Near Matlock, Derbyshire.	Darley Dale. Pale brown to white.	Sandstone. Fine grained. Very strong and weathers well. Largely used.	Yes
	Rowsley. SE of Bakewell, Derbyshire.	Dukes. Deep Red.	Limestone. Fine grained.	No
	Stanton Lees, Derbyshire.	Endcliffe. Pink to brown.	Sandstone. Fine to coarse texture. Very durable. One of the 'Dale' stones.	No
	Birchover, NW of Matlock, Derbyshire.	Dungeons. Pink.	Sandstone. Fine to medium texture. Very durable.	No
	Wirksworth, SW of Matlock, Derbyshire.	Hadene. Cream.	Limestone. Fine grained. Said to be indistinguishable from Hopton Wood.	No
	Hopton, Wirksworth, SW of Matlock, Derbyshire.	Hopton Wood. Cream.	Limestone. Fine grained. Takes a good polish and more used for decorative work.	No
	Near Matlock, Derbyshire.	Hall Dale. Pink.	Sandstone. Fine grained. One of the 'Dale' stones.	Yes
	NW of Matlock, Derbyshire.	Stancliffe. Honey to very light drab.	Sandstone. Close grained, uniform texture, good working properties. Very hard and durable. One of the 'Dale' stones. St. George's Hall, Liverpool, restoration at 32 St. Mary's Street and 35 Heriot Roco (1993 and 1998), Dynamic Eearth Building, Holyrood Road, (1999) Edinburgh. The Crescent, Buxton.	Yes
	Birchover, NW of Matlock, Derbyshire.	Stanton Park. Pink and brown.	Sandstone. Fine to coarse grained. Very durable.	No
	Matlock. Derbyshire.	Whatstandwell. Light brown to pink.	Sandstone. One of the coarser 'Dale' stones.	Yes
	Matlock, Derbyshire.	Lumshill. Light buff.	Sandstone. Medium grained.	Yes
Nottinghamshire	Worksop. Nottinghamshire.	Steetley.	Magnesian Limestone.	No
	Mansfield, Woodhouse, Nottinghamshire.	Mansfield Woodhouse. Warm yellow.	Magnesian Limestone. Fine grained, compact.	No
	Mansfield, Nottinghamshire.	White Mansfield. Creamy yellow.	Magnesian Sandstone. Fine, even grained. Not durable in polluted atmospheres. Southwell Minster, Newark Town Hall.	Yes
	Linby, S of Mansfield, Nottinghamshire.	Linby. Brownish yellow.	Magnesian Limestone. Coarse grained but free working. Not durable in polluted atmospheres. Newstead Abbey.	No
	Two quarries (Lindley's and Sill's) at Mansfield, Nottinghamshire.	Red Mansfield. Warm red colour.	Magnesian Sandstone. Fine even grained. Not durable in polluted atmospheres.	No

County (Old)	Location of Quarry/Mine	Name and Colour	Type, characteristics and locations of use	Available Yes/No
Lincolnshire	Ancaster and Grantham, Lincolnshire.	Ancaster, Weatherbed. Brownish yellow. Freestone Creamy white.	Limestone. Coarse grained. Freestone has a finer, even texture. Both free working and good for ashlar. Not durable in polluted atmospheres. Newark castle, Harlaxton Manor.	Yes
	Great Casterton, Near Stamford, Lincolnshire.	Casterton (also known as Stamford) Pale brownish cream.	Limestone. Medium grained with small fine shell fragments. Free working good for ashlar. Not suitable for polluted atmospheres. Much of Stamford and the locality with some use at Cambridge and elsewhere in the Fenlands and East Anglia.	No
Caernarvon and Merioneth	The following quarries are shown on the 2001 BGS map as currently active in the production of Welsh slate in shades of deep blue, blue, dark grey, blue grey, green bronze and heather mainly for roofing purposes but also for cladding: Porthmadoc and Llechwedd, Cut-y-Bugail, Gloddfa Ganol all near Blaenau Ffestiniog, Penrhyn Light Blue, near Bathesda, Peny Orsed, near Nantile and Aberilefeni, near Corris all in Gwynedd with Berwyn, east of Corwen near Llangollen in Conwy.			
Shropshire	Clive, near Shrewsbury, Shropshire.	Grinshill. White through cream to orange and deep red.	Sandstone. Fine grained, durable. Resistant to pollution.	Yes
Staffordshire	Hollington near Uttoxeter, between Stoke and Derby, Staffordshire.	Hollington. White, red, salmon and mottled.	Sandstone. Even grained, fine to medium, hardens on exposure. The pinkish stone chosen for the rebuilding of Coventry Cathedral and work on Hereford and Birmingham cathedrals and Ludlow Church, Upper part BHS, Eastgate, Gloucester.	No
Rutland	Clipsham, near Oakham, Rutland.	Clipsham. Pale cream, buff to light brown.	Some blue patches which weather to grey. Limestone. Medium to coarse grained with shell fragments. Good for ashlar. A little more durable in polluted atmospheres than other limestones. Many local churches. Used for restoration of Houses of Parliament, York, Canterbury, Ripon and Salisbury Cathedrals. Examination Schools, 'Tom Tower', Christ Church, St Aldates (1909 rebuild) Oxford. Eltham Palace, London (1935).	Yes
	Ketton, near Stamford, Rutland.	Ketton. Yellow, buff to pink.	Limestone. Medium even grained regular texture free of shell fragments. Ideal for the ashlared walls of dwellings in the classical style. Good weathering except in polluted atmospheres. Much used in Cambridge and some use in Stamford and at Audley End.	Yes
Norfolk	Snettisham, north of Kings Lynn, Norfolk.	Carstone. Light brown.	Sandstone. Strongly impregnated with iron. Coarse grained and hard. Turns darker on exposure. Stables at Houghton Hall.	Yes
Warwickshire	Edge Hill, NW of Banbury, Warwickshire.	Hornton. Brown, greenish blue, sometimes mixed.	Ironstone. Close grained, easily worked and suitable for ashlar. Not durable in polluted atmospheres. Broughton Castle.	Yes

County (Old)	Location of Quarry/Mine	Name and Colour	Type, characteristics and locations of use	Available Yes/No
Northamptonshire	Collyweston, near Stamford, Northamptonshire.	Collyweston.	Limestone. Sandy texture. Fissile and used for roofing purposes.	Yes
	Duston, west of Northampton, Northamptonshire.	Duston. Warm russet brown.	Ironstone. Sandy texture.	No
	Moulton, NE of Northampton, Northamptonshire.	Moulton. Warm russet reddish, yellowy brown.	Ironstone. Sandy texture. Much used in Northampton and surrounding area. Classed as sandstone on BGS map.	No
	Weldon, near Corby, Northamptonshire.	Weldon. Pinkish brown to buff.	Limestone. Fine even grained with some shell matter. Open textured but notably frost resistant. Easily worked and suitable for ashlar but not durable in polluted atmospheres. Kirby Hall, Castle Ashby, Ruston IHall, Boughton House, Haunt Hill House.	No
Cambridgeshire	Burwell, NE of Cambridge, Cambridgeshire.	Burwell. White.	Limestone. Chalk often called 'clunch'. Still available to SW of Cambridge at Barrington Quarry.	No
Gloucestershire	Coleford, Gloucestershire.	Forest of Dean. Dark blue grey	Sandsone. Fine grained, hard and durable. Nat West, Eastgate, Gloucester.	Yes
	Fishponds, Bristol.	Pennant. Dark blue and grey. and red.	Sandstone. Fine grained. Hard and difficult to work but much used locally. Very durable. Shire Hall, Gloucester (1816).	No
	Guiting. East of Cheltenham, Gloucestershire.	Guiting. 'Yellow', warm brown to yellow and 'White' white to cream available.	Limestone. Coarse grained with fossil content. Royal Oxford Hotel ('yellow').	Yes
	Forest of Dean. Gloucestershire.	Red Wilderness.	Sandstone.	Yes
	Farmington, Northleach, Gloucestershire.	Painswick. Cream.	Limestone. A 'Cotswold' building stone.	No

A plethora of quarries which are or were at one time producing 'Cotswold' Limestone for building are shown on the BGS map. Among those listed as currently active in 2001 are Campden and Chipping Campden, Stanley, Cotswold Hill, Flick, Happylands, Coscombe, Kineton Thorns, Huntsmans, Swellwold, Brockhill and Soundborough. Among the buildings where the stone has been used are Gloucester Cathedral, south wall, base ragstone, St. Michael's Church, tower, St. Mary's de Crypt, Southgate, St. John's Church (1734), Church House, wall and dressings, all in Gloucester, Buttermarket, High Town, (1861), Museum and Library. Dressings to windows, Broad Street both in Hereford and St. John the Baptist Church, Cirencester.

Oxfordshire	Stonesfield, Near Woodstock, NW of Oxford, Oxfordshire.	Stonesfield. Light brown.	Limestone. Used mainly for roofing.	No

County (Old)	Location of Quarry/Mine	Name and Colour	Type, characteristics and locations of use	Available Yes/No
Oxfordshire. (continued)	Bladon, near Oxford, Oxfordshire.	Bladon. Light brown with traces of blue.	Limestone. Coarse grained and used mainly for rubble walling, sandy and fissile. New Bodleian Library, random coursed and hammer dressed (1939). Sometimes used as an inferior substitute for Clipsham.	No
	Taynton, NW of Burford, Oxfordshire.	Taynton. Orange-yellow or white, honey shade.	Limestone. Coarse grained containing shells. Radcliffe camera, Radcliffe Square, upper part, Oxford.	No
	Headington, suburb of Oxford,	Headington.	Limestone. Freestone, soft, fairly porous, easily weathered. Fully worked by about 1700. Oxfordshire. Radcliffe Camera, Radcliffe Square, lower part, Oxford.	No
Bedfordshire	Dunstable	Tottenhoe. Greenish grey.	Limestone. Gritty feel from shell fragments.	Yes
Somerset	Doulting, East of Shepton Mallet, Somerset.	Doulting. Pale buff to light brown.	Limestone. Coarse grained with crinoid fragments. Not suitable for polluted atmospheres. Wells Cathedral, Corpus Christi College (older part), Oxford.	Yes
	Norton-sub-Hamden, West of Yeovil, Somerset.	Ham Hill. Light yellow to light brown.	Limestone. Coarse grained with shell fragments. Not suitable for polluted atmospheres.	Yes
	Charlton Mackrell, S of Glastonbury, N of Yeovil, Somerset.	Blue Lias.	Limestone.	No
	Bishop Sutton, S of Bristol W of Bath, Bath and North East Somerset.	Stowey.	Limestone	Yes
	Combe Down, E of Bath. Bath and North East Somerset.	Combe Down. Pale brown to light cream.	Limestone. Even grained, easy working. Can be porous. Not suitable for polluted atmospheres. A 'Bath' stone of which much of the city was built. Other Somerset quarries in area shown on BGS map produced similar stone. See also Wiltshire.	Yes
Wiltshire	Tisbury. Vale of Wardour, W of Salisbury, Wiltshire.	Tisbury. Greenish grey.	Limestone. Sandy texture.	No
	Chilmark, Vale of Wardour, W of Salisbury, Wiltshire.	Chilmark. Greeny grey.	Limestone. Sandy texture similar to Tisbury. Salisbury Cathedral, Romsey Abbey, Wilton House. The BGS map shows it as quarried with a sandstone called Teffont.	Yes
	Bradford-on-Avon, Wiltshire.	Bradford.	Limestone. A 'Bath' stone.	No
	Limpley Stoke, SE of Bath. On border of Wiltshire and Bath and North East Somerset.	Stoke Ground. Cream to buff.	Limestone. Fine grained. Very suitable for ashlar and details. A 'Bath' stone.	Yes

County (Old)	Location of Quarry/Mine	Name and Colour	Type, characteristics and locations of use	Available Yes/No
Wiltshire (continued)	Winsley. W of Bradford-on-Avon, Wiltshire.	Winsley Ground.	Limestone. A 'Bath' stone.	No
	Corsham, between Bath and Chippenham, Wiltshire.	Monks Park. Pale cream bordering on white.	Limestone. One of the 'Bath' stones but finer and more even grained than the others. Free working. Not suitable for polluted atmospheres, but durable otherwise.	Yes
	Box Hill, N of Bradford-on-Avon, Wiltshire.	Box Ground, also known as St. Aldhelm. Cream to light brown.	Limestone. A 'Bath' stone, no longer worked. Malmesbury Abbey, Wilts.	No
	Corsham, between Bath and Chippenham, Wiltshire.	Corngrit. Light cream.	Limestone. One of the 'Bath' stones but coarser grained than Corsham Down. Durable except in polluted atmospheres.	No
	Corsham, between Bath and Chippenham, Wiltshire.	Corsham Down. Light cream.	Limestone. One of the 'Bath' stones. Fine grained and free working.	No

'Bath' stone was mined at other locations in the area among them Hartham Park near Corsham and Hazelbury near Box and Farley Down, Monkton Farley, north west of Bradford-on-Avon. Also at Westwood, south west of Bradford-on-Avon. Apart from much use in Bath, Bristol and the surrounding area, 'Bath' stone was extensively used elsewhere. In Oxford the Sheldonian Theatre was refaced in 1830, the Baliol College Broad Street frontage, in 1850 with very fine joints and iron cramps and, mixed with Portland stone, the Ashmolean Museum, Beaumont Street. In Gloucester, the 1856 frontage to the Eastgate Shopping Centre is in 'Bath' stone as are the dressings to the Guildhall in the High Street at Worcester.

Surrey	Near Godalming, SW of Guildford, Surrey.	Bargate (Burgate). Brown to yellow.	Sandstone. Coarse in texture and durable. Sometimes iron staining. Charterhouse School.	No
	Reigate, Surrey.	Reigate. Greenish.	Sandstone. Although a freestone and capable of being ashlared, its use for Westminster Abbey and Southwark Cathedral among other locations in London proved unsatisfactory due to porosity.	No

Other quarries for 'Reigate' stone were at Bletchingley, Godstone and Limpsfield all to the east of and between Reigate and Sevenoaks and at Merstham and Gatton to the north of Redhill. Stone variants of both Reigate and Bargate were also quarried at Woolmer Forest and Selborne to the south east of Alton, both in Hampshire.

Kent	Bethersden, West of Ashford, Kent.	Bethersden.	Limestone. Known as a 'paludina' marble because of containing remains of snail shells.	No
	Aylesford, NE of Maidstone, Kent.	Kentish Ragstone. White to greenish grey.	Sandstone, from Lower Greens and Hythe beds. Tough, hard and relatively intractable. Chiefly used for rubble walling. Keep of Rochester Castle, White Tower, London, West Gate, Caterbury, Knole House, Igtham Moat, Cooling Castle, Maidstone Gaol, Archbishop's Palace Maidstone. Eltham Palace, London, Great Hall.	No

County (Old)	Location of Quarry/Mine	Name and Colour	Type, characteristics and locations of use	Available Yes/No
Kent (continued)	Tunbridge Wells, Kent.	Calverley (or Tilgate). Variegated brown.	Sandstone. Its use much in evidence in Tunbridge Wells.	No
Devonshire	Near Tavistock, Devonshire.	Merrivale. Silver grey.	Granite. Medium grained with feldspar crystals.	Yes
	Beer, East Devon, Devonshire.	Beer. White to pale grey or cream.	Limestone. Fine grained, light in weight and compact. Soft, easy to work, hence much used for interior work. Not suitable for polluted atmosphere and though not considered suitable used locally for external work. Upper level Guildhall, Exeter (1592) and repairs to Cathedral. Essentially chalk.	Yes
	Oreston. Near Plymouth, Devonshire.	Radford. Grey to red.	Limestone. Fine grained, free working.	No
Dorset	Isle of Portland, S of Weymouth, Dorset.	Portland. White to grey. Dorset.	Limestone. The 2001 BGS map indicates that the Basebed, Roach and Whitbed levels are available. Much used generally since Wren chose it for the rebuilding of St. Paul's Cathedral.	Yes
	Isle of Purbeck, W of Swanage, Dorset.	Purbeck. Grey to greeny grey.	Limestone. Used as building stone but also much used when polished as 'Purbeck marble' for interior columns such as those at Salisbury Cathedral and many other churches. The Isle of Purbeck also produces a good building stone called Purbeck-Portland which was available in 2001.	Yes
Hampshire	Quarr Abbey. Near Ryde, Isle of Wight.	Quarr.	Limestone. Contained shell fragments.	No
	Ventnor, Isle of Wight.	St. Boniface and Green Ventnor. Blue grey to green grey.	Sandstone. Fine grained. Much used on Island and at Winchester and Chichester (detached bell tower) Cathedrals and for repairs at Herstmonceux Castle.	No
Sussex	Quarries to S and SW of East Grinstead, West Sussex.	Wealden. Blend of dark grey and pale fawn.	Sandstone. Fine grained freestone. Good for ashlar. Wakehurst Place, Bodiam Castle.	Yes
	West Hoathley, West Sussex.	Sussex Sandstone. West Sussex.	Sandstone. Quarry locations as per Stone Federation's Natural Stone Glossary of early 1990s. The 1991 BGS map shows an entirely different location.	No
	Horsham, West Sussex.	Horsham. Brown to yellow.	Sandstone. Fine grained, easily split into slabs for roofing and paving.	No

County (Old)	Location of Quarry/Mine	Name and Colour	Type, characteristics and locations of use	Available Yes/No
Cornwall	St Breward, Bodmin, Cornwall.	Tor Brake, Tor Down De Lank and Hantergantick. Silvery to greenish grey.	Granites. Medium grained. Mainly used for engineering work.	Yes
	Penryn, NW of Falmouth, Cornwall.	Penryn. Medium grey.	Granite. Medium to coarse grained. Used for both engineering and general building work. *Others active to west of Falmouth*	No*
	Delabole, NE of Port Isaac, Cornwall.	Delabole. Green, grey green, green and rustic red.	Slate. Other active quarries in the area are producing similar slate called Trevillet Merryfield and Trecame, Tyne's and Prince of Wales.	Yes

Index

Note:The paged list of contents at the beginning of each Section provides the principle references.

The Index extends the coverage to Court cases, organisations, trade associations, individuals, British Standards and other publications. Where a photograph is referenced it is to the page on which it appears.

It is recommended, that whole sub-sections be read in preference to paragraphs being considered in isolation.

C

U

V

W

Z